Studies in Polymer Science 9

Polymer Solutions

Studies in Polymer Science

Other titles in the series

Studies in Polymer Science 9

Polymer Solutions

by

Hiroshi Fujita

Emeritus Professor, Department of Macromolecular Science, Osaka University, Toyonaka, Osaka 560, Japan

ELSEVIER
Amsterdam — Oxford — New York — Tokyo   1990

ELSEVIER SCIENCE PUBLISHERS B.V.
Sara Burgerhartstraat 25
P.O. Box 211, 1000 AE Amsterdam, The Netherlands

Distributors for the United States and Canada:

ELSEVIER SCIENCE PUBLISHING COMPANY INC.
655, Avenue of the Americas
New York, NY 10010, U.S.A.

ISBN 0-444-88339-8

Printed in The Netherlands

INTRODUCTION

Purposes

Polymer solution study dates back as early as the late 1920s when Staudinger utilized the quantity now called the intrinsic viscosity in his attempt to prove the existence of macromolecular compounds. However, it was shortly after World War II that its real progress commenced, and in this early stage of development Flory played a decisive role in bringing the study of dilute polymer solutions to the subject of academic interest. Recognizing for the first time that excluded–volume repulsion between the monomers or structural units (hereafter called segments) of polymers is the major determining factor for both static and dynamic properties of polymer molecules in dilute solution, he constructed a molecular theory of these properties and confirmed his fundamental ideas by comparison of the predictions therefrom with experimental results. His pioneering achievements are elegantly summarized in his classic volume *Principles of Polymer Chemistry* published in 1953. An explosive development in both theory and experiment on dilute polymer solutions has since taken place, with excluded–volume effects as the central theme. In a book entitled *Modern Theory of Polymer Solutions* published in 1971, Yamakawa summarized and discussed in a scholarly fashion most of the important contributions made to polymer solutions by the end of the 1960s. However, his book was mainly concerned with solutions at infinite dilution. In fact, at that time, the theory of polymer solutions was almost synonymous with that of infinitely dilute ones.

The situation began to change about the middle of the 1960s when there occurred an interest in expanding the physics of polymer solutions to finite dilutions, and shortly came a second stage of remarkable development in polymer solution study. It was the semi–dilute solution that became the main focus of research at this stage. Here by semi–dilute solution we mean a system in which the polymer concentration is sufficiently low, yet polymer chains are not isolated but overlap with one another. The chain overlap, however, is not as extensive as in more concentrated solutions. As a result there occurs a spatial correlation of density fluctuations which extends over a range comparable to the size of a single polymer coil. Thus, semi–dilute solutions may be compared to solutions of small molecules near the critical point. This analogy has attracted the interest of polymer–oriented physicists, and many important theoretical predictions have been deduced. It is probably fair to say that the polymer community owes much to theoretical physicists with regard to the recent enhancement of knowledge on concentrated polymer solutions.

At present, the discipline of polymer solutions has grown to the state which covers the whole range of concentration, from dilute solutions to undiluted

V

polymers (melts), though it still leaves much to be desired in the region of high concentration. This book aims to summarize typical theoretical and experimental material which is hopefully of use in grasping the state of our current understanding of dilute and concentrated polymer solutions. In so doing, the author does not intend to prepare a well-balanced review but is mainly interested in clarifying what has essentially been solved and what apparently remains unsolved or unsettled in both theory and experiment. Thus he tries to be critical and selective rather than to enumerate a large body of recent information with no bias. For example, the data to be compared with theoretical predictions are often chosen on his own criteria and experiences, and "typical" theories which are referred to are within the limits accessible by his knowledge and ability. The author also tries to interpret and translate some sophisticated concepts mainly invoked by physicists into language intelligible to general polymer physical chemists.

Scope

This book is divided into Part I and Part II, the former concerned with infinite dilution and the latter with finite dilutions. It is not intended to be a textbook nor a reference book for specialists. Differing from Yamakawa's monograph quoted above, it virtually omits going deeply into the mathematical derivations of chosen theories except for a few cases. Furthermore, the technical backgrounds of quoted experimental results are skipped. Emphasis is placed on showing the scope and limitations of the "typical" theories in comparison with the "best" experimental data available to date.

The discussion is almost entirely limited to monodisperse linear polymers (except in two sections dealing with macrorings) dissolved in pure solvents. Hence, practically important problems relating to polydispersity, branching, chemical heterogeneity (copolymers), mixed solvents, and so forth are placed outside the scope of the book. Equally important, only "global" polymer properties are discussed. Here by global properties we mean ones which are essentially determined by the basic character of the polymer molecule, i.e., it is a long chain having an almost infinite degree of freedom. Thus this book does not deal with solution properties which sensitively reflect the microscopic features (the chemistry) of individual polymer molecules, but focuses on the properties of grossly coarse-grained polymer models.

As the center of gravity shifted to concentrated systems in the 1970s, the study of dilute polymer solutions began to lose the popularity and activity that it had maintained until the late 1960s, and there arose a notion that little of significance remained to be investigated in the field of dilute solutions. One of the main purposes of Part I is to present a variety of evidence showing that this field is by no means closed.

Following Chapter 1 which mentions some basic concepts in chain statistics, we deal in Chapter 2 with recent advances in the so-called two-parameter theory of dilute polymer solutions. This chapter is by and large designed to be a continued part of Yamakawa's book. One of the recent newcomers in the polymer community is the renormalization group theory. Whether agreeable or not, it has made a great impact on those who are interested in theoretical aspects of polymer solutions. Unfortunately, the basic concepts involved in it are not always familiar in polymer science. For this reason an introductory account of its elementary parts is presented in Chapter 3. Chapter 4 is concerned with some special topics which hopefully illustrate the recent trends in dilute solution studies. Up to this chapter the discussion is limited to ideally flexible polymer chains. Effects of chain stiffness are considered in Chapter 5, the last chapter in Part I. Marked progress in theory and the availability of various semi-flexible polymers in recent years have made it possible to undertake more detailed and systematic work on stiff chains in solution than before. The main findings from such work are summarized in this chapter.

Part II starts with Chapter 6 which explains fundamental concepts needed for reading recent contributions to the physics of concentrated polymer solutions. The most important of them are the correlation length and the screening effects of chain overlap on intrachain segment-segment interactions. As mentioned above, the semi-dilute solution has become one of the central objects in recent studies of polymer concentrates. Chapter 7 is devoted to a discussion on the static and transport properties of this special type of solution. The blob concept and scaling laws are central in this discussion. Polymer self-diffusion in concentrated solutions and melts has become one of the most up-to-date subjects in polymer physics. Motivated by the reptation theory proposed by de Gennes and formulated elegantly by Doi and Edwards, many polymer scientists have put much effort into elucidating this Brownian motion from both theoretical and experimental sides. However, it is still controversial whether the reptative mode dominates the chain motion in polymer concentrates. In Chapter 8, we talk about this situation by referring to typical theories and experimental data for self-diffusion and tracer diffusion coefficients. The following chapter turns to a classic subject which, however, still leaves much to be worked out. It is the thermodynamics of phase separation in polymer solutions. Since the pioneering work of Shultz and Flory in the early 1950s, this phenomenon has been explored intensively by many workers, notably Koningsveld and his collaborators. The main aim in such efforts has been to establish, either theoretically or empirically, the free energy of mixing which enables us to predict phase relations in polymer solutions quantitatively. If we look over the existing literature somewhat carefully, we will find that no satisfactory form of this energy is as yet established even for binary systems consisting of a monodisperse polymer and a pure solvent. The situation is much worse for systems containing more than one polymer component. In Chapter 9, we summarize typical efforts toward accurate predictions of phase equilibria in binary and quasibinary solutions and point out what difficulties have to be overcome in order to reach them. Part II ends with Chapter 10 which discusses some other topics currently calling attention in the study of condensed polymer systems.

This book deals almost exclusively with equilibrium and steady transport properties. Time-dependent or dynamic processes are discussed only fragmentarily in connection with polymer self-diffusion (Chapter 8) and spinodal decomposition (Chapter 10). It far exceeds the ability and experiences of the author to write about them. However, we have to remark that the center of gravity in polymer solution study is rapidly moving toward dynamic processes and also surface phenomena.

The author has tried to include as many "typical" experimental data as possible in the manuscript, thereby complementing the discussion written in his limited command of English (which is desperately remote in grammar and syntax from his mother language). Because they have been chosen largely on his preference and prejudice, someone may find some of them "atypical" or irrelevant. The references collected at the end of each chapter are neither extensive nor complete and essentially limited to those published by the end of 1988. Many of them have been chosen from the author's own library and collections.

Acknowledgments

I wrote this book with the cooperation of Professor T. Norisuye of the Department of Macromolecular Science, Osaka University, who wrote the initial draft for chapters 3 and 5 and suggested the revision of the manuscript in many points. Two former colleagues of mine, Professor Y. Einaga of the Department of Polymer Chemistry, Kyoto University and Dr. T. Sato of the Department of Macromolecular Science, Osaka University, read several chapters and drew some of the graphs cited. I am greatly indebted not only to them but also to my friends Professor K. Kajiwara and Dr. H. Urakawa of the Kyoto Institute of Technology, who kindly and effectively prepared the camera-ready copy of the manuscript with a Macintosh laser printer. Finally, my deep appreciation goes to Elsevier Science Publishers, who made it possible for me to pour the last energy of my scientific career into a book.

CONTENTS

PART I   DILUTE SOLUTIONS

PART II CONCENTRATED SOLUTIONS

Chapter 6 Basic Concepts 179

XV

PART I

DILUTE SOLUTIONS

Chapter 1   Some Fundamentals

1.   Chain Models

1.1   Random Flight Chain

The molecule usually referred to as a linear homopolymer is an entity consisting of many identical chemical residues (called monomers) covalently bonded together in a linear fashion. Thus it can be visualized as a long chain of monomer units. In general, each monomer unit is given a certain freedom of rotation about the axis of its preceding monomer unit, and this freedom allows the chain to kink randomly and hence to take on a huge number of different conformations. For this feature the chain as a whole behaves like a flexible string. However, internal rotations of successive monomers cannot take place independently because their constituent atoms or atomic groups interact with each other when they come close. Owing to this correlated internal rotation the chain has some resistance against bending or twisting. The chain thus cannot be perfectly flexible. In actual polymers usually considered "flexible", the correlation concerned persists only over the range of a few successive monomers, which is very short in comparison with the entire chain length of typical polymer molecules. Then we come to the idea that when we are interested in the events occurring on chain length scales, flexible polymer molecules may be modeled by a hypothetical chain which is made up of rigid thin rods of equal length connected linearly by universal joints. In this model, which is called the **random flight chain** or **freely jointed chain**, each rod can take any direction with equal probability. Each rod constituting a random flight chain is referred to as the **segment** or bond.

We can introduce into a random flight chain a constraint that the angle between neighboring segments, i.e., the bond angle, is held constant. Then the model chain is called the **freely rotating chain**. However, this model is not explicitly treated below, since, in many cases, the bond angle constraint can be accounted for by changing the segment length accordingly. This operation is a kind of coarse–graining which will be mentioned in Section 3.3.

A pair of monomer units of a polymer chain that are separated by more than a chain segment can come close as a result of deformation of the intervening chain between them. Then they tend to interact. This interaction, which Flory referred to as a long–range interference of monomer units, decidedly affects the number of conformations that the chain can take up; for example, those in which two monomers occupy the same point in space are not realizable. In order to take this effect into account when the polymer is modeled by a random flight chain we usually place small beads at the ends of each segment (i.e., at

2

the universal joints) and assume that intrachain interactions (interchain ones as well) act only between the beads. Thus, each segment, i.e., the rod between a pair of adjacent beads, is treated as a phantom.

The random flight chain no longer retains information about the primary chemical structure and local interactions of the actual polymer molecule. It is therefore useful only for formulating its so-called global or large-scale behavior, i.e., one which is essentially determined by its entire length (or molecular weight) as well as its overall flexible nature and also by the long-range interference of its monomers. In fact, polymers in solution exhibit many properties which can be categorized as global, and some of them can be measured by experiment. Part I of this book deals with typical global behavior of linear homopolymers in solution at the limit of infinite dilution. The main purpose is to clarify how such behavior is related quantitatively to the length and flexibility of the chain and the strength of bead–bead interactions. Up to end of Chapter 4, unless otherwise stated, the chain is assumed to be ideally flexible.

1.2 Spring–bead Chain

The random flight chain is not always convenient for theoretical formulation of global polymer properties, because it is subject to the constraint that the length of each segment is fixed (the freely rotating chain is even more inconveneient owing to a further constraint associated with a fixed bond angle). Therefore, theorists favor a less constrained chain model called the **spring–bead chain**. This is generated by replacing each segment of a random flight chain with a spring whose end-to-end distance fluctuates following the Gaussian distribution function (see Section 2.3 of Chapter 1) and thus which is referred to as the Gaussian spring. Again, each spring is treated as a phantom, and it is assumed that interactions occur between small beads attached to its ends.

1.3 Continuous Chains

When moving along a random flight chain or a spring–bead chain, we encounter a discontinuous change in direction at each bead. In these chains, the interactions act between the beads discretely distributed along the chain. Such features retaining the zig–zag nature of actual polymer molecules, both random–flight and spring–bead chains have long been liked and used by polymer chemists in theoretical formulations of polymer systems. However, some theorists in physics have preferred a more abstract model in which the microscopic discreteness of actual polymer chains is washed out. In that, the polymer chain is coarsened to the extreme: a smooth space curve of certain length. This curve is called the **continuous chain model**.

3

A continuous chain can be generated by a limiting process in which the segment of a freely rotating chain is shortened to zero, with the total contour length of the chain held constant, and at the same time the bond angle approaches 180°. This was first proposed by Kratky and Porod [1] and the generated chain is referred to as the **wormlike chain** or KP chain. Its distinct feature is that it is no longer a flexible chain but has resistance against bending. For this feature the wormlike chain has been used as a model in formulating global behavior of stiff or semi–flexible polymer molecules, as discussed in Chapter 5.

A continuous chain can also be derived by making all springs in a spring–bead chain vanishingly short, with its total contour length held fixed. The resulting chain has no stiffness but is completely flexible. The Edwards chain discussed later in this chapter is an example.

We may constrain a freely rotating chain in such a way that the internal rotation of each segment is allowed to occur only over a limited range of rotational angle. If the limiting process leading to the wormlike chain is carried out with this additional constraint taken into account, we obtain a continuous chain model first considered by Miyake and Hoshino [2] and also by Burchard [3]. Yet another continuous chain model called the **helical wormlike chain** was proposed by Yamakawa and Fujii [4]. This chain is derived from a discrete chain somewhat different from the random flight chain [5].

2. Basic Assumptions and Approximations

2.1 Excluded–volume Effect

We suppose a very dilute gas of random flight chains. As a result of thermal rotation of chain segments each chain will take up a great number of conformations, with a very short interval of time being spent for the passage from one conformation to another. In so doing, it automatically avoids taking those conformations in which any pair of its beads overlap, because it is physically impossible for them to occupy the same volume element in space at the same time. When a pair of beads come close they exert a repulsive force F on each other. The strength of this force depends on the separation between the beads and also on the chemistry of each bead as well as the temperature and pressure of the gas. Now we introduce a liquid solvent into the system to make a dilute solution of random flight chains. Each bead has a chance to interact with solvent molecules as well as other beads. As a result the force that acts between a pair of beads becomes no longer equal to the vacuum value F. If the bead–solvent interaction favors bead–solvent contact over the bead–bead one, the solvent is said to be good (for the polymer considered), while if the reverse is the case, the solvent is said to be poor or bad. Thus, a good solvent tends to prevent a pair of beads from approaching or tries to pull them apart. This suggests that the bead–solvent interaction has the effect equivalent to inducing some additional force between a pair of beads.

In fact, according to the statistical mechanical theory formulated by McMillan and Mayer [6] and by Saito [7], the solution of a polymer in a pure solvent should be equivalent in thermodynamic behavior to a hypothetical gas of the same polymer whose segments repel each other with a force given by the sum of the vacuum value F and some additional force F'. The latter summarizes the effect due to segment–solvent interactions and is called the solvent–mediated force. Thus, in the McMillan–Mayer–Saito theory, the segment–solvent interactions are lumped into an unknown F', and the thermodynamics of polymer solutions can be formulated by applying the well–established theory of gases. However, we note that, strictly speaking, this remarkable advantage can be used at a fixed chemical potential of the solvent, i.e., for the solution at osmotic equilibrium with the solvent.

In a good solvent, the segment–solvent interaction tends to pull a pair of segments apart, so that the solvent–mediated force F' should be repulsive as is F. On the other hand, F' should be attractive in poor solvents. Hence, as the solvent is made poorer by changing either solvent species or temperature, the situation should be reached in which the attractive F' cancels or suppresses the repulsive F so that the net force $F + F'$ (which is usually called the mean force) becomes zero or even negative.

What we is concerned with in the physics of polymer systems is not their physical properties for individual polymer conformations but those averaged over the ensemble of such conformations under given conditions of polymer and solvent. For flexible polymers the averages depend primarily on the strength of the net interaction force $F + F'$ and the number of segments contained in the chain. Knowledge of F and F' is thus essential for the understanding of polymers in solution.

The above discussion is concerned with the interactions between a pair of beads, i.e., binary cluster interactions. However, there are chances for beads to interact forming clusters higher than the binary one. Flory was the first to reach a very important recognition that polymer conformations and hence global polymer properties are significantly influenced by the potential energy stored in the polymer chain as a result of formation of binary, ternary, and higher–order clusters of segments. The term **excluded–volume effect** is used to describe any effect arising from intrachain or interchain segment–segment interactions. As mentioned in the INTRODUCTION, Flory's recognition of this effect triggered the development of polymer solution studies for the last four decades.

2.2 Potential Energy of the Chain

We begin with formulating the potential energy stored in an isolated polymer molecule in a pure solvent. To do this we adopt the spring–bead chain model because it is easier for mathematical treatment. When it is used, the potential energy possessed by the springs must be added to that arising from mean force interactions among the beads of the chain.

The spring–bead chain considered is assumed to consist of $N + 1$ beads, two of which are situated at the chain ends. Tacitly, it is assumed that N is sufficiently large in comparison with unity. Thus, in what follows, $N + 1$ is often approximated by N with no special notice. Anyway this approximation implies that we are implicitly considering a sufficiently long polymer chain. We assign numbers $0, 1, 2, \ldots, N$ to the beads, with 0 for the bead at one chain end. Placing bead 0 at the space–fixed coordinate origin, we denote the vector drawn from bead 0 to bead j by $\mathbf{r}_j$. Then the vector $\mathbf{R}_j$ from bead j−1 to bead j is expressed by $\mathbf{r}_j - \mathbf{r}_{j-1}$ and is here called the spring bond vector. Each spring has an elastic energy denoted here by $w k_B T$, where k_B and T are the Boltzmann constant and the absolute temperature, respectively. We assume that w depends only on R_j, the magnitude of $\mathbf{R}_j$. Then the total elastic energy W of the chain is expressed by

$$W = k_B T \sum_{j=1}^{N} w(R_j) \tag{2.1}$$

Next, the mean force potential associated with a binary cluster of beads i and j is denoted by u_{ij}, that with a ternary cluster of beads i, j, and m by u_{ijm}, and so on. We introduce Δu_{ijm} defined by

$$u_{ijm} = u_{ij} + u_{jm} + u_{mi} + \Delta u_{ijm} \tag{2.2}$$

Similar expressions can be written for the mean force potentials of clusters higher than the ternary one, though not displayed here. The quantity u_{ij} is referred to as the binary cluster (interaction) energy, and Δu_{ijm} and Δu_{ijmn} as the residual (or irreducible) potentials of ternary and quaternary clusters, respectively.

The total mean force potential U of the chain can be expressed as follows:

$$U = U_2 + \Delta U_3 + \Delta U_4 + \ldots \tag{2.3}$$

where

$$U_2 = \sum_{i=0, j>i}^{N} u_{ij} \tag{2.4}$$

$$\Delta U_3 = \sum_{i=0,m>j>i}^{N} \Delta u_{ijm} \qquad (2.5)$$

$$\Delta U_4 = \sum_{i=0,n>m>j>i}^{N} \Delta u_{ijmn} \qquad (2.6)$$

In most of the previous calculations the terms ΔU_3, $\Delta U_4, \ldots$ have been neglected in comparison with U_2. This is called the **binary cluster approximation**, and is equivalent to Kirkwood's superposition approximation in the theory of liquids. As another approximation it is reasonable to assume that u_{ij} depends only on r_{ij}, i.e., $u_{ij} = u(r_{ij})$. With these approximations we obtain

$$U = U_2 = \sum_{i=0,j>i}^{N} u(r_{ij}) \qquad (2.7)$$

The function $u(r_{ij})$ depends on the combination of polymer and solvent as well as T, but has nothing to do with the bead number and hence N. In this book, we confine ourselves to non-ionic polymers, so that all the interactions involved in the system are short-ranged. Thus, as has been done by most previous authors, $u(r_{ij})$ may be approximated by

$$u(r_{ij}) = k_B T \beta \delta(\mathbf{r}_{ij}) \qquad (2.8)$$

where $\delta(\mathbf{r})$ is a three-dimensional delta function and β is a parameter which has the dimension of volume and is referred to as the **excluded-volume strength** throughout this book. Sometimes β is called the binary cluster integral, but such naming is not used below. This parameter is not constant for a given polymer but varies with solvent as well as temperature. In fact, it can be negative when the mean force is attractive. Hence, the convention regarding β as an effective core volume of one bead sometimes leads to a conceptional difficulty.

Summarizing, under the binary cluster approximation, we obtain

$$H = W + U = k_B T \sum_{k=1}^{N} w(R_k) + k_B T \beta \sum_{i=0,j>i}^{N} \delta(\mathbf{r}_{ij}) \qquad (2.9)$$

for the total potential energy H (often called the Hamiltonian) of a "neutral" spring-bead chain. Until the effect of ΔU_3 is discussed in Chapter 4 we will proceed with H given by eq 2.9 as the basis of our formulation of polymer behavior in dilute solutions.

2.3 Spring Bond Probability

The probability $\tau(R)$ that the spring bond vector has an end–to–end distance of R is expressed in terms of $w(R)$ as

$$\tau(R) = \exp[-w(R)] / \int \exp[-w(R)] \, d\mathbf{R} \qquad (2.10)$$

because this bond can take any direction with equal probability. Here the differential $d\mathbf{R}$ actually stands for $4\pi R^2 dR$, and the integral extends from zero to infinity of the variable R. The quantity a defined by

$$a^2 = \int \tau(R) R^2 \, dR \qquad (2.11)$$

gives the mean length of the spring bond vector, and is called the spring length in the subsequent presentation. As mentioned before, each spring in the spring–bead chain is assumed to be Gaussian, which means that $\tau(R)$ is expressed by

$$\tau(R) = (3/2\pi a^2)^{3/2} \exp[-3R^2/(2a^2)] \qquad (2.12)$$

Advantages of this assumption will be made clear in a later section.

2.4 Theta Condition

The osmotic pressure Π of a dilute solution of a monodisperse polymer with molecular weight M is expressed as a power series of polymer mass concentration c (weight of polymer per unit volume of solution) as

$$\Pi/RT = c/M + A_2 c^2 + A_3 c^3 + \dots \qquad (2.13)$$

where R is the gas constant (not to be confused with the magnitude of a vector $\mathbf{R}$), T the absolute temperature, and M the molecular weight of the polymer. This series is usually called the osmotic virial expansion, with $A_i \, (i = 2, 3, \dots)$ being referred to as the i-th virial coefficient of the solution. Experimentally, A_2 can be evaluated by determining the initial slope of $\Pi/(RTc)$ plotted against c. However, the estimation of A_3 and the higher virial coefficients is not a simple matter for experimentalists. Thus, available experimental information about the virial coefficients is largely limited to A_2.

For a series of homologous polymers A_2 depends on M as well as T and the solvent species. Experimental studies have repeatedly shown that for a given polymer there is a combination of poor solvent and temperature θ for which A_2 vanishes regardless of M (except for oligomers and similar low–molecular–weight polymers). This special poor solvent at θ is called the theta solvent, and θ the theta temperature. However, in what follows, we use the term theta condition synonymously with the former. For polystyrene the theta condition is cyclohexane at $34.5(\pm 0.5)°$C and for poly(isobutylene) it is benzene at $24.0°$C. Sometimes there exists more than one θ solvent for one polymer species. Such θ solvents include binary mixtures of good and poor solvents. In this book, we will not concern ourselves with mixed solvents.

Molecular theory of the second virial coefficient (see Section 2 of Chapter 2) shows that if the binary cluster approximation is valid, this coefficient vanishes under the solvent condition (specified by solvent species and temperature) in which the potential energy U_2 associated with the binary cluster interaction happens to be zero. According to eq 2.7 and 2.8, $U_2 = 0$ is equivalent to $\beta = 0$. Since β depends only on temperature for a given combination of polymer and solvent it follows that the θ temperature does not depend on M in the binary cluster approximation. This conclusion is consistent with most experimental results reported to date. It is mainly for this reason that we proceed with the Hamiltonian H based on the binary cluster approximation, i.e., eq 2.9. Thus, unless otherwise stated below, $\beta = 0$ is taken as the condition specifying a θ solvent or a θ temperature. There is an argument by some theorists [8] that, in poor solvents encompassing the θ temperature, the binary cluster approximation is inadequate and at least ΔU_3 must be added to H. This seems reasonable, because U_2 is supposed to be very small in such solvents, but, as will be discussed in Chapter 4, the inclusion of ΔU_3 brings about some yet unsolved difficulties.

3. Distribution Functions

3.1 Spring–bead Chain

In applying statistical mechanics to the spring–bead chain, it is usual to ignore the conformation of the springs and to describe individual conformations of the chain in terms of the positions of the beads, which are specified by $\{\mathbf{r}\} \equiv (\mathbf{r}_1, \mathbf{r}_2, \ldots, \mathbf{r}_N)$. We note that bead 0 is fixed at the coordinate origin so that its position does not enter the set $\{\mathbf{r}\}$. Denoting by $P(\{\mathbf{r}\})d\{\mathbf{r}\}$ the probability of finding the chain conformation in an infinitesimal range between $\{\mathbf{r}\}$ and $\{\mathbf{r} + d\mathbf{r}\}$, we obtain

$$P(\{\mathbf{r}\}) = (1/Z)\exp\left[-H(\{\mathbf{r}\})/k_\mathrm{B}T\right] \tag{3.1}$$

where H is the Hamiltonian of the chain and Z is defined by

$$Z = \int \exp\left[-H(\{\mathbf{r}\})/k_\mathrm{B}T\right]d\{\mathbf{r}\} \tag{3.2}$$

9

the latter being the partition function (the sum of states) of the chain. With eq 2.9 substituted into eq 3.1 and consideration of eq 2.10, we find

$$P(\{\mathbf{r}\}) = (1/Z_N) \prod_{k=1}^{N} \tau(R_k) \exp\left[-U_2(\{\mathbf{r}\})/k_B T\right] \tag{3.3}$$

with

$$Z_N = Z/[\int \exp\left[-w(R)\right]d\mathbf{R}]^N \tag{3.4}$$

$$U_2(\{\mathbf{r}\}) = k_B T \beta \sum_{j>i}^{N} \delta(\mathbf{r}_{ij}) \tag{3.5}$$

Equation 3.3 is the basis of the theoretical calculations of global polymer properties which depend on chain conformation. For example, the probability density (normalized distribution function) $G(\mathbf{R})$ that the chain has a specified end–to–end distance $\mathbf{R}$ is expressed in terms of $P(\{\mathbf{r}\})$ as follows:

$$G(\mathbf{R}) = \int P(\{\mathbf{r}\})\delta(\mathbf{R} - \sum_{j=1}^{N} \mathbf{R}_j)d\mathbf{R}_k \tag{3.6}$$

The mean–square end–to–end distance $\langle R^2 \rangle$ can then be calculated from

$$\langle R^2 \rangle = \int \mathbf{R}^2 G(\mathbf{R})d\mathbf{R}/\int G(\mathbf{R})d\mathbf{R} \tag{3.7}$$

In a θ solvent, $\beta = 0$ (under the binary cluster approximation) so that eq 3.6 gives

$$G(\mathbf{R})_\theta = (1/Z_N) \int \prod_{k=1}^{N} \tau(R_k)\delta(\mathbf{R} - \sum_{j=1}^{N} \mathbf{R}_j)d\mathbf{R}_k \tag{3.8}$$

where the subscript θ signifies the value under the θ condition (this convention will be used throughout the ensuing presentation). Since the spring is Gaussian we can substitute eq 2.12 into $\tau(R_k)$, and then by application of the Wang–Uhlenbeck method (see [9] for details) we obtain

$$G(\mathbf{R})_\theta = (3/2\pi\langle R^2 \rangle_\theta)^{3/2} \exp\left(-3\mathbf{R}^2/2\langle R^2 \rangle_\theta\right) \tag{3.9}$$

where

$$\langle R^2 \rangle_\theta = N a^2 \tag{3.10}$$

Equation 3.9 shows that when not perturbed by intrachain interactions an isolated spring–bead chain consisting of Gaussian springs behaves as Gaussian regardless of the value of N. This feature explains why the spring–bead chain is favored for theoretical formulations of global polymer behavior. It is well known [10] that if N is sufficiently large, $G(\mathbf{R})_\theta$ and $\langle R^2 \rangle_\theta$ of a random flight chain are also expressed by eq 3.9 and 3.10, respectively, with a being reinterpreted as the length of one segment, and also that these equations hold asymptotically (i.e., for $N \gg 1$) for chains involving the bond angle constraint and a short-ranged correlation among the internal rotations of successive segments, though a can no longer be regarded as the segment length. Thus we can expect that whatever their local chemical structure, linear polymers (flexible as a whole) under the θ condition asymptotically exhibit global behavior characteristic of the Gaussian spring–bead chain.

3.2 Edwards Continuous Chain

Transformation of the relations for the spring–bead chain to those for a continuous chain invokes a sophisticated mathematical technique. Here only its outline and final consequences are presented, because the author does not find himself qualified to expound it with any rigor. The reader is advised to consult Freed's book [11].

We consider a spring–bead chain consisting of N Gaussian springs. Its contour length L is expressed by

$$L = N \Delta s \tag{3.11}$$

where Δs denotes the contour length of one spring. When the chain is not perturbed by intrachain interactions, eq 3.10 holds and can be rewritten with the aid of eq 3.11 as

$$\langle R^2 \rangle_\theta = L(a^2 / \Delta s) \tag{3.12}$$

We let N increase infinitely and Δs diminish to zero in order to transform the spring–bead chain into a continuous chain. In so doing, we impose the condition that L and $\langle R^2 \rangle_\theta$ remain unchanged. Then it follows from eq 3.12 that a should be diminished so as to make b' defined by

$$b' = a^2 / \Delta s \tag{3.13}$$

converge to a certain finite value b as Δs tends to zero. Thus

$$b = \lim_{\Delta s \to 0} a^2 / \Delta s \tag{3.14}$$

and we obtain the important expression

$$\langle R^2 \rangle_\theta = Lb \tag{3.15}$$

11

in the continuous chain limit. The length parameter b is hereafter called the Kuhn segment length for continuous chains. It is the basic scale for specifying the length of a continuous chain.

In eq 3.3 we put U_2 equal to zero, substitute eq 2.12 for $\tau(R_k)$, and replace a^2 by $b'\Delta s$. The result reads

$$P(\{\mathbf{r}\})_\theta = \Xi \exp\left[-3 \sum_{k=1}^{N} (\mathbf{r}_k - \mathbf{r}_{k-1})^2 / (2b'\Delta s)\right] \tag{3.16}$$

where Ξ is a proportionality factor which is to be determined from the normalization condition for $P(\{\mathbf{r}\})_\theta$. In the limit to a continuous chain, $\{\mathbf{r}\}$ tends to a function $\mathbf{r}(s)$ describing the shape of a continuous space curve. Here s is the contour length measured along the curve from its one end, and $\mathbf{r}(s)$ represents the vector connecting the curve point specified by s to that for $s = 0$. In this limit, $\Delta s \to ds$ and $\mathbf{r}_k - \mathbf{r}_{k-1} \to \mathbf{r}(s) - \mathbf{r}(s-ds)$. Furthermore, $b' \to b$. Thus, the continuous chain version of eq 3.16 is found to be

$$P[\mathbf{r}(s)]_\theta \delta\mathbf{r}(s) = D[\mathbf{r}(s)] \exp\left\{-(3/2b) \int_0^L [d\mathbf{r}(s)/ds]^2 ds\right\} \tag{3.17}$$

Here, $P[\mathbf{r}(s)]_\theta$ is a functional, and its product with $\delta\mathbf{r}(s)$ indicates the probability that the shape of the curve is found between those specified by $\mathbf{r}(s)$ and $\mathbf{r}(s) + \delta\mathbf{r}(s)$, with $\delta\mathbf{r}(s)$ denoting a function of s which is infinitesimally small for any s in the range $0 < s < L$. $D[\mathbf{r}(s)]$ is another functional equal to the continuous chain limit of Ξ multiplied by $\delta\mathbf{r}(s)$. The precise meaning of this functional need not be known in the present context.

Next it can be shown that eq 3.3 with $U_2 \neq 0$ is transformed in the continuous chain limit into

$$P[\mathbf{r}(s)] = P[\mathbf{r}(s)]_\theta \exp\left(-U_2[\mathbf{r}(s)]/k_B T\right) \tag{3.18}$$

where $U_2[\mathbf{r}(s)]$ is the continuous version of $U_2(\{\mathbf{r}\})$. When N is increased infinitely, β must be diminished to zero because otherwise U_2 diverges. The way to prevent U_2 from the divergence is to make $\beta/(\Delta s)^2$ remain finite as $\Delta s \to 0$. With this in mind we define β_c by

$$\beta_c = \lim_{\Delta s \to 0} \beta(b'/\Delta s)^2 \tag{3.19}$$

The quantity β_c has the same dimension as β and can be regarded as the excluded-volume strength of a continuous chain.

Transforming eq 3.5 taking eq 3.19 into account, we obtain

$$U_2[\mathbf{r}(s)] = k_B T \left(\beta_c/2b^2\right) \int_0^L \int_0^L \delta(\mathbf{r}(s) - \mathbf{r}(s'))\, ds\, ds' \qquad (3.20)$$

It appears that substitution of eq 3.20 into eq 3.18 completes the derivation of $P[\mathbf{r}(s)]$ for continuous chains. But this is not the case for the following reason.

The restriction $j > i$ to the sum in eq 3.5 is imposed to avoid the untenable condition that each bead in the spring–bead chain interacts with itself. However, this bead self–interaction creeps into eq 3.20 as the integrand corresponding to $s = s'$. Thus eq 3.20 is physically unacceptable, and must be modified so as to be free from self-interaction. Edwards [12] then proposed replacing eq 3.20 by

$$U_2[\mathbf{r}(s)] = k_B T \left(\beta_c/2b^2\right) \int_0^L \int_{\substack{0 \\ |s-s'|>\lambda}}^L \delta(\mathbf{r}(s) - \mathbf{r}(s'))\, ds\, ds' \qquad (3.21)$$

The quantity λ is referred to as the **cut–off length** or simply the cut–off. With the introduction of this length the contact of paired contour points s and s' satisfying $|s - s'| < \lambda$ is prohibited. In other words, the potential energy arising from such contacts is removed from U_2.

Substitution of eq 3.17 and 3.21 into eq 3.18 yields

$$P[\mathbf{r}(s)]\delta\mathbf{r}(s) = D[\mathbf{r}(s)] \exp\left\{-H[\mathbf{r}(s)]/k_B T\right\} \qquad (3.22)$$

where $H[\mathbf{r}(s)]$ is a "Hamiltonian" defined by

$$H[\mathbf{r}(s)]/k_B T = (3/2b) \int_0^L [d\mathbf{r}(s)/ds]^2\, ds$$

$$+ (\beta_c/2b^2) \int_0^L \int_{\substack{0 \\ |s-s'|>\lambda}}^L \delta(\mathbf{r}(s) - \mathbf{r}(s'))\, ds\, ds' \qquad (3.23)$$

which is due originally to Edwards [12]. The model chain characterized by this Hamiltonian is hereafter referred to as the **Edwards continuous chain**. It is specified by four parameters L, b, β_c, and λ, but b can be hidden by the transformations

$$\mathbf{c}(s) = (3/b)^{1/2}\, \mathbf{r}(s) \qquad (3.24)$$

$$v = (3)^{3/2}\, \beta_c\, b^{-7/2} \qquad (3.25)$$

13

The resulting H is

$$H[\mathbf{c}(s)]/k_{\mathrm{B}}T = (1/2)\int_0^L [d\mathbf{c}(s)/ds]^2\, ds$$

$$+ (v/2)\int_0^L \int_{\substack{0 \\ |s-s'|>\lambda}}^L \delta(\mathbf{c}(s)-\mathbf{c}(s'))\, ds\, ds' \qquad (3.26)$$

The parameter v is also named the excluded-volume strength in the subsequent presentation, though it no longer has the dimension of volume.

Finally, we note that eq 3.26 holds as it stands for the Edwards continuous chain in the space of $d\ (\leq 4)$ dimensions if $\mathbf{c}(s)$ and v are redefined as

$$\mathbf{c}(s) = (d/b)^{1/2}\mathbf{r}(s) \qquad (3.27)$$

$$v = (d)^{d/2}\beta_c b^{-(4+d)/2} \qquad (3.28)$$

In what follows, H defined by either eq 3.23 or 3.26 is called the **Edwards Hamiltonian**.

3.3 Coarse–graining

Actual polymer molecules are microscopically discrete in structure. Introduction of a cut–off length λ gives the continuous chain model some kind of discreteness. Hence it is not only necessary to remove the self–interaction but also serves to make the model somewhat realistic. The point is that there exists no guiding principle for the choice of λ. In fact, we may assign any positive value to this length as long as it is smaller than the entire chain length. This flexibility is not characteristic of continuous chains but also involved in either the random flight chain or the spring–bead chain. In the latter the segment length or the spring length can be chosen at will.

We call a continuous chain with no cut–off the primary chain (some authors [11], [13] use the term bare instead of primary). If a cut–off λ is introduced, only pairs of contour points of the primary chain which are separated by more than λ are allowed to interact. In other words, the chain conformations containing loops shorter than λ are removed from those realizable by the primary chain. This smearing of chain loops below a certain size is an example of what is usually called **coarse–graining** in physics. With increasing λ, the chain is more coarse–grained and intrachain interactions can occur only between contour points more separated along the chain.

The Hamiltonian $H[\mathbf{r}(s)]$ changes as the primary chain is coarse–grained, because it depends on λ. Global properties of a continuous chain being governed by $H[\mathbf{r}(s)]$, they should vary with λ. An ingenuous formulation of this variation is the key of the renormalization group theory, which has recently opened a novel way to approach global polymer behavior in solution. It is noteworthy that the flexible choice of λ finds its advantage in this technique. An elementary account of the normalization group theory as applied to dilute polymer solutions is given in Chapter 3.

Coarse–graining is a well–known maneuver in theoretical physics. Generally it is successfully used to extract the behavior of many–body systems which reflects the large scale structure or order formed by their constituent units (atoms, molecules, etc.). Interestingly, a single polymer molecule can be treated as a many–body system unless the degree of polymerization is too low. This fact makes it possible to approach its global behavior by coarse–graining. However, coarse–graining is not a new concept in the polymer community. It was already introduced as early as the 1930s when Kuhn replaced actual polymer molecules by random flight chains, but polymer physical chemists paid little attention to what happens to chain behavior when the segment length of such a model chain is varied.

References

1. O. Kratky and G. Porod, Rec. Trav. Chim. Pay–Bas **68**, 1106 (1948).
2. A. Miyake and Y. Hoshino, Rep. Prog. Polym. Phys. Jpn **18**, 69 (1975).
3. W. Burchard, Brit. Polym. J. **3**, 209 (1971).
4. H. Yamakawa and M. Fujii, J. Chem. Phys. **64**, 5222 (1976).
5. H. Yamakawa and J. Shimada, J. Chem. Phys. **68**, 4722 (1978).
6. W. G. McMillan and J. E. Mayer, J. Chem. Phys. **13**, 276 (1945).
7. N. Saito, "Kobunshi Butsuri (Polymer Physics)," (Rev. Ed.), Syokabo, Tokyo, 1967, Chap. 4.
8. Y. Oono, J. Phys. Soc. Jpn **41**, 228 (1976); T. Oyama and Y. Oono, J. Phys. Soc. Jpn **42**, 1348 (1977); Y. Oono and K. F. Freed, J. Phys. A: Math. Gen. **15**, 1931 (1982).
9. H. Yamakawa, "Modern Theory of Polymer Solutions," Harper & Row, New York, 1971, Chap. 2.
10. P. J. Flory, "Statistical Mechanics of Chain Molecules," John Wiley & Sons, New York, 1969.
11. K. F. Freed, "Renormalization Group Theory of Macromolecules," John Wiley & Sons, New York, 1987.
12. S. F. Edwards, Proc. Phys. Soc. (London) **85**, 613 (1965); see also K. F. Freed, Adv. Chem. Phys. **22**, 1 (1972).
13. Y. Oono, Adv. Chem. Phys. **61**, 301 (1985).

Chapter 2   Two–Parameter Theory

1.   Chain Dimensions

1.1   End Distance Expansion Factor

We begin with discussing excluded–volume effects on $\langle R^2 \rangle$, the mean–square end–to–end distance of an isolated polymer molecule. As usual we introduce a dimensionless quantity α_R defined by

$$\alpha_R{}^2 = \langle R^2 \rangle / \langle R^2 \rangle_\theta \qquad (1.1)$$

and called the **end distance expansion factor.** Here the subscript θ signifies the θ condition defined in Chapter 1. If the molecule is coarse–grained to a spring–bead chain model, $\langle R^2 \rangle_\theta$ is represented by (see eq 1–3.10)[1]

$$\langle R^2 \rangle_\theta = N a^2 \qquad (1.2)$$

where N is the total number of springs and a the mean spring length.

The general expression for $\langle R^2 \rangle$ can be written by using eq 1–3.3 through eq 1–3.7, yielding

$$\langle R^2 \rangle = Q^{-1} \iint \mathbf{R}^2 \, \delta(\mathbf{R} - \mathbf{r}_N)$$
$$\times \{ \exp [-3 \sum_{k=1}^{N} \mathbf{R}_k{}^2 / (2a^2) - \beta \sum_{n>m}^{N} \sum^{N} \delta(\mathbf{r}_m - \mathbf{r}_n)] \} \, d\mathbf{R}_k \, d\mathbf{R} \qquad (1.3)$$

where Q is a normalization factor (not explicitly written here for simplicity) and

$$\mathbf{r}_N = \sum_{j=1}^{N} \mathbf{R}_j \qquad (1.4)$$

A variety of approximations have been proposed to calculate eq 1.3. All the analytical calculations so far reported have actually dealt not with discrete chains but with continuous ones in the sense that the sums in eq 1.3 were replaced by appropriate integrals. The Edwards continuous chain version of eq 1.3 is shown to become

---

[1] In quoting eq a of another chapter X, we hereafter denote it by eq X–a.

17

$$\langle R^2 \rangle = Q^{-1} \iint \mathbf{R}^2 \delta(\mathbf{R} - \mathbf{r}(L) + \mathbf{r}(0)) \exp\left[-(3/2b)\int_0^L (d\mathbf{r}/ds)^2\, ds\right.$$

$$\left.- (\beta_c/2b^2)\int_0^L \int_{\substack{0 \\ |s-s'|>\lambda}}^L \delta(\mathbf{r}(s) - \mathbf{r}(s'))\, ds\, ds'\right] D[\mathbf{r}]\, d\mathbf{R} \tag{1.5}$$

with

$$Q = \iint \delta(\mathbf{R} - \mathbf{r}(L) + \mathbf{r}(0)) \exp\left[-(3/2b)\int_0^L (d\mathbf{r}/ds)^2\, ds\right.$$

$$\left.- (\beta_c/2b^2)\int_0^L \int_{\substack{0 \\ |s-s'|>\lambda}}^L \delta(\mathbf{r}(s) - \mathbf{r}(s'))\, ds\, ds'\right] D[\mathbf{r}]\, d\mathbf{R} \tag{1.6}$$

Now we make the substitutions

$$s/L = x, \qquad \mathbf{r}(s)/(Lb)^{1/2} = \hat{\mathbf{r}}(x) \tag{1.7}$$

and use the relation

$$\delta(\mathbf{r}(s) - \mathbf{r}(s')) = (Lb)^{-3/2}\delta(\hat{\mathbf{r}}(x) - \hat{\mathbf{r}}(x')) \tag{1.8}$$

Then, from eq 1.5 and 1.6 and the fact that the continuous chain version of eq 1.2 is given by eq 1–3.15, we find

$$\langle R^2 \rangle = \langle R^2 \rangle_\theta\, F(z, \lambda/L) \tag{1.9}$$

or by eq 1.1

$$\alpha_R{}^2 = F(z, \lambda/L) \tag{1.10}$$

Here, F is a universal function and the variable z is a dimensionless quantity defined by

$$z = (1/2\pi)^{3/2} v L^{1/2} \tag{1.11}$$

or

$$z = (3/2\pi)^{3/2} \beta_c b^{-7/2} L^{1/2} \tag{1.12}$$

with v given by eq 1–3.25. In this book, z is called the **excluded–volume variable**. As will be seen below, this variable plays the most fundamental role in polymer solution theory.

1.2 Two–parameter Theory

As mentioned in Section 3.2 of Chapter 1, the Hamiltonian H of the Edwards continuous chain depends on L, b, β_c, and λ. Hence, any equilibrium property of this chain in dilute solutions ought to be expressed as a function of these four parameters. Interestingly, eq 1.9 indicates that actually $\langle R^2 \rangle$ is governed by three combined parameters Lb, z, and λ/L formed from the four (for continuous chains $\langle R^2 \rangle_\theta = Lb$ according to eq 1-3.15). Hence, if $\lambda/L \ll 1$, $\langle R^2 \rangle$ essentially depends on Lb and z (or α_R depends only on z). We can expect that the same holds for any equilibrium behavior of the Edwards continuous chain in dilute solutions. Any theory of dilute polymer solutions fitting to this expectation is called the **two–parameter theory**. We may say that it is a theory of the Edwards continuous chain subject to the condition $\lambda/L \ll 1$. The use of the Edwards Hamiltonian combined with this condition is hereafter referred to as the **two–parameter approximation**.

The initial prescription of the two–parameter theory was made by Flory [1], and in his 1971 book Yamakawa [2] summarized a variety of efforts made to develop it by the end of the 1960s. The present chapter describes more recent contributions to this theory and related experimental studies.

1.3 Perturbation Theory

Assuming that $F(z, \lambda/L)$ can be expanded in powers of z in the vicinity of $z = 0$, we write eq 1.10 as

$$\alpha_R{}^2 = 1 + C_1 z + C_2 z^2 + C_3 z^3 + \cdots \tag{1.13}$$

where the C_m are functions of λ/L. Reported calculations of these coefficients went to the limit $\lambda/L \to 0$. Although, as pointed out by Muthukumar and Nickel [3], the process to this limit needs a deliberate treatment, they paid little attention to it and yet derived correct results.

Teramoto [4] in 1951 was the first to report $C_1 = 4/3$, and this finding was confirmed independently by several subsequent authors [5–8]. In 1955, Fixman [9] found $C_2 = (28\pi/27) - (16/3) = -2.075$, but it was in 1967 that C_3 was reported to be 6.459 by Yamakawa and Tanaka [10]. Calculations of C_m become increasingly laborious so that these coefficients for $m \geq 4$ remained unknown until recently. By a very sophisticated analysis Muthukumar and Nickel [3] in 1984 succeeded in obtaining the following values:

$$
\begin{array}{ll}
C_1 = 1.333, & C_2 = -2.075 \\
C_3 = 6.2969, & C_4 = -25.057 \\
C_5 = 116.13, & C_6 = -594.72
\end{array}
\tag{1.14}
$$

19

A small difference in this C_3 from that of Yamakawa and Tanaka [10] had already been pointed out by Barrett [11].

The absolute values of C_m increase so rapidly with m that the z series in eq 1.13 would converge only for very small values of z or might be only asymptotic. This finding seems tragic if we consider the great amount of computational work needed for obtaining the above numerical results.

The C_m values of Muthukumar and Nickel correspond to the limit $\lambda/L \to 0$. Some authors [12, 13] suspected that the C_m for $m > 1$ might diverge as $\ln(\lambda/L)$ in this limit, but, importantly, Muthukumar and Nickel [3] proved that the logarithmic terms cancel with one another and do not affect the final results.

1.4 Asymptotic Forms

Early workers already felt that the z series for $\alpha_R{}^2$ might be of use only in the very vicinity of $z = 0$. It was then natural that a considerable interest arose in deriving closed approximate expressions which may be used to describe α_R up to larger values of z. Various methods have been proposed, and those published by the end of the 1960s are summarized and discussed in detail by Yamakawa [2]. The earliest one is the historic mean–field theory of Flory [14],[2] which gives

$$\alpha_R{}^5 - \alpha_R{}^3 = 2.60\,z \tag{1.15}$$

This relation predicts that $\alpha_R{}^5$ is asymptotically proportional to z for large z. Expressions for $\alpha_R(z)$ (and also $\alpha_S(z)$ which appears below) which give this type of asymptotic behavior are often called **fifth–power type**. Another feature of Flory's theory is that it permits evaluation of the proportionality constant (called the prefactor or front factor) in the asymptotic relation $\alpha_R{}^5 \sim z$. In a paper of 1960, Kurata et al. [15] criticized eq 1.15 by deriving an expression of third–power type on the basis of an ellipsoid polymer coil model. Although their conclusion was not ultimately accepted, this paper played an important role because it triggered intensive debates over the chain length dependence of polymer dimensions in the years to come.

Assuming the asymptotic form of α_R for $z \to \infty$ to be

$$\alpha_R{}^\gamma = A\,z \tag{1.16}$$

we ask ourselves what would be the correct values of the exponent γ and the front factor A. First we note that Flory's equation 1.15 predicts $\gamma = 5$ and $A = 2.60$. When eq 1.15 is expanded in powers of z we find 2.60 for the

---

[2]In mean–field approaches, the intrachain interaction energy is estimated neglecting segment density fluctuations due to chain connectivity. Thus, each segment sees other segments as if they were distributed uniformly in the polymer coil.

coefficient C_1 in eq 1.13, in disagreement with the exact value $4/3$. To make up for this defect in Flory's equation Stockmayer [16] proposed modifying 2.60 in eq 1.15 to $4/3$. Equation 1.15 so changed is called the modified Flory equation.

Extensive work in the 1960s seeking closed approximate expressions for $\alpha_R(z)$, as summarized in Yamakawa's book [2], led to confusion, predicting three values 3, 4, and 5 for γ. In this connection we must note that $\langle R^2 \rangle$ is not directly measurable so that we were unable to judge by experiment which of these values was closer to the truth. Reiss [17], Edwards [18], and Yamakawa [19] attempted a more exact evaluation of γ and all arrived at $\gamma = 5$ (see Ref. [2] for a review). However, this conclusion of the 1960s did not come to be accepted widely.

A breakthrough came in the 1970s when de Gennes [20] suggested that the renormalization group theory should be powerful for elucidating global polymer behavior in dilute solutions. In fact, this theory opened the door to an analytical approach to γ and has led to a number of interesting results. For example, using a highly sophisticated technique, le Guillou and Zinn–Justin [21] showed γ to be 5.68, though they were unable to calculate the front factor A. This γ value, now considered the most accurate, importantly indicated that the correct asymptotic behavior of α_R was not fifth–power type. It should be noted that none of the previous theories had predicted γ larger than 5. The renormalization group theory more recently developed by Douglas and Freed[22] (explained in Chapter 3) allows A as well as γ to be estimated approximately. Thus, they obtained 5.45 for γ, which is also larger than 5 and can be compared favorably with the value of le Guillou and Zinn–Justin.

Though these new theoretical values of γ are clearly different from 5, the differences are only about 10%, indicating that the mean–field theory of Flory nearly hit the target. This means that eq 1.15 is by no means an unreasonable approximation for a discussion of excluded–volume effects in polymer solutions. The problem is whether 2.60 in it is adequate or not. It is also apparent that the various closed approximate equations of third or fourth–power type derived in the 1960s now have lost their significance.

In view of the overwhelming difficulties involved in analytical approaches to chain dimensions, many computer simulations of chain conformations have been undertaken since the pioneering work of Wall and collaborators [23] in the 1950s. Excellent review articles summarizing advances in this type of "experiment" have been published. Here we only touch upon the results which are considered essential for the context of our discussion.

As for α_R, which is the quantity most thoroughly investigated by computer simulations, it is pertinent to mention the results obtained by Domb and collaborators [24–26] using the Domb–Joyce discrete chain model [27] (which is a random flight chain with a bond angle constraint and some special bead–bead interaction). They enumerated the total possible chain conformations generated on various three–dimensional lattices (simple cubic, face–centered cubic, diamond, etc.), computed α_R for the chains with N up to 20, and inferred its N dependence for larger N by extrapolation. In this way they concluded that $\gamma = 5$ would be most probable regardless of the type of lattice. With this result, eq 1.16 can be rewritten

$$\alpha_R{}^2 = A' z^{2/5} \sim N^{1/5} \tag{1.17}$$

Domb et al. [25] found $A' = 1.64$ common to almost all the lattices studied, in support of the expectation (by the two–parameter theory) that α_R should be a universal function of z. This value of A' gives $A = 3.44$, which is about 30% larger than Flory's original value of 2.60 and indicates that Stockmayer's modification [16] of A from 2.60 to 4/3 is meaningless. Some may be reluctant to accept the finding of Domb's school because it is an inference by long extrapolation from data on relatively short chains. However, it should be noted that more recent Monte Carlo simulations on much longer chains by Alexandrowicz [28] yielded a result consistent with it. The author believes that eq 1.17 with $A' = 1.64$ is one of the most important findings in polymer physics, because, as mentioned below, it established a way for evaluating the excluded–volume strength β by experiment.

1.5   Interpolation Formulas

With the known expression for $z \gg 1$ from the above–mentioned computer simulations and that for $z \ll 1$ from perturbation calculations, Domb and Barrett [29] made an interpolation formula for α_R. The latest one [30] proposed after some modifications of the original reads

$$\alpha_R{}^2 = (1 + 20z + 155.54z^2$$
$$+ 591.86z^3 + 325z^4 + 1670z^6)^{1/15} \tag{1.18}$$

which agrees with the perturbation theory up to z^3 when expanded in powers of z. Two more remarks may be in order. Equation 1.18 contains no term with z^5, and it is dominated by the term with z^6 for $z > 1$. The latter means that $\alpha_R{}^2$ as a function of z essentially enters the asymptotic region when z reaches the order of unity. Tanaka [31] illustrated that eq 1.18 fits quite closely typical "experimental" results from computer simulations of previous authors.

Taking for granted that γ is equal to 5 and using the perturbation theory up to z^3, Stockmayer [32] applied the Padé approximant method to obtain

$$\alpha_R{}^5 = 1 + \left(\frac{10z}{3}\right)\frac{1 + 2.953z}{1 + 3.509z} \tag{1.19}$$

22

This gives $A' = 1.51\,(A = 2.81)$, which is not very different from the computer simulation value of 1.64 due to Domb et al. With the C_m values now available up to $m = 6$ we can make a more accurate Padé approximant for $\alpha_R{}^5$. It reads [33]

$$\alpha_R{}^5 = \frac{1 + 11.367z + 36.992z^2 + 31.411z^3}{1 + 8.034z + 12.069z^2} \tag{1.20}$$

This gives $A' = 1.466\,(A = 2.604)$, which happens to be very close to the original Flory value and is further from the value 1.64 of Domb et al. than is Stockmayer's. Recently, Suzuki [34] pointed out that when a closed approximate expression

$$\alpha_R{}^2 = 0.572 + 0.428(1 + 6.23z)^{1/2} \tag{1.21}$$

which was derived by Yamakawa and Tanaka [10] in the 1960s was expanded in powers of z, the coefficients came very close to the values listed in eq 1.14 all up to $m = 6 : C_1 = 1.333, C_2 = -2.075, C_3 = 6.459, C_4 = -25.13, C_5 = 109.5, C_6 = -511.2$. However, eq 1.21 predicts the fourth-power asymptotic behavior, in disagreement with the finding of Domb et al.

1.6 Radius Expansion Factor

The mean–square radius of gyration $\langle S^2 \rangle$ of a spring–bead chain consisting of $N + 1$ beads is expressed by

$$\langle S^2 \rangle = \frac{1}{N+1} \sum_{j=0}^{N} \langle S_j{}^2 \rangle \tag{1.22}$$

where $\langle S_j{}^2 \rangle$ is the mean–square distance between bead j and the center of mass of the chain. Practically, $\langle S^2 \rangle$ is more interesting than $\langle R^2 \rangle$ as a measure of chain dimensions, because it can be determined by scattering techniques. We note that eq 1.22 can be rewritten as [35]

$$\langle S^2 \rangle = \frac{1}{(N+1)^2} \sum_{j>i}^{N} \langle \mathbf{r}_{ij}{}^2 \rangle \tag{1.23}$$

where $\mathbf{r}_{ij}$ is the distance between beads i and j.

We introduce a dimensionless quantity α_S defined by

$$\alpha_S{}^2 = \langle S^2 \rangle / \langle S^2 \rangle_\theta \tag{1.24}$$

and call it the **radius expansion factor**. In the two–parameter approximation this factor becomes a universal function of the excluded–volume variable z, as does the end distance expansion factor. As is well known, the following relation holds for Gaussian chains:

$$\langle S^2 \rangle_\theta / \langle R^2 \rangle_\theta = 1/6 \tag{1.25}$$

On the other hand, typical simulation experiments [23, 26, 36] found that $\langle S^2 \rangle / \langle R^2 \rangle$ for different lattice chains approached values near 0.155 as z increased. Thus, we see that α_S^2 / α_R^2 decreases from unity to about 0.933 as z increases from zero to infinity.

Perturbation calculations show [2] that α_S^2 in the vicinity of $z = 0$ is expressed by

$$\alpha_S^2 = 1 + (134/105)z - 2.082z^2 + \ldots \tag{1.26}$$

The coefficients of z and z^2 are due to Zimm et al. [37] and Yamakawa et al. [38], respectively, but those of the higher z terms still remain unknown. Combining eq 1.26 with the corresponding z series for α_R^2 mentioned earlier, we obtain

$$\alpha_S^2 / \alpha_R^2 = 1 - 0.057z + 0.069z^2 + \ldots \tag{1.27}$$

Domb and Barrett [29] made an interpolation formula for α_S^2 / α_R^2 which agrees with eq 1.27 for small z and converges to the limiting value of 0.933 for large z. It was corrected by Suzuki [34] to give

$$\alpha_S^2 / \alpha_R^2 = 0.933 + 0.067 \exp(-0.85z - 0.67z^2) \tag{1.28}$$

If eq 1.18 is used with eq 1.28, it is possible to calculate α_S as a function of z over the entire range of positive z. We call the combined equation (not written explicitly here) the Domb–Barrett interpolation formula for the radius expansion factor.

In Figure 2–1, in the form of α_S^2 plotted against $z^{2/5}$, this formula is compared with the original Flory equation (eq 1.15 with the subscript R replaced by S), the Yamakawa–Tanaka equation ($\alpha_S^2 = 0.541 + 0.459(1 + 6.04z)^{0.46}$ [10]), and the data from simulation experiments on four different lattices [23, 26, 39–42]. It can be seen that the D–B curve gives the closest agreement with the plotted points. This curve becomes essentially linear, i.e., enters the asymptotic region, for $z > 2$. It is interesting that the asymptotic relation between α_S and z is reached at such a relatively small value of z. Equally interesting, the asymptotic linear part of the D–B curve, indicated by a dashed line in Figure 2–1, passes through the coordinate origin. This feature is due to the lack of the term with z^5 in eq 1.18. The dashed line can be expressed by

$$\alpha_S^2 = 1.53z^{2/5} \tag{1.29}$$

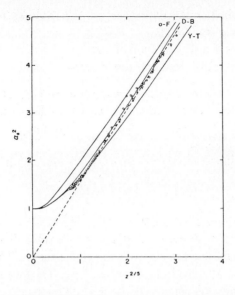

Fig. 2–1. Relations between α_S^2 and $z^{2/5}$ for four different lattice chains and by three approximate formulas. The plotted points are for diamond lattice (Wall and Erpenbeck [23], Suzuki [41], Wall and Hioe [39], Jurs and Reissner [40]), simple cubic lattice (Kron and Ptitsyn [42], Domb and Hioe [26]), face–centered cubic lattice (Domb and Hioe [26]), and body–centered cubic lattice (Domb and Hioe [26]). Solid curves, original Flory equation (o–F), Yamakawa–Tanaka equation (Y–T), and Domb–Barrett interpolation equation (D–B). Dashed line, fitting the asymptotic linear part of the D–B curve. It is important to observe that the dashed line passes through the coordinate origin.

The absence of an additional constant in this asymptotic relation provides an important criterion in testing the validity of eq 1.29 with experimental data.

1.7 Evaluation of Excluded–volume Strength

1.7.1 Excluded–volume Variable for the Spring–bead Chain

The excluded–volume variable z introduced in eq 1.11 has been defined for the Edwards continuous chain with length L and excluded–volume strength v. Since all the theoretical expressions presented above for α_R and α_S are concerned with such chains, the variable z appearing in them is the one so defined. However, though identical notation was used, the z that appeared in

25

the discussion on computer simulations is not the same as that, because it is concerned with discrete lattice chains. For clarity we should have designated the discrete chain z by a different symbol, say z'.

For the spring–bead chain the usual definition of z' is

$$z' = (3/2\pi a^2)^{3/2} \beta \, (M/m)^{1/2} \tag{1.30}$$

where M is the molecular weight of the chain, m the molar mass of one bead, and a the spring length. With eq 1–3.13, we may rewrite eq 1.30 as

$$z' = (3/2\pi)^{3/2} (b')^{-7/2} \beta \, (b'/\Delta s)^2 L^{1/2} \tag{1.31}$$

where we have used the approximation $L = \Delta s \, [(M/m) - 1] \sim \Delta s \, (M/m)$ (Δs is the contour length of one spring), which is valid for large N. In the limit $\Delta s \to 0$ with L fixed constant, with the aid of eq 1–3.19 and 1–3.25, the right–hand side of eq 1.31 reduces to that of eq 1.11. Thus we see that z defined by eq 1.11 is the continuous chain version of z'. It is customary to equate z in the theoretical expressions for continuous chains to z' when experimental data on actual polymers are analyzed by use of those expressions. This convention is adopted throughout the subsequent discussion. In other words, no distinction is made between z and z'.

In terms of a combined parameter B defined by

$$B = \beta/(m^{1/2} a^3) \tag{1.32}$$

eq 1.30 can be written

$$z = (3/2\pi)^{3/2} B \, M^{1/2} \tag{1.33}$$

In this expression, M is a measurable quantity (we tacitly consider that a given polymer and its model chain should have the same molar mass), but B is not so since m, a, and β are model constants (i.e., these depend on the extent to which the molecule is coarse–grained). Thus, z is also model–dependent. This means that either B or z cannot be evaluated directly by experiment. Their estimation requires theoretical expressions which relate experimentally measurable quantities to z. Especially when we are concerned with good solvent systems, we need to have them valid up to sufficiently large z. At present, the best available to us is the asymptotic relation given by eq 1.29, though it is not exact.

1.7.2 Methods

As mentioned above, eq 1.29 holds for $z > 2$ and hence for $\alpha_S{}^2 > 2$ at least. Thus, when measured values of α_S are plotted against $M^{1/5}$ (note that $z \sim M^{1/2}$), the data points should fall on a straight line passing through the coordinate origin in the region of $\alpha_S{}^2 > 2$. The slope K of the line can be used to evaluate B from

$$B = (2\pi/3)^{1.5}(K/1.53)^{2.5} \tag{1.34}$$

which follows from eq 1.29 and 1.33.

The above method cannot be applied when the available data for α_S are limited to those for $\alpha_S{}^2 < 2$, as in the case where we deal with low–molecular–weight polymers or poor solvent systems. In such a case, only curve–fitting is practical for the estimation of B. Thus we prepare two log–log graphs, drawing on one the $\alpha_S{}^2$ vs. z curve graphing the Domb–Barrett interpolation formula, and plotting on the other the measured data (under fixed solvent conditions) of $\alpha_S{}^2$ against $M^{1/2}$. Then, we shift the data points on the latter horizontally until they are most satisfactorily superimposed on the curve of the former. By eq 1.33 the amount of shift is equal to $(3/2)\log(3/2\pi) + \log B$. This relation allows B to be estimated.

Miyaki et al. [43] and Miyaki [44] tested these methods with data on several polymer + solvent systems [43, 45–48]. Figure 2-2 illustrates the plots of $\alpha_S{}^2$ against $M^{1/5}$ constructed from these data. As expected, the data points for $\alpha_S{}^2 > 2$ are fitted by solid lines which pass through the coordinate origin. Miyaki [44] (see also Ref. [49]) also showed that his data on polystyrene (PS) in cyclohexane (CH) near the θ temperature can be analyzed by the above–mentioned curve–fitting method. In Figure 2-3, the $\alpha_S{}^2$ data for PS in benzene and CH are plotted against z calculated by using the B values determined by Miyaki [44]. The solid, dashed, and dot–dash lines show the Domb–Barrett interpolation equation, the Yamakawa–Tanaka equation, and the original Flory equation, respectively. Of the three, the solid line fits the data points most accurately. Suzuki [34] showed that his empirical equation

$$\alpha_S{}^2 = 0.609 + 0.391(1 + 6.53z)^{1/2} \tag{1.35}$$

gives an equally good fit to the data of Figure 2-3, but this equation is not the fifth–power type. Anyway, the result shown in Figure 2-3 may be taken as excellent evidence for the two–parameter approximation to chain dimensions of flexible linear polymers and also as a triumph of the simulation studies by Domb's school.

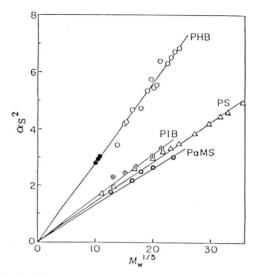

Fig. 2–2. Check of asymptotic relation 1.29 with typical data. $\triangle$, PS in benzene [43, 44]; $\circ$, $\bullet$, PHB in TFE [45, 46]; $\oplus$, PIB in CH [47]; $\odot$, PαMS in toluene [48]. PS = polystyrene, PHB = poly (D-β-hydroxybutyrate), PIB = poly (isobutylene), PαMS = poly (α-methyl-styrene). TFE = trifluoroethanol, CH = cyclohexane. The slope of each indicated line is proportional to $B^{2/5}$. The value of B for PHB in THF is the largest of those reported so far for polymer + solvent systems.

1.7.3 Remarks

As noted above, B is a model parameter associated with the spring–bead chain. Its constituting parameters a, m, and β vary with the degree of coarse-graining of the actual polymer, but β depends also on solvent quality. Hence it is meaningful to compare the values of B for a given polymer species under different solvent conditions, but not to compare those for different polymers in a given solvent. Being model–dependent, it is misleading to expect that B may give quantitative information about the monomer–monomer (solvent–mediated) interaction in actual polymers or to attempt to calculate B from knowledge on the structure of the polymer molecule and the potential between its repeat units and the solvent.

The excluded–volume variable z for a given polymer can be varied experimentally by changing either M or solvent condition. The zero of z can be obtained in a θ solvent, while a larger z is made available by choosing a better solvent (which gives a larger B) as well as a higher–molecular–weight polymer

28

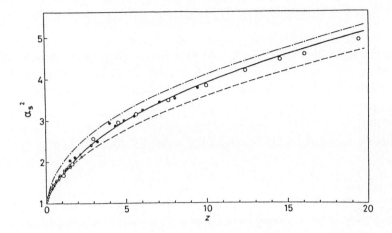

Fig. 2–3. $\alpha_S{}^2$ for polystyrene as a function of z. Points, experimental data on benzene and CH solutions [44]. Solid line, D–B interpolation formula. Dashed line, Yamakawa–Tanaka equation. Dot–dash line, original Flory equation.

sample. Miyaki et al. [43] extracted a polystyrene fraction with M_w as high as 6×10^7 and measured its $\langle S^2 \rangle$ in benzene and CH. This is probably the highest molecular weight that could be measured directly on a synthetic polymer. Usually, polymers with M of the order of 10^7 may be called ultrahigh. As z increases, the beads of a spring–bead chain exclude more strongly with one another, and in the limit $z \to \infty$, they virtually no chance of forming binary clusters. This extreme state is sometimes referred to as self–avoiding.

1.8   Mazur Distribution Function for End Distance

Because of the great difficulty in analytical calculation of $G(\mathbf{R})$, the distribution function for the end distance $\mathbf{R}$ of a polymer chain, many attempts have been made to find closed approximate expressions for this function. Typical of such expressions is the Mazur function (or the generalized Domb–Gillis-Wilmers function) deduced from computer simulations [50]. Its form is

$$W(\mathbf{R}) \equiv 4\pi \mathbf{R}^2 G(\mathbf{R}) = C\mathbf{R}^s \exp[-(\mathbf{R}/\sigma)^t] \tag{1.36}$$

where C is a normalized factor, and s, t, and σ are adjustable parameters. It can be shown that C is given by

$$C^{-1} = (\sigma^{s+1}/t)\Gamma((s+1)/t) \tag{1.37}$$

where Γ denotes the gamma function. The parameter σ is related to $\langle R^2 \rangle$ by

$$\sigma^2 = \langle R^2 \rangle \, \Gamma((s+1)/t)/\Gamma((s+3)/t) \tag{1.38}$$

Hence the right–hand side of eq 1.36 may be regarded as a function of three parameters $s, t,$ and $\langle R^2 \rangle$. Domb et al. [51] reduced the number of parameters to two by putting $s = t$, but Mazur [50] treated s and t as independent. However, as explained below, there is a theoretically expected relation between these two parameters.

Equation 1.16 can be put into a more frequently used form as

$$\langle R^2 \rangle \sim N^{2\nu} \quad (z \gg 1) \tag{1.39}$$

where the exponent $\nu (= (1+\gamma)/2\gamma)$ is 0.60 in the mean–field theory of Flory [14] and 0.588 in the renormalization group theory of le Guillou and Zinn–Justin [21]. Fisher [52] proved that ν is related to the parameter t by

$$t = 1/(1-\nu) \tag{1.40}$$

Moreover, des Cloizeaux [53] derived the relation

$$s = 2 + 0.168 \, (1/\nu) \tag{1.41}$$

Thus we are not allowed to treat t and s as independent, and we get t = 2.43 and s = 2.29 for $\nu = 0.588$.

Summarizing, we may conclude that the Mazur function for $G(\mathbf{R})$ depends only on a single parameter $\langle R^2 \rangle$, because there is little allowance for the value assignable to ν. In this connection it is instructive to recall that $W(\mathbf{R})$ for Gaussian chains depends only on $\langle R^2 \rangle_\theta$.

The Mazur function can be used to calculate $\langle R^{-1} \rangle$ and $\langle R^{2n} \rangle$ $(n = 2, 3, \dots)$, but its usefulness extends no more than that. Calculation of such important global polymer properties as mean–square radius of gyration, particle scattering function, intrinsic viscosity, and diffusion coefficient needs an analytical expression for $W_{ij}(\mathbf{R})$, the distribution function for the distance $\mathbf{R}$ between specified beads i and j. We find no a priori reason for this function to be approximated by the Mazur function. Knowledge about W_{ij} has increased in recent years mainly on the basis of computer simulation experiments. Some of the typical results from such work are described below.

1.9 Distribution of Intrachain Distance

For simplicity the function $W_{ij}(\mathbf{R})$ is referred to below as the distribution of intrachain distance $\mathbf{R}$. We consider the following three cases:

Case I: Beads i and j are at the ends of the chain

Case II: Bead i is at one end of the chain and bead j at any point inside the chain

Case III: Beads i and j are both at any points inside the chain

Part of a chain between a pair of its beads is called the central chain, and each of the remaining two subchains the end chain. Thus we have only a central chain in case I, one end chain in case II, and two end chains in case III.

From computer simulation experiments with as short a random flight chain as $N = 15$ Domb et al. [51] suggested that the same form as eq 1.36 should be valid for the distribution of intrachain distance. Baumgärtner [54] performed Monte Carlo calculations on a much longer random flight chain of $N = 160$, though limited to case II where bead j is situated at the midpoint of the chain and to case III where beads i and j are at one third of the chain from its ends. He confirmed the suggestion of Domb et al., with the following values obtained for s and t:

Case	s	t
I	2.270 ± 0.006	2.44 ± 0.05
II	2.55 ± 0.06	2.6 ± 0.15
III	2.9 ± 0.1	2.48 ± 0.05

These values for case I are very close to the above–mentioned "theoretical" ones for $\nu = 0.588$ ($s = 2.29, t = 2.43$). The t values for the three cases are nearly coincident and close to 2.50 which can be obtained from Fisher's relation (eq. 1.40) and Flory's ν value of 0.6. On the other hand, the s values indicate a significant dependence on the location of the central chain, being larger for the central chain situated in a more internal part of the entire chain. Summarizing, we see that the distribution of intrachain distance $\mathbf{R}$ is described by the same form as the Mazur function but varies with the location of the central chain approximately through the position–dependent parameter s.

The central chains treated by Baumgärtner are relatively long. Redner [55] did computer simulation experiments to investigate the behavior of shorter central chains, but his random flight chain was as short as $N = 15$.

Finally, we touch upon des Cloizeaux's theory [53], which dealt with $W_{ij}(\mathbf{R})$ in the limit of $\mathbf{R} \rightarrow 0$ for subchains in an infinitely long chain. It derived $s = 2.273, 2.46, 2.71$ for cases I, II, III, respectively. The s value for case I is very close to, but those for cases II and III are somewhat smaller than

Baumgärtner's values. It is splendid that des Cloizeaux was able to predict the increase in s accompanying the change I → II → III.

1.10 Mean Square and Mean Reciprocal of Intrachain Distance

1.10.1 Theory of Barrett

For an unperturbed (Gaussian) spring–bead chain the mean square $\langle r_{ij}^2 \rangle$ and mean reciprocal $\langle r_{ij}^{-1} \rangle$ of the distance r_{ij} between its beads i and j are equal to $a^2|i-j|$ and $(6/\pi a^2)^{1/2}|i-j|^{-1/2}$. These are well–known facts and are cited here without proof (see Ref. [2]). How are these averages expressed when the chain is perturbed by intrachain excluded–volume interactions ? This problem was first brought us by Peterlin [56] in 1955. Solution of it needs information about $W_{ij}(\mathbf{R})$, but, as mentioned above, work on this function is as yet in the process of development.

Very recently, Barrett [57] has derived closed approximate expressions for the above two averages on the basis of Monte Carlo simulation data. His idea is as follows.

We define dimensionless factors α_{ij}^2 and α_{ij}^{-1} by

$$\alpha_{ij}^2 = \langle r_{ij}^2 \rangle / a^2 |i-j| \tag{1.42}$$

$$\alpha_{ij}^{-1} = \langle r_{ij}^{-1} \rangle (\pi/6)^{1/2} a \, |i-j|^{1/2} \tag{1.43}$$

These factors are unity for Gaussian chains and depend on the total chain length as well as where beads i and j are situated on the chain. In place of i and j as well as N we may use the new variables n, x, and x' defined by

$$n = j - i \quad (j > i), \quad i = nx', \quad N = n(1 + x + x') \tag{1.44}$$

Then, the length of the central chain is given by na, and the lengths of the two end chains by nxa and $nx'a$. Further, we introduce y defined by

$$y = x'/x \tag{1.45}$$

and assume $0 < y < 1$ without loss of generality. The three cases considered in the preceding section are described in terms of n, x, and x' as follows. Case I: $x = x' = 0$, Case II: $x' = 0$, and Case III: $x \neq 0, x' \neq 0$.

In what follows, α_{ij}^m ($m = -1, 2$) is denoted by $\alpha(m; x, y)$, and the excluded–volume variable for the entire chain by z as before and that for the central chain by Z. Then by eq 1.33 we have

$$Z = (3/2\pi)^{3/2} B M'^{1/2} \tag{1.46}$$

32

where M' is the molecular weight of the central chain.

Long ago Teramoto et al. [58] calculated $C_1(2)$ in the expansion

$$\alpha(2; x, y) = 1 + C_1(2)Z + \ldots \qquad (1.47)$$

and recently Barrett [57] succeeded in calculating $C_1(-1)$ in the expansion

$$\alpha(-1; x, y) = 1 + C_1(-1)Z + \ldots \qquad (1.48)$$

These coefficients are functions of x and y, and the reader who is interested in their actual forms is advised to consult the original papers.

The average $\langle r_{ij}{}^m \rangle$ is regarded as a function of m, n, x, and y. Barrett [57] assumed that when n is increased with x and y fixed constant, i.e., the total chain length is increased with a fixed partition of the central and end chains, this function $X(m; n, x, y)$ behaves as

$$X(m; n, x, y) = A(m; x, y)n^{\delta(m)} \quad (n \to \infty) \qquad (1.49)$$

If this assumption is accepted, it turns out that $\delta(m)$ can be obtained from the end distance distribution function $W(\mathbf{R})$ for an isolated central chain. Barrett chose the Domb–Gillis–Wilmers function, i.e., the Mazur function for the case $s = t$, for $W(\mathbf{R})$. This choice gives

$$\delta(m) = m\nu \qquad (1.50)$$

where ν is the parameter defined by eq 1.39. Barrett took Flory's value of 0.6 for ν. Then $\delta(2) = 1.2$ and $\delta(-1) = -0.6$.

Barrett carried out Monte Carlo simulations on chains with $y = 0$ and 1 and analyzed the data by eq 1.49 to evaluate $A(m; x, 0)$ and $A(m; x, 1)$ for $m = 2$ and -1. The results were independent of the lattices studied as illustrated in Figure 2–4. It can be seen from the figure that for either m both $A(m; x, 0)$ and $A(m; x, 1)$ become constant for $x > 2$, suggesting that when the central chain is long, its X is not affected by the segments of the flanking end chain which are separated by more than twice the central chain from its ends.

On the basis of the data shown in Figure 2–4 combined with the perturbation calculations for $\alpha(2; x, y)$ and $\alpha(-1; x, y)$ Barrett made the following interpolation formulas:

$$\alpha(2; x, y) = \left\{ 1 + 5C_1(2)Z \left[1 + \frac{1.89 + 2x(1 + y)}{1 + x(1 + y)} Z \right] \right\}^{0.2} \qquad (1.51)$$

$$\alpha(-1; x, y) = \left\{ 1 - 10C_1(-1)Z \left[1 + \frac{2.1 + 4x(1 + y)}{1 + x(1 + y)} Z \right] \right\}^{-0.1} \qquad (1.52)$$

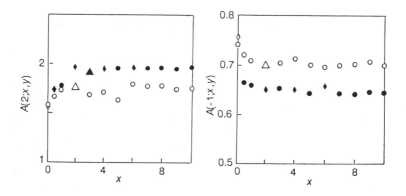

Fig. 2–4. Dependence of $A(2; x, y)$ and $A(-1; x, y)$ on x in the two limiting cases $y = 0$ and $y = 1$. $\circ, \bullet$: sc lattice; $\diamond, \blacklozenge$: bcc lattice; $\triangle, \blacktriangle$: fcc lattice. Filled and unfilled points are for $y = 1$ and $y = 0$, respectively.

Figure 2–5 shows the Z dependence of $\alpha(2; x, y)$ and $\alpha(-1; x, y)$ computed from these equations in the limit of large x for cases I, II, and III. The curves suggest that when the end chains are much longer than the central chain, the excluded–volume effect on the latter may not undergo a significant influence from the former. In other words, subchains in a long chain which themselves are relatively long behave approximately as if they are not flanked by end chains. Although this conclusion probably breaks down for short subchains, the important work of Barrett seems to justify to some extent the sometimes used approximation of replacing $W_{ij}(\mathbf{R})$ by the Mazur function $W(\mathbf{r}_{ij})$.

1.10.2   Applications

Another approximation that has often been employed in previous calculations is to replace $\alpha(-1; x, y)$ by $[\alpha(2; x, y)]^{-1/2}$. In fact, this approximation has already been used in writing eq 1.49 with respect to the exponent but not to the prefactor. Using eq 1.51 and 1.52, Barrett [57] compared $P^{-1/2}$ and Q in the asymptotic expressions $\alpha(2; x, y) = PZ^{0.4}$ and $\alpha(-1; x, y) = QZ^{-0.2}$ valid for $Z \gg 1$. The first, second, and third rows of Table 2–1 correspond, respectively, to the limiting cases of no end chains, one end chain of infinite length, and two end chains of infinite length. In each, we see $Q < P^{-1/2}$, but the difference stays less than 10 %, suggesting that the above-mentioned approximation is not so bad for chains with large Z.

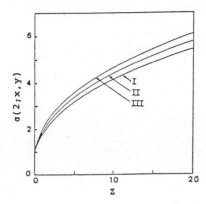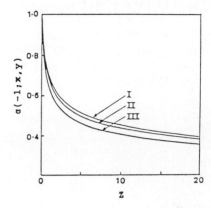

Fig. 2–5. Approximations to $\alpha(2; x, y)$ and $\alpha(-1; x, y)$ in the limit $x \gg 1$ for cases I, II, and III.

The continuous chain version of eq 1.23 gives for the radius expansion factor α_S

$$\alpha_S{}^2 = (3/8) \int_{-1}^{1} \int_{-1}^{1} \alpha(2; x, y) |\xi - \eta| \, d\xi \, d\eta \tag{1.53}$$

where ξ and η are related to x and y by

$$x = (1 - \eta)/|\xi - \eta| \tag{1.54}$$

$$y = \xi/(1 - \eta) \tag{1.55}$$

Substitution of eq 1.51 followed by numerical integration gave the values presented in the second column of Table 2-2. Comparison with the third column computed by the Domb–Barrett equation shows differences less than 5% for $z > 2$, suggesting that eq 1.51 may be fairly accurate unless Z is close to 1.

Table 2–1. Comparison of $P^{-1/2}$ and Q

x'	x	P	$P^{-1/2}$	Q
0	0	1.659	0.776	0.749
∞	0	1.78	0.750	0.678
∞	∞	2.04	0.700	0.674

35

Table 2–2. Test of eq 1.51

z	$\alpha_S{}^2$ (eq 1.51)	$\alpha_S{}^2$ (DB eq.)
0	1	1
1	1.73	1.64
2	2.15	2.06
5	2.99	2.92
10	3.89	3.85
20	5.10	5.07
40	6.70	6.69
100	9.65	9.65

2.   Second Virial Coefficient

2.1   Introduction

The second virial coefficient A_2 of a monodisperse polymer solution is defined by eq 1–2.13. Historically, it has been one of the central subjects in polymer solution studies. Yet, the theories available at present are not self–contained for typical experimental data. In this section, we give an account highlighting the gaps remaining between theory and experiment on the A_2 of solutions of linear flexible polymers.

In a solvent in which mean forces between segments are repulsive, polymer chains cannot freely interpenetrate with one another. This means that in such a solvent there appear correlations between the relative positions and conformations of a pair of chains which tend to overlap. The second virial coefficient is a measure of the effect which these correlations exert on the thermodynamic behavior of very dilute solutions. Thus it should depend on the strength of segment–segment interactions. Though not readily obvious, it also depends on the chain length of the polymer. The chain length dependence on A_2 is the subject of prime interest in the discussion that follows. In the binary cluster approximation, the mean force potential $U(1,2)$ of a pair of identical spring–bead chains 1 and 2 (each consisting of $N+1$ beads) in a pure solvent is expressed by

$$U(1,2) = \sum_{j_1=0}^{N} \sum_{k_2=0}^{N} u(r_{j_1 k_2}) \tag{2.1}$$

36

where $u(r)$ is the interaction energy of two beads separated by a distance r, and j_1 and k_2 the running indices for the beads of chains 1 and 2, respectively. If $u(r)$ is approximated by $k_B T \delta(\mathbf{r})$, the statistical theory of liquid solutions allows A_2 to be represented as [2]

$$A_2 = -(N_A/2VM^2) \iint P(\{\mathbf{r}_1\})P(\{\mathbf{r}_2\})$$

$$\times \{\exp[-\beta \sum_{j_1=0}^{N} \sum_{k_2=0}^{N} \delta(\mathbf{r}_{j_1 k_2})] - 1\} \, d\{\mathbf{r}_1\} \, d\{\mathbf{r}_2\} \qquad (2.2)$$

where N_A is the Avogadro constant, V the volume of the solution, $P(\{\mathbf{r}_p\})$ the one–body distribution function of chain p, which is actually given by eq 1–3.3, and β the excluded–volume strength.

2.2 Penetration Function

It follows from eq 2.2 that A_2 becomes proportional to β in the limit $\beta \to \infty$. As mentioned in Section 2.4 of Chapter 1, the θ condition is a combination of solvent species and temperature for which A_2 vanishes. Hence, in the binary cluster approximation, $\beta = 0$ is equivalent to the θ condition, as has been repeatedly referred to in the preceding discussions. However, we note that this equivalence is not obtained if the (residual) ternary cluster interaction is taken into account (see Section 2 of Chapter 4).

Theorists prefer a dimensionless quantity h defined by

$$A_2 = (N_A \beta / 2m^2)h \qquad (2.3)$$

where m is the molar weight of one bead. This quantity does not vanish but tends to unity as β approaches zero. Sometimes it is called the reduced second virial coefficient. Transforming eq 2.2 to the Edwards continuous chain and letteing λ/L tend to zero, we find that h becomes a universal function of the excluded–volume variable z. Hence, in the framework of the two–parameter theory, h depends only on z, as do the expnasion factors α_R and α_S. Since, as eq 1.33 indicates, z is proportional to $M^{1/2}$, it turns out that A_2 depends on M through the z dependence of h. This fact was first pointed out by Zimm [59] as early as 1946. Experimental results show, almost without exception, that A_2 decreases monotonically as the molecular weight gets higher. But it is not always easy to predict such molecular weight dependence intuitively.

Unfortunately, h cannot be estimated from measurement of A_2, because β is not directly measurable. Thus experimentalists prefer another dimensionless quantity Ψ called the **penetration function**. It is defined by

$$\Psi = zh/\alpha_S^3 \qquad (2.4)$$

where α_S is the radius expansion factor. In the two–parameter theory Ψ is a universal function of z.

It is a simple matter to show that

$$\Psi = \frac{A_2 M^2}{4\pi^{3/2} N_A \langle S^2 \rangle^{3/2}} \tag{2.5}$$

The quantities on the right–hand side are all experimentally measurable and so Ψ is a measurable quantity, the reason that experimentalists prefer Ψ to h. But it should be noted that Ψ contains excluded–volume effects on chain dimensions as well as on A_2.

As seen from eq 2.4, Ψ vanishes at $z = 0$, i.e., under the θ condition. On the other hand, if Ψ asymptotically converges to a finite value, designated below by $\Psi(\infty)$, as z increases indefinitely, $h(z)$ should vary as $z^{-0.4}$, i.e., as $M^{-0.2}$, in the limit of large z, because eq 1.29 indicates $\alpha_S{}^3 \sim z^{0.6}$ for $z \gg 1$. This limiting behavior of $h(z)$ immediately leads to the prediction $A_2 \sim M^{-0.2}$ ($M \to \infty$).

2.3 Perturbation Calculations

Because exact analytical calculation of $h(z)$ is hopelessly difficult, attempts have been made to evaluate by perturbation the coefficients D_m in the z expansion of this function:

$$h(z) = 1 - D_1 z + D_2 z^2 + \ldots \tag{2.6}$$

Such work was initiated by Zimm [59] and taken over by Albrecht [60], and Tagami and Casassa [61]. However, the calculations of these authors were approximate because they neglected intrachain interactions, i.e., approximated $P(\{\mathbf{r}_p\})$ by the value for $\beta = 0$. Analyses allowing for both intra and interchain interactions were first made by Kurata and Yamakawa [62] and continued by Kurata et al. [63] and Tagami and Casassa [61]. Recently, Tanaka and Šolc [64] corrected these previous calculations for errors and arrived at

$$h(z) = 1 - 2.865z + 13.928z^2 + \ldots \tag{2.7}$$

No evaluation of the term with z^3 is as yet reported, but as may be judged from the coefficients in eq 2.7, its absolute value would be quite large.

2.4 Approximate Closed Expressions

Various methods have been proposed to find approximate closed expressions for $h(z)$ which are hopefully of practical use for analyzing experimental data.

Those reported by the end of the 1960s are discussed in Yamakawa's book [2]. The most important in this period is the Kurata–Yamakawa equation [63, 65], which reads

$$h(z) = \frac{[1 - (1 + K\bar{z})^{-(2D_1 - K)/K}]}{(2D_1 - K)\bar{z}}$$ (2.8)

where

$$\bar{z} = z/\alpha_S{}^3$$ (2.9)

$$K = (3D_2/D_1) - (2D_1 + 5.313)$$ (2.10)

We note that the numerical term 5.313 is associated with the expansion of one chain in the presence of another [61, 63]. Substitution of the correct values for D_1 and D_2 into eq 2.8 gives [64]

$$h(z) = \frac{[1 - (1 + 3.537\bar{z})^{-0.620}]}{2.193\bar{z}}$$ (2.11)

If K is treated as an adjustable parameter, most of the closed expressions reported by the end of the 1960s can be derived as special cases of eq 2.8 [2]. This fact is quite intriguing. Equation 2.8 predicts that $h(z)$ varies as $z^{-0.4}$ in the limit of large z.

The line TS in Figure 2–6 shows Ψ as a function of $\alpha_S{}^3$, computed from eq 2.8 and the Domb–Barrett interpolation formula for α_S. It gives a curve which is convex upward and approaches a $\Psi(\infty)$ value of 0.456 from below. This limiting value is twice as large as the experimental estimates of $0.22 - 0.25$ (see the next section).

Recently, by combining the perturbation expansion with Monte Carlo simulation data, Barrett [66] constructed the following interpolation formula:

$$h(z) = (1 + 14.3z + 57.3z^2)^{-0.2}$$ (2.12)

Expansion on the right–hand side yields $D_1 = 2.86$ and $D_2 = 13.1$, in good agreement with the correct values. For $z \gg 1$ this equation gives $h = 4.45z^{-0.4}$. The line B in Figure 2–6 has been calculated from eq 2.12 and the Domb–Barrett interpolation formula for $\alpha_S{}^3$. It levels off more rapidly than the line TS does and approaches 0.235 from below. This limiting value is in excellent accord with the experimental values quoted above.

2.5 Comparison with Experiment – Penetration Function

The prediction by two–parameter theory that Ψ should be a universal function of α_S can be tested experimentally with no assumption because these two

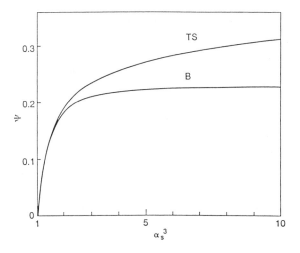

Fig. 2–6. Theoretical curves for Ψ as a function of $\alpha_S{}^3$. TS, eq 2.11 combined with the Domb–Barrett (DB) equation for α_S. B, eq 2.12 combined with the DB equation.

quantities are directly measurable. However, the test is not always easy because Ψ is sensitive to errors in the measurements of A_2 and $\langle S^2 \rangle$.

Most previous studies at α_S close to unity, i.e., at small values of z, were performed by changing both M and T in poor solvents near the θ temperature. The polymer samples used had high M since otherwise no reliable measurement of $\langle S^2 \rangle$ could be made. Figure 2–7, which illustrates some typical data obtained in this way, substantially confirms the prediction concerned, while the line B fits the data points only at $\alpha_S{}^3 < 2$.

In order to swell a polymer coil to $\alpha_S{}^3 > 3$ it is necessary to use a good solvent and a high molecular weight sample. In this connection we note that A_2 and chain dimensions in good solvents are practically independent of temperature. In other words, β in such solvents undergoes little change when T is varied. Although Daoud and Jannink [70] took the reduced temperature $(T - \theta)/\theta$ or $(T - \theta)/T$ as a measure of the "goodness" of a solvent, i.e., as proportional to β, this fact implies that their idea is not valid in good solvent systems. Probably the proportionality holds only in poor solvents near the θ temperature. In fact, it has been assumed by many authors who analyzed experimental data for such solvent systems.

Figure 2–8 shows some typical data for Ψ in the region of large $\alpha_S{}^3$. Though somewhat badly scattered, the plotted points may be fitted by a single curve as indicated, thus conforming to the prediction that Ψ depends only on α_S. A guess estimate gives $\Psi(\infty) = 0.22 - 0.25$. More importantly, Ψ decreases to $\Psi(\infty)$

40

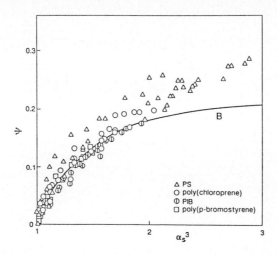

Fig. 2–7. Data for Ψ in the region of $\alpha_S{}^3$ relatively close to unity. Different symbols mark poly(chloroprene) [67], PS [68], poly(p–bromostyrene) [69], and PIB [47] in various poor solvents. Solid curve, reproduction of the line B in Fig. 2–6.

with increasing $\alpha_S{}^3$. This decreasing trend of Ψ is opposite to the behavior of the line TS or B in Figure 2–6. The discrepancy is serious, and the following remarks may be in order.

Firstly, none of the existing theories for Ψ, except for one due to Gobush et al. [73], predict the decrease of Ψ to $\Psi(\infty)$. Secondly, this behavior of Ψ in good solvents was revealed in recent years when polymer samples with ultrahigh M and systems with very large β [45, 72] became available. Thirdly, the determination of Ψ at large $\alpha_S{}^3$ is sensitive to the polydispersity of polymer samples. Thus some may doubt the reliability of the data points displayed in Figure 2–8. For example, Douglas and Freed [74] suspected that the polystyrene samples used happened to be systematically more polydisperse at higher M, but Fujita and Norisuye [75] commented that this was less likely. In section 2.7, we will show that the decreasing trend of Ψ in good solvent systems has to be accepted as the fact, even if conflicting with the current theories.

2.6 Breakdown of the Two–parameter Theory

Figure 2–9 shows Miyaki's data for Ψ for polystyrene in benzene and cyclohexane [44]. In the range of $\alpha_S{}^3$ from 2 to 4, the Ψ values in the two solvents do not superimpose on a single curve, showing that Ψ is not a universal func-

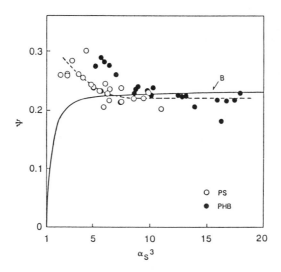

Fig. 2-8. Data for Ψ in the region of high $\alpha_S{}^3$. Different marks are for PS [43, 44, 68, 71] and PHB [45, 72] in good solvents. The line B is the same as that in Fig. 2-6, and the dashed line is an eye guide to the trend of plotted points.

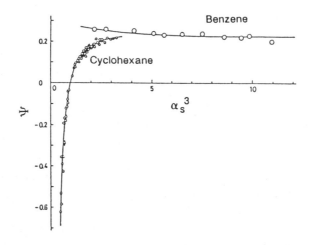

Fig. 2-9. Data for Ψ for PS in benzene (◯) and in CH (◦), showing the breakdown of the two-parameter theory [44].

tion of α_S, i.e., this graph demonstrates a breakdown of the two-parameter approximation.

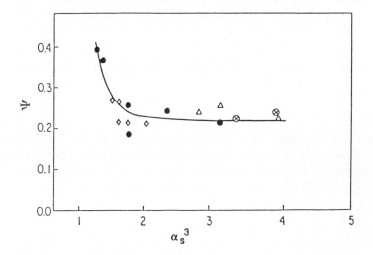

Fig. 2–10. Recent data on Ψ for PS in toluene in the region of $\alpha_S{}^3$ near unity. $\bullet$, SANS [77]; $\triangle$, LS [78]; $\diamond$, SANS [79]; $\otimes$, SANS [80].

   In connection with Figure 9–2, it is interesting to ask whether Ψ continues to rise or eventually fall to zero after passing a maximum as $\alpha_S{}^3$ approaches unity (i.e., $z \to 0$). The latter seems a mandate, because, according to eq 2.4, Ψ should vanish at $z = 0$. To check this prediction scattering experiments have to be extended to low M, since the change in temperature has little effect on $\langle S^2 \rangle$ in good solvents. Use of light scattering, however, is not advantageous, since its accuracy for the determination of $\langle S^2 \rangle$ and hence of Ψ and α_S sharply drops as M is lowered. Very recently, using small–angle neutron scattering (SANS), Huber et al. [76] succeeded in measuring A_2 and $\langle S^2 \rangle$ for polystyrene in toluene down to $\alpha_S \sim 1.1$ (which corresponded to $M = 1200$). The filled circles in Figure 2–10 show Ψ calculated by Huber and Stockmayer [77] from the data of Huber et al. Interestingly and contrary to the prediction, they keep rising first gradually and then sharply as $\alpha_S{}^3$ approaches unity. When this novel finding is combined with the earlier data displayed in Figures 2–8 and 2–9 we may conclude rather safely that Ψ for polystyrene in a good solvent declines monotonically toward a limiting value as α_S increases from unity to infinity. Whether this behavior is general or specific to polystyrene remains to be seen by future experiments. Anyway it is surprising that the upswing of Ψ in Figure 2–10 disagrees with the prediction that $\Psi = 0$ at $z = 0$ (see eq 2.4). Huber and Stockmayer [77] attributed the finding of Huber et al. to chain stiffness.

43

However, the effect of chain stiffness is unlikely to persist up to very long chains, so that it is preferable to seek another mechanism for the observed decline in Ψ in the region of high α_S.

2.7   Comparison with Experiment — Molecular Weight Dependence of A_2

As noted above, Zimm [59] was the first to point out on a theoretical consideration that A_2 should depend on M. Flory and Krigbaum [81] in 1950 formulated a mean–field theory concluding that A_2 should decrease with increasing M. But the predicted M dependence was weaker than that observed in those days. The discrepancy motivated extensive literature on theoretical and experimental studies of A_2 which are well summarized in Yamakawa's book [2].

The experimental data reported to date for linear polymers in good and marginal solvents indicate, almost without exception, that A_2 is a decreasing function of M and that when the range of M is relatively narrow (say less than two decades), this function is empirically described by a power law as

$$A_2 \sim M^{-\delta} \tag{2.13}$$

with δ usually found in the range 0.2–0.30. However, as exemplified by the extensive data [82] on poly(methyl methacrylate) shown in Figure 2–11, plots of $\log A_2$ vs. $\log M$ over a wider range of M follow a curve which is weakly convex–downward. Importantly, no curve of opposite curvature has ever been reported. Thus, δ is not constant but decreases with increasing M. In general, this decrease is so slow that it is not easy to estimate $\delta(\infty)$, the asymptotic limit of $\delta(M)$. At present, $\delta(\infty)$ is inferred to be in the range 0.20–0.22, regardless of the kind of solvent (good and marginal). Little is as yet explored on $\delta(M)$ in poor solvents. We note that $\delta(\infty) = 0.2$ is consistent with the theoretical prediction made in Section 2.2.

The data of Figure 2–11 extend down to the region of oligomers and show that a fairly sharp rise of A_2 occurs in this region. Similar behavior of the M dependence of A_2 can be seen in the recent data of Huber et al. [76] and Zhang et al. [83] on polystyrene in toluene. More work is needed for A_2 of low–molecular–weight polymers in various solvents.

Now, using the fact that eq 1.29 is good enough for $z > 2$, we obtain from eq 2.5

$$\Psi = CM^{0.2}A_2 \quad (z > 2) \tag{2.14}$$

where C is a constant. From this it follows that

$$\frac{d^2 \log \Psi}{d(\log \alpha_S{}^3)^2} = \left(\frac{10}{3}\right)^2 \frac{d^2 \log A_2}{d(\log M)^2} \tag{2.15}$$

44

holds for $\alpha_S{}^3 > 2$ (equivalent to $z > 2$) in a fixed solvent. If this relation is combined with the observed fact that $\log A_2$ vs. $\log M$ relations are always convex downward we see that $d^2 \log \Psi / d(\log \alpha_S{}^3)^2$ is positive, which means that if Ψ is positive as in good solvents, it should decrease monotonically toward $\Psi(\infty)$ with increasing $\alpha_S{}^3$. This behavior is indeed what we have illustrated above with some recent experimental data. Thus, it is certain that the current theories predicting an up–going approach of Ψ to $\Psi(\infty)$ are not adequate.

The second virial coefficient of polymer solutions has been the subject of theoretical study for decades. Nonetheless, no persuasive explanation is as yet available concerning its molecular weight dependnece in good solvent systems. To be important, the defect of the existing theories of A_2 is not limited to this point. They reveal another discrepancy with experiment when we look at the behavior of A_2 in poor solvents below the θ temperature.

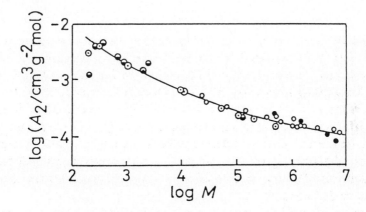

Fig. 2-11. Typical A_2 data for poly (methyl methacrylate) in acetone, a good solvent [82].

2.8   Second Virial Coefficient below the θ Temperature

Though not specially mentioned, the discussion up to this point has been limited to the polymer in solvents in which A_2 is non–negative, so that β and z are zero or positive in the binary cluster approximation. Available data on A_2 for polymer solutions below θ are still scant and fragmentary. This is mainly due to the technical difficulties explained in Section 2.3 of Chapter 4. Recently, Tong et al. [84] undertook fairly systematic measurements of A_2 on polystyrene in cyclohexane, and Takano et al. [85] did similar work on poly(isoprene) in dioxane. The results obtained were essentially similar for both systems and revealed quite unexpected behavior. In Figure 2-12 are displayed the data for the system polystyrene + cyclohexane.

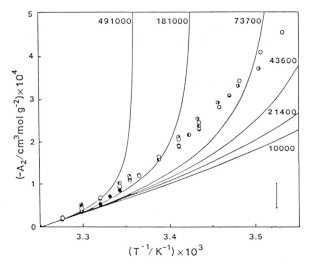

Fig. 2-12. Data on A_2 for PS in CH below θ. Different marks are for different M. Solid curves, calculated by the Tanaka–Šolc equation 2.16 for indicated M values.

Interestingly, the data points for different M nearly fall on a single curve, indicating that A_2 is virtually independent of M and varies only with temperature. This finding does not conform to the two–parameter theory which predicts the indicated solid curves for different M when, for example, the following approximate expression due to Tanaka and Šolc [64] is used for $h(z)$:

$$h(z) = (1 + 6.585z)^{-0.418} \tag{2.16}$$

Use of other available expressions for $h(z)$ gives curves similarly divergent for different M. The discrepancy observed here between theory and experiment is so pronounced that one may be tempted to attribute it to a gross error in experimental measurements. In this connection it is worth noting that very recently Perzynski et al. [86] have reported A_2 data on polystyrene in cyclohexane which substantiated the finding of Tong etal. At present, there appears no theory that can explain these findings adequately.

46

2.9 Third Virial Coefficient

The third virial coefficient A_3 is defined as the coefficient of c^3 in eq 1–2.13 for osmotic pressure Π. Theoretically, it reflects the excess interaction of ternary bead clusters, while experimentally its knowledge is important for accurate determination of A_2 by analysis of osmotic pressure, light scattering, and sedimentation equilibrium data. However, accurate measurement of A_3 is not a simple task, and the available data are as yet neither abundant nor systematic.

In discussing A_3 we often use a reduced third virial coefficient g defined by

$$g = A_3/(MA_2{}^2) \tag{2.17}$$

In the framework of the two–parameter theory this dimensionless quantity becomes a universal function of the excluded–volume variable z. Yamakawa's book [2] summarizes some representative theories of $g(z)$ presented by the end of the 1960s. Unfortunately, no substantial progress in the theory of g has since evolved.

Long ago Flory [1] advocated the choice $g = 0.25$ for analysis of osmotic pressure data, and later Berry [68] proposed changing 0.25 to 0.333 when light scattering data are analyzed. Though these g values have been routinely used, a naive question is whether g is constant or M–dependent. Using a renormalization group method, des Cloizeaux [87] derived $g \sim 0.44$ in the good solvent limit, but this is considerably higher than 0.277 recently obtained by Douglas and Freed [88] in the same limit. The g value for hard sphere solutions is known to be 5/8 (= 0.625) for any M [2].

As is the case for A_2, A_3 should depend on temperature and solvent quality as well as M. Probably, the M dependence is of prime interest for polymer physical chemists, but its experimental determination has long remained unattacked except for Vink's work [89] on polystyrene in cyclohexane near the θ temperature. Very recently, Kniewske and Kulicke [90] and Sato et al. [91] estimated A_3 as a function of M for polystyrene in toluene (25°C) and benzene (25°C), respectively. Their data are shown in Figure 2–13. The data points for either solvent can be fitted by

$$A_3 \sim M^{0.59} \tag{2.18}$$

According to the two–parameter theory [2], g is expected to converge to a finite value at the limit of large z, while according to what we have learned in previous sections A_2 is likely to vary with $z^{-0.4}$ at the same limit. Hence it follows from eq 2.17 that asymptotically $A_3 \sim z^{1.2} \sim M^{0.6}$ in good solvent systems. This consequence conforms to eq 2.18. Thus we are tempted to say that g reached an asymptotic value at molecular weights studied by the above-mentioned authors. In fact, Kniewske and Kulicke found a constant g value of

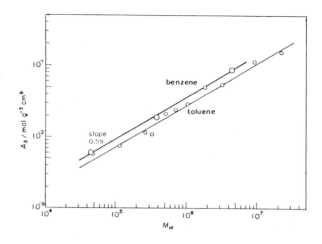

Fig.2–13. Molecular weight dependence of A_3 for PS in toluene (o) [90] and in benzene (◯) [91].

0.333 for all of their polystyrene samples (actually, they assigned this value to g in estimating A_3 by curve-fitting analysis of light scattering data). However, Sato et al. measured A_3 more directly and found that g increased with M as shown in Figure 2–14. Thus g seems to be M–dependent.

The g curve in Figure 2–14 indicates that Flory's g value of 0.25 is adequate for $M < 2 \times 10^5$ and Berry's of 0.333 for $5 \times 10^5 < M < 1.5 \times 10^6$, but assigning a constant g is no longer good for higher M. The theories of g described in Yamakawa's book all predict that g tends to zero as M is lowered. This prediction arises from the use of the binary cluster approximation. The data of Sato et al. suggest that g remains at about 0.20 even at low M. The discrepancy suggests that much remains to be investigated on short polymer chains in dilute solution.

In the binary cluster approximation, A_3 as well as A_2 must vanish simultaneously under the θ condition. This prediction, however, does not agree with the osmotic pressure data of Flory and Daoust [92] on poly(isobutylene) in benzene, which gave positive A_3 at the θ temperature (24°C). More definite evidence for non–vanishing A_3 under the θ condition can be seen from the osmotic pressure data of Vink [89] on polystyrene in cyclohexane. Thus it seems mandatory to abandon the binary cluster approximation in the region near the θ condition.

However, the inclusion of the (residual) ternary cluster interaction gives rise to various difficulties as discussed in Section 2 of Chapter 4. This dilemma is one of the most serious problems that the current theory of dilute polymer solutions faces.

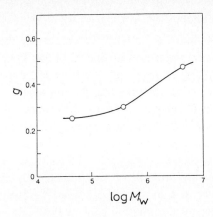

Fig. 2–14. Molecular weight dependence of the reduced third virial coefficient g for PS in benzene [91].

3.   Steady Transport Coefficients

3.1   Intrinsic Viscosity and Diffusion Coefficient

The intrinsic viscosity $[\eta]$ and the diffusion coefficient D_0 at infinite dilution of a polymer molecule are typical dynamic global properties that have attracted many polymer chemists since the beginning of the development of polymer science. These physical properties are sometimes called **steady (-state) transport coefficients**, since each is concerned with the rate at which energy is dissipated as heat from a steadily rotating or translating polymer molecule to its surrounding solvent. What has captured the prime interest of polymer researchers is their dependence on the molecular weight M of the molecule. A huge amount of experimental data on this dependence has been accumulated and empirically analyzed, while many theories have been proposed for the interpretation of those data in terms of molecular concepts and parameters. Yet there exist significant problems that wait adequate theoretical explanations. The main purpose of Section 3 is to discuss some typical of them. For simplicity, as up to this point, we deal only with monodisperse linear flexible polymers in pure solvents, in both theory and experiment. Furthermore, when $[\eta]$ is referred to, it is limited to the value at the limit of zero shear rate. In other words, the shear rate effect on $[\eta]$ will not be considered, though it is a very interesting subject. We also use the term diffusion coefficient to mean its value at infinite dilution. Polymer diffusion at finite dilutions is discussed in Chapters 7 and 8.

49

3.2 Remarks on Theory

We do not intend to make a detailed exposition of the typical theories of transport coefficients, as written in Yamakawa's book [2], but focus on a comparison of their consequences with well–documented experimental results. In this connection, the following remarks may be in order.

The classic papers published in 1948 by Kirkwood and Riseman [93] and by Debye and Bueche [94] marked the beginning of extensive literature on the modern theories of $[\eta]$ and D_0, but the Kirkwood–Riseman theory has turned out to dominate the subsequent progress. The theories of these authors were concerned with steady slow motion of an isolated polymer molecule under a constant external force or in a constant shear flow of solvent, and formulated by applying hydrodynamics to perturbative solvent flows caused by the motion of the polymer. In a paper of 1954, Kirkwood [95] presented a more general theory, called the diffusion equation approach, which can treat time–dependent external forces or imposed solvent flow as well as thermal fluctuations of the polymer chain. It also allows the incorporation of microscopic constraints such as fixed bond length and bond angle into the theory. When the polymer is replaced by the spring–bead chain model so that only chain connectivity is retained as the structural constraint, the diffusion equation approach can be developed to a very satisfactory state. The clue to this development was prescribed by Rouse [96], Bueche [97], and Zimm [98] in the 1950s and has grown to be the central part of current polymer dynamics, which includes quasi–elastic light (and neutron) scattering, viscoelastic responses, dielectric dispersion, and so forth. One of the recent problems of theoretical interest is to work out the corresponding dynamic theory for more realistic chain models whose repeat units are subject to various structural constraints. For steady transport coefficients of the spring–bead chain the dynamic theories of Zimm and others predict essentially the same behavior as does the Kirkwood–Riseman theory. We thus make no discrimination between them in the subsequent comparisons of theory and experiment.

3.3 Experimental Information

3.3.1 Houwink–Mark–Sakurada Relation for Intrinsic Viscosity

One of the most surprising generalities in the world of polymers is that $[\eta]$ for a series of homologous polymers under a fixed solvent condition (solvent species and temperature) follows a simple power law as

$$[\eta] = KM^\nu \tag{3.1}$$

over an extended range of M. Here, K and ν are constants for the polymer + solvent system considered. Equation 3.1 is referred to as the **Houwink–Mark–Sakurada (HMS) relation** in this book. Since it was discovered half

50

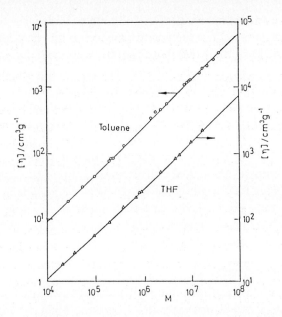

Fig. 2–15. HMS relations for PS in toluene and THF [100].

a century ago the HMS relation has been confirmed invariably on numerous combinations of polymer and solvent, though the documented range of M vary with experimentalists as well as systems investigated.

The main experimental facts that have to be explained theoretically are as follows:

(1) When $[\eta]$ is plotted against M on a log–log graph paper, it gives a straight line over a wide range of M;

(2) The slope ν of the line for linear flexible polymers in non-θ solvents in which the second virial coefficient A_2 is positive is in the range $0.5 < \nu < 0.8$. Polymers giving ν larger than 0.8 are usually suspected to be semi-flexible;

(3) In general, ν is larger for a better solvent, i.e., for a larger A_2;

(4) Under the θ condition where A_2 vanishes, ν for flexible linear polymers is always 0.5. Thus we have

$$[\eta]_\theta = K_\theta M^{0.5} \tag{3.2}$$

where the subscript θ signifies the value under the θ condition. For a given polymer K_θ is nearly independent of solvent species. Einaga et al. [99] showed eq 3.2 to hold up to M values as high as 6×10^7 for polystyrene in cyclohexane at the θ temperature $34.5°$C.

The range of M available to experimentalists is limited, which implies that it is meaningless to ask whether the HMS relation holds or not up to infinitely large M. However, recent studies on "ultrahigh" molecular–weight polymer samples give reason to believe the positive answer to this question. Figure 2–15 shows the data of Meyerhoff and Appelt [100], who made measurements up to $M = 4 \times 10^7$. The data points in either solvent exhibit no discernible curvature, giving 0.724 and 0.714 for the ν values for toluene and THF, respectively. For polystyrene in benzene Einaga et al. [99] found a linear relation with $\nu = 0.75 \pm 0.01$ even up to a higher M. It seems difficult to suspect that the linear relations found in these studies would begin to bend at a molecular weight of the order of 10^8 or higher. If it happens, then it would become virtually impossible to compare theoretical calculations for infinitely long chains with experiment and thus they would have no practical value. The author maintains the opinion that once established over a wide range of M, more than, say, two decades, the HMS relation should be valid up to any higher molecular weight. In what follows, any theory of $[\eta]$ is discussed on the basis of this view.

3.3.2 Diffusion Coefficient

Measurement of D_0 requires more skill and more expensive equipment than that of $[\eta]$, and its accuracy decreases as M becomes higher. Thus, the available data on D_0 are not as abundant as those of $[\eta]$ and few of them reach very high molecular weights. However, we may deduce the following empirical rules from them.

(1) The M dependence of D_0 in a fixed solvent follows a power law as

$$D_0 = K'M^{-\mu} \tag{3.3}$$

over a wide range of M (actually the entire range of M studied in each experiment). Here K' and μ are constants. No special name is given to this relation;
(2) In θ solvents, μ is always equal to 0.5 within experimental errors. Thus we have

$$D_{0\theta} = K'_\theta M^{-0.5} \tag{3.4}$$

(3) In non–θ solvents in which A_2 is positive, μ is found in the range between 0.5 and 0.6. This range is so narrow that somewhat different μ values are reported by different authors for the same polymer + solvent systems.

3.4 Comparison with Experiment—Molecular Weight Dependence

3.4.1 θ Solvents

Theory of flexible polymers in dilute solutions is greatly simplified when it is concerned with the θ condition, because intrachain interactions need not be considered and the chain can be treated as Gaussian. In fact, the original versions of existing theories on dynamic polymer behavior started with Gaussian chains. One of the most important results from them is that $[\eta]_\theta$ increases with increasing M in such a way that ν in eq 3.1 decreases monotonically from unity to 0.5. Therefore, the theoretically predicted relation between $\log[\eta]_\theta$ and $\log M$ is not linear but convex upward, in disagreement with eq 3.2. The curvature of this relation arises from the hydrodynamic phenomenon that part of the polymer domain allows the solvent to permeate through and this fraction ξ diminishes toward zero as the chain becomes longer. This phenomenon is usually called the partial draining of solvent. If it happens that $\xi = 0$ at any chain length, it follows that ν is 0.5 over the entire range of M, and eq 3.2 results.

As early as the 1930s Kuhn [101] had suggested that the solvent inside the polymer coil is immobilized, which means $\xi = 0$. Flory [14] adopted it as a postulate in formulating his pioneering theory of polymer solutions summarized in his 1953 book [1]. However, unless the hydrodynamic mechanism responsible for this postulate is clarified, we cannot say that eq 3.2 has been explained by molecular theory. It is amazing that we are still unable to explain why actual polymer molecules behave as if non-draining for any chain length.

Since a substantial part of a polymer coil is occupied by solvent molecules it seems difficult to preclude the possibility that partial solvent draining takes place, i.e., $\xi \neq 0$, especially when the chain is relatively short. There must be an as yet unknown mechanism that happens to make the polymer behave as impermeable to the solvent. Yamakawa and Fujii [102] showed that chain stiffness gives rise to an effect on $[\eta]$ which opposes the effect of solvent draining. In fact, when an appropriate value is assigned to d/q, where q is the persistence length expressing chain stiffness (see Chapter 5) and d is the diameter of the chain, their theory predicts $[\eta]_\theta$ which varies linearly with $M^{0.5}$ over a much wider range of M than does the theory for perfectly flexible Gaussian chains. However, it is unlikely that the value of d/q happens to be nearly the same for any combination of polymer and θ solvent. This ratio may vary from system to system.

The Kirkwood-Riseman theory [93] and the Zimm theory [98] as well predict that μ in eq 3.3 under the θ condition decreases monotonically from 1 to 0.5 with increasing M, and thus fail to derive eq 3.4. Again, we have to invoke complete immobilization of solvent inside the polymer coil to explain this relation by these theories. The Yamakawa-Fujii theory shows that chain stiffness also counteracts the draining effect on D_0, but a d/q value different from that needed for $[\eta]_\theta$ has to be used in order to obtain a maximum suppression of the draining effect.

3.4.2   Non-θ Solvents

Owing to the intrachain interaction the theoretical treatment of $[\eta]$ and D_0 in non-θ solvents is far more difficult than that in θ solvents. Usually, though still unjustified theoretically, it is assumed that the polymer coil in non-θ solvents is also impermeable to solvent molecules. Making this assumption and going to

53

the Edwards continuous chain, we find that the **viscosity expansion factor** α_η defined by

$$\alpha_\eta{}^3 = [\eta]/[\eta]_\theta \tag{3.5}$$

becomes a universal function of the excluded–volume variable z, as do the expansion factors α_R and α_S, and that the same thing holds for the **friction expansion factor** α_f defined by

$$\alpha_f = f/f_\theta \tag{3.6}$$

Here, f denotes the friction coefficient of the polymer chain, which is defined as the force required to have the center of mass of the chain translate at unit velocity in a solvent at rest. This quantity is related to D_0 by the Einstein equation

$$D_0 = k_B T/f \tag{3.7}$$

where k_B is the Boltzmann constant and T the absolute temperature.It also defines the Stokes radius R_H, the hydrodynamic size of the polymer coil, by

$$f = 6\pi\eta_0 R_H \tag{3.8}$$

where η_0 is the viscosity coefficient of the solvent. Thus we have

$$\alpha_f = D_{0\theta}/D_0 = R_H/R_{H\theta} \tag{3.9}$$

3.4.3 Approximate Expressions

For small values of z we may expand $\alpha_\eta{}^3(z)$ and $\alpha_f(z)$ in powers of z as

$$\alpha_\eta{}^3 = 1 + E_1 z + E_2 z^2 + \ldots \tag{3.10}$$

$$\alpha_f = 1 + F_1 z + F_2 z^2 + \ldots \tag{3.11}$$

Fujita et al. [103] tried to compute E_1 exactly except for the preaveraging approximation (see Ref.[2]) and obtained $E_1 = 1.29$. But Shimada and Yamakawa [104] noticed a numerical error in the computation of Fujita et al. and found the correct E_1 value to be 1.14. Calculation of E_2 seems almost prohibitive. The exact value of F_1 was obtained by Stockmayer and Albrecht [105] to be 0.609, but the value of F_2 still remains unknown.

Some attempts have been made to obtain approximate expressions which describe α_η and α_f over an extended range of z. Weill and des Cloizeaux [106] derived the relation

$$\alpha_\eta{}^3 = \alpha_R{}^2 \alpha_f \tag{3.12}$$

for a polymer coil with no solvent draining (the non–draining limit). If this relation is correct, α_η can be obtained by calculating α_R and α_f as functions of z. Weill and des Cloizeaux used their theory as described in Section 1 of Chapter 4 to evaluate the latter two. However, the logical process leading to eq 3.12 is not clear to the author. Tanaka [107] proposed the empirical equations

$$\alpha_\eta{}^5 = 1 + 1.90z \tag{3.13}$$

$$\alpha_f{}^5 = 1 + 3.045z \tag{3.14}$$

and showed that these fit some typical experimental data closely, though they are not consistent with the empirical relations 3.1 and 3.3. Equation 3.13 predicts that $[\eta]$ tends to be proportional to $M^{0.8}$ at large M regardless of solvent species with $\beta > 0$. This asymptotic behavior, however, does not conform to the observed results (see Section 3.3.1).

Barrett [57,108] calculated $[\eta]$ and R_H by numerically solving the Kirkwood-Riseman hydrodynamic equations for non–draining coils with the use of his approximate expression for $\langle 1/R_{ij}\rangle$ given in eq.1.52. Here, R_{ij} denotes the distance between bead i and j. Though the preaveraging approximation was used, his calculations are significant, because they are the direct approaches in the framework of the Kirkwood–Riseman theory to $[\eta]$ and R_H of linear flexible chains perturbed in any degree by volume exclusion. Combining the numerical results obtained with the first–order perturbation calculations, Barrett [57, 108] constructed the following interpolation formulas valid in the non–draining limit:

$$\alpha_\eta{}^3 = (1 + 3.8z + 1.9z^2)^{0.3} \tag{3.15}$$

$$\alpha_f = (1 + 6.09z + 3.59z^2)^{0.1} \tag{3.16}$$

These can be rewritten

$$[\eta] = C_\eta M^{1/2}(1 + 3.8z + 1.9z^2)^{0.3} \tag{3.17}$$

$$f = C_f M^{1/2}(1 + 6.09z + 3.59z^2)^{0.1} \tag{3.18}$$

by taking into account that $[\eta]_\theta$ and f_θ for non–draining polymers vary as $M^{1/2}$. Both C_η and C_f are constants for a given polymer. The lines in Figure 2-16 show computed values of $\log[\eta]$ and $\log f$ plotted against $\log z^2$ ($\sim \log M$). They are distinctly curved, with the slope changing from 0.5 to 0.8 for $[\eta]$ and from 0.5 to 0.6 for f as M increases, thus failing to explain the empirical relations 3.1 and 3.3. It should be noted that even if the non–draining condition invoked in the calculations of Barrett is relaxed, the asymptotic slopes for $[\eta]$ and f remain unchanged, being 0.8 and 0.6, respectively.

55

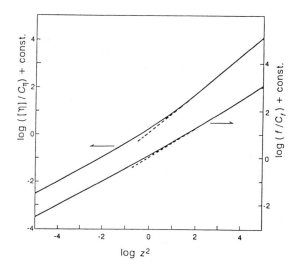

Fig. 2-16. Molecular weight dependences of $[\eta]$ and f predicted by Barrett's equations 3.17 and 3.18; note that $z^2 \sim M$. Dashed lines, asymptotes at large z.

3.4.4 Remarks

Some authors (for example, Weill and des Cloizeaux [106]) argue that because of the long–range nature of hydrodynamic interactions between paired beads the theoretically predicted asymptotic slope 0.8 of the relation between $\log [\eta]$ and $\log M$ will be approached so slowly that it may not be observable experimentally. This implies that actually observed HMS relations with $\nu < 0.8$ are mere segments of a non–linear dependence of $\log[\eta]$ on $\log M$.

Barrett's calculated curve for $[\eta]$ in Figure 2-16 appears to enter the asymptotic region at z^2 above 40. According to the estimate by Einaga et al. [43, 44], $z^2 = 40$ for polystyrene in benzene corresponds to $M \sim 6 \times 10^6$. Thus if the experiment is done with ultrahigh molecular weight polystyrene now available, it should be possible to check the above–mentioned argument. For example, Einaga et al. found for polystyrene in benzene that eq 3.1 held accurately up to $M \sim 60 \times 10^6$, but the ν value was 0.75 instead of 0.8. The critical M of polystyrene in toluene corresponding to $z^2 = 40$ is probably not very different from that in benzene. The straight line for polystyrene in toluene shown in Figure 2-15 extends to $M \sim 4 \times 10^7$, but its slope 0.724 is even smaller than that in benzene. These findings suggest that the predicted limiting slope of 0.8 is not experimentally observable but theoretically incorrect for an as yet unknown reason.

As mentioned above, in the non–draining limit, both the viscosity and friction expansion factors for the Edwards continuous chain become universal functions of z. If this is the case, we can expect that plots of α_η (or α_f) against $\log M$ for homologous polymers in different solvents should be superimposable on a single curve by horizontal shifting of the data points for each solvent. Moreover, the "master curves" so determined for different polymers should be brought to a single curve by further horizontal displacement. However, it can be easily proved that such superimposability is not obtained when $[\eta]$ (or D_0) follows the HMS relation with ν which varies with solvent species. This is illustrated with actual data, along with the related discussion, in Section 3.6.

3.5 Comparison with Theory—Hydrodynamic Factors

3.5.1 Experimental Information

As explained in Section 3.4.1, the observed M dependence of $[\eta]_\theta$ and $D_{0\theta}$ can be deduced from the Kirkwood–Riseman theory (the Zimm theory as well) if it is postulated that the polymer coil is in the non–draining state at any chain length. Then we ask how accurately this theory can predict the prefactors K_θ in eq 3.2 and K'_θ in eq 3.4. To discuss this problem it is usual to use the "hydrodynamic factors" Φ_θ and ρ_θ defined as follows:

$$\Phi_\theta = [\eta]_\theta M/(6\langle S^2\rangle_\theta)^{3/2} \tag{3.19}$$

$$\rho_\theta = \langle S^2\rangle_\theta^{1/2}/R_{\mathrm{H}\theta} \tag{3.20}$$

Some authors use in place of ρ_θ a hydrodynamic factor P_θ defined by

$$P_\theta = k_{\mathrm{B}}\Theta/[(6\langle S^2\rangle_\theta)^{1/2}\eta_0 D_{0\theta}] \tag{3.21}$$

where Θ denotes the theta temperature in degrees kelvin. In passing, Φ_θ and P_θ were first introduced by Flory and Fox [109] and by Mandelkern and Flory [110], respectively.

Since, as we have learned, $\langle S^2\rangle_\theta = K''_\theta M$, substitution of eq 3.2 into eq 3.19 gives

$$\Phi_\theta = K_\theta/(6K''_\theta)^{3/2} \tag{3.22}$$

Similarly, we obtain

$$\rho_\theta = 6\pi\eta_0 K'_\theta(K''_\theta)^{1/2}/k_{\mathrm{B}}\Theta \tag{3.23}$$

$$P_\theta = 6^{1/2}\pi/\rho_\theta \tag{3.24}$$

These relations indicate that Φ_θ, ρ_θ, and P_θ are independent of M and that P_θ and $1/\rho_\theta$ differ by a constant $6^{1/2}\pi (= 7.695)$.

Experimental values of Φ_θ and ρ_θ are considerably influenced by the polydispersity of polymer samples used. Hence, those for a rigorous comparison with theoretical predictions have to be determined with sufficiently narrow-distribution samples. However, even literature data chosen on this criterion are badly scattered over a considerable range. This is especially true of Φ_θ. Nonetheless, considering the experimental difficulties, we may conclude that both Φ_θ and ρ_θ are universal to a first approximation, the former being mostly found in the range $(2.0 - 2.7) \times 10^{23}$ mol^{-1} and the latter in the range $1.25 - 1.35 (6.15 - 5.70$ for $P_\theta)$. For the most thoroughly investigated system polystyrene + cyclohexane the best values of Φ_θ and ρ_θ seem to be $2.5_5 \times 10^{23}$ mol^{-1} by Miyaki et al. [111] and 1.27 by Schmidt and Burchard [112], respectively. It is highly desirable that more experimental work be undertaken on a variety of polymer + θ solvents to enrich our knowledge about the hydrodynamic factors.

3.5.2   Calculated Values

If the spring–bead chain model is used, eq 1.2 combined with eq 1.25 gives for K_θ''

$$K_\theta'' = a^2/(6m) \tag{3.25}$$

If K_θ and K_θ' in the non–draining limit are expressed as

$$K_\theta = N_A (\pi/6)^{3/2} a^3 m^{-3/2} \Gamma \tag{3.26}$$

$$K_\theta' = k_B \Theta m^{1/2} (\eta_0 a)^{-1} \Gamma' \tag{3.27}$$

Γ and Γ' are dimensionless constants whose values depend on theory. Substitution of eq 3.25, 3.26, and 3.27 into eq 3.22 and 3.23 yields

$$\Phi_\theta = N_A (\pi/6)^{3/2} \Gamma \tag{3.28}$$

$$\rho_\theta = 6^{1/2} \pi \Gamma' \tag{3.29}$$

We see that for theoretical purposes a dimensionless quantity Φ_θ/N_A is more suitable than Φ_θ itself.

According to the Kirkwood–Riseman theory, we have $\Gamma = 1.259$ and $\Gamma' = 0.192$ [2], which yield $\Phi_\theta = 2.87 \times 10^{23}$ mol^{-1} and $\rho_\theta = 1.48$. On the other hand, the diffusion equation approach due to Zimm [98] gives the same value for ρ_θ and a somewhat smaller value of 2.84×10^{23} mol^{-1} for Φ_θ. A more elaborate evaluation of Yoshizaki and Yamakawa [113] using the diffusion equation approach led to 2.856×10^{23} mol^{-1} for the latter. Having been derived by use of the preaveraging approximation to the Oseen interaction tensor (see Ref.[2]), these numerical results are not exact. The following round figures may be taken as the best for Φ_θ, ρ_θ, and P_θ under this approximation:

$$\Phi_\theta = 2.85 \times 10^{23} \text{ mol}^{-1} \qquad (3.30)$$

$$\rho_\theta = 1.50 \qquad (3.31)$$

$$P_\theta = 5.20 \qquad (3.32)$$

These values are higher than the corresponding "best" experimental ones quoted in Section 3.5.1, about 12% for Φ_θ and about 15% for ρ_θ. Though small, the differences cannot simply be attributed to the inaccuracy of the experimental values, and have indeed sparked more precise investigations of polymer hydrodynamics in recent years.

3.5.3 Problem of the Preaveraging Approximation

The preaveraging approximation replaces a tensor function of bead coordinates by a set of fixed numbers and thus greatly simplifies solving the basic hydrodynamic equations. Many have suspected that it was mainly responsible for the above-mentioned differences between calculated and experimental hydrodynamic factors. In the mid-1960s Pyun and Fixman [114] developed a perturbation theory which allows us to approach polymer hydrodynamics without invoking this approximation, and obtained 2.66×10^{23} mol^{-1} as an approximate value of Φ_θ. The details can be seen in Yamakawa's book [2]. By improving the Pyun–Fixman theory Bixon and Zwanzig [115] in 1978 found 2.76 in place of 2.66. This new value is rather close to the value given in eq 3.30, suggesting that the preaveraging approximation is not as serious as one might anticipate. If so, we must look for something else to approach the experimental values of Φ_θ.

Zimm [116] in 1980 reported Monte Carlo solutions to the non-preaveraged Kirkwood–Riseman equations, giving in the non-draining limit

$$\Phi_\theta = 2.51 \times 10^{23} \text{ mol}^{-1} \qquad (3.33)$$

$$P_\theta = 5.99 \quad (\rho_\theta = 1.28) \qquad (3.34)$$

Interestingly, these values are favorably compared with the "best" experimental ones quoted above. de la Torre et al. [117] carried out similar computer calculations for chains subject to bond length, bond angle, and internal rotation constraints, and reached conclusions in support of Zimm's. Thus it appeared that the long-suspended question about the gap between theory and experiment

59

concerning Φ_θ and ρ_θ was almost resolved. But the story did not end with these numerical studies.

The Kirkwood–Riseman theory is based on the assumption (rigid–body approximation) that the polymer molecule would rotate or translate maintaining its instantaneous conformation. Thus it neglects thermal fluctuations of chain segments. Though Zimm wrote a justification of this assumption, Fixman [118] criticized it and performed computer calculations called the dynamic simulation of chain motion. He was unable to draw definite conclusions owing to computational difficulties, but made a guess estimate that Φ_θ and P_θ calculated with the fluctuating non–averaged Oseen interaction tensor would be about 8% lower and only about 1% higher, respectively, than those with the non–fluctuating preaveraged Oseen tensor. If these values are correct, it turns out that Zimm and de la Torre et al. had overestimated errors due to the preaveraging approximation. Whether this conclusion is right or not is one of the most important unsolved problems in polymer physics.

3.5.4 Remarks

Recently, Oono [119] and Oono and Kohmoto [120] applied the renormalization group theory to polymer hydrodynamics of the Kirkwood–Riseman scheme. They computed Φ_θ and P_θ to first order in ϵ, where $\epsilon = 4 - d$ with d being the dimensionality of space, and obtained in three dimensions

$$\Phi_\theta = 2.36 \times 10^{23} \text{ mol}^{-1} \tag{3.35}$$

$$P_\theta = 6.20 \tag{3.36}$$

which compare rather favorably with the experimental values. It should be noted that the preaveraging approximation is not needed for the analysis to first order in ϵ. Wang et al. [121] extended the calculation of P_θ to second order in ϵ without invoking the preaveraging approximation and found 6.55 for that in three dimensions. This value is further from experiment than is the first order value in eq 3.36, illustrating that renormalization group calculations to a higher order in ϵ do not always yield more accurate results. In passing, we note that the preaveraging approximation becomes imperative for calculating Φ_θ to second order in ϵ.

The theories of the hydrodynamic factors referred to above all use the binary cluster approximation. However, when we are concerned with polymer solutions near the θ condition, at least (residual) ternary cluster interactions will have to be taken into consideration. Whether such additional interactions may account, if in part, for the above–mentioned gap between theory and experiment is yet to be investigated.

3.6 Tests of the Two–Parameter Approximation

As mentioned repeatedly, according to the two–parameter theory, the expansion factor α_S, the penetration function Ψ, and the hydrodynamic expansion factors α_η and α_f in the non–draining limit should become universal functions of a single variable z. These non–dimensional quantities are experimentally determinable without any assumption. Thus, the validity of the two–parameter theory can be tested directly by looking at whether a single curve independent of polymer and solvent condition (solvent species and temperature) is obtained or not when any of them is plotted against the other. Such tests were made by many authors (for example, see Ref. [2] and [119]). Here we refer to a recent one by Miyaki and Fujita [49] (and also Miyaki [44]), who used the following criteria A and B.

Criterion A. If α_S is a universal function of z, where $z = (3/2\pi)^{3/2} BM^{1/2}$ as defined by eq 1.33, the following should hold:

(i)   For a given polymer we choose any solvent condition S as the reference. Then, for any other solvent condition S' there exists a constant value of a dimensionless quantity a_M such that the plot of $\log \alpha_S^2$ against $\log M$ for S' can be superimposed on that for S when the abscissa for the former is changed to $\log(M/a_M)$. The resulting composite curve is called the master curve of α_S;

(ii)   The master curves of α_S for different polymers can be brought to a single curve by relevant horizontal shifting.

If the values of B for S and S' are denoted by B_S and $B_{S'}$, respectively, it can be shown that a_M is related to these values by

$$a_M = (B_S/B_{S'})^2 \tag{3.37}$$

Criterion B. If α_η is a universal function of z, the following should hold:

(i)   Plots of $\log \alpha_\eta^3$ vs. $\log M$ for a given polymer in different solvent conditions are brought to a master curve of α_η when each is replotted against $\log(M/a_M)$, with a_M being the same as that determined from α_S;

(ii)   The master curves of α_η for different polymers are reduced to a single curve by relevant horizontal displacement.

Figure 2–17 shows that α_S data on narrow–distribution polystyrene in benzene (25, 30°C), methyl ethyl ketone (35°C), and cyclohexane (different temperatures above θ) obey criterion A (i). Here, $M^\star$ denotes M/a_M, and benzene at 25°C is chosen as the reference state S. The data points in Figure 2–18 have been obtained by shifting horizontally the data for poly(isobutylene) fractions in cyclohexane (25°C), n–heptane (25°C), and isoamyl isovalerate (IAIV) (different temperatures above θ) until they matched the master curve of α_S for polystyrene. The good matching is consistent with criterion A (ii).

61

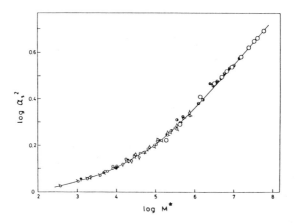

Fig. 2–17. Verification of criterion A (i). $M^\star = M/a_M$. Polymer: PS. Different marks are for benzene, MEK, and CH. The solid curve gives an eye guide to the trend of the plotted points.

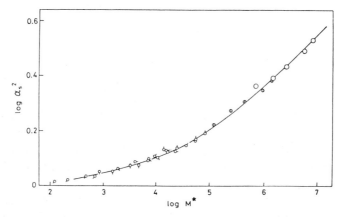

Fig. 2–18. Verification of criterion A (ii). Polymer: PIB. Different marks are for CH, n–heptane, and IAIV. Solid line is the reproduction of the curve in Fig. 2–17.

Figure 2–19 displays $\log \alpha_\eta^3$ plotted against $\log M^\star$ for polystyrene in the same three solvents as in Figure 2–17. Apparently, these data do not conform to criterion B (i). It was found that the α_η data for poly (isobutylene) in isoamyl isovalerate, n–heptane, and cyclohexane exhibited a similar breakdown of criterion B (i) [49]. However, we have to remark that the master curves of α_η

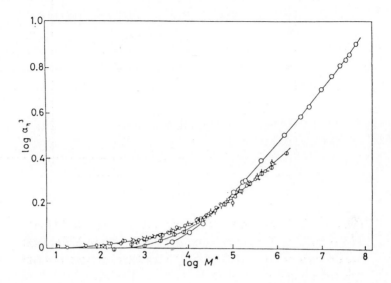

Fig. 2–19. Breakdown of criterion B (i). Polymer: PS. ○, benzene; ⦶, MEK; ○, CH.

for polystyrene + cyclohexane and poly (isobutylene) + isoamyl isovalerate are superimposable, thus obeying criterion B (ii). Summarizing, we can conclude that the two–parameter theory for $[\eta]$ does not always hold in actual polymer solutions. More recently, Varma et al. [78] working on polystyrene in various solvents have shown that the same conclusion applies for α_f as well as α_η.

If criteria A and B hold, α_η should be a universal function of α_S. Figure 2–20 displays a test of this prediction with the polystyrene data shown in Figures 2–17 and 2–19. The plotted points are not fitted by a single line but are split into two groups, one consisting of the benzene data and the other of the cyclohexane and methyl ethyl ketone data. This result can be associated with the breakdown of criterion B (i), because, as shown above, polystyrene in these three solvents obeys criterion A (i). Since the two split lines lie quite close, they may not be discerned unless measurements are made with high accuracy, and the conclusion in support of criterion B (i) may be erroneously drawn.

A momentary consideration reveals that the breakdown of criterion B is a consequence of the fact that $[\eta]$ follows the HMS relation with the index ν varying with solvent quality. Thus, after all, we go back to the basic question of why this relation holds so generally.

One of the basic approximations in two–parameter approaches to the steady

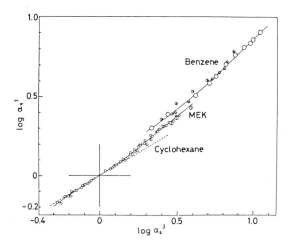

Fig. 2–20. $\alpha_\eta{}^3$ as a function of $\alpha_S{}^3$ for PS in CH, MEK, and benzene. The dashed line is calculated from eq 1.26 for $\alpha_S{}^2$ and eq 3.10 with $E_1 = 1.14$ [104] for $\alpha_\eta{}^3$.

transport coefficients is the neglect of partial drainage of the polymer coil. However, the author is not aware of any previous theory which has shown a better agreement with experimental results by the incorporation of the draining effect. In their recent renormalization group calculations, Wang et al. [122] had to allow the draining parameter h (see Ref.[2]) to decrease with increasing M in order to explain the HMS power law. According to its definition, h is a linearly increasing function of $M^{1/2}$, and at present we find no justification for the maneuver of Wang et al.

According to hydrodynamics, in the limit where the spring–bead chain reduces to a single bead, $[\eta]$ tends to the Einstein value ($2.5 v_p$, where v_p is the specific volume of the bead) if the bead is modeled by a rigid sphere. Thus, when extrapolated to zero molecular weight, the $[\eta]$ vs. M relation for any actual polymer should have a non–zero intercept. This prediction indeed agrees with experiment, but the Kirkwood–Riseman theory does not, yielding the zero intercept at $M = 0$. The discrepancy can be attributed to the approximation that the Kirkwood–Riseman formalism treats the bead as a point force. When the bead has a finite size, it receives a moment from the surrounding solvent flow and dissipates an additional energy which increases viscosity.

Recently, Yoshizaki et al. [123] worked out a theory for $[\eta]$ by approximately calculating this moment and derived the expression

$$[\eta] = [\eta]_P + [\eta]_E \tag{3.38}$$

where $[\eta]_\mathrm{P}$ is the intrinsic viscosity calculated by the point–force approximation and $[\eta]_\mathrm{E}$ that of the Einstein sphere. We note that the latter is independent of M. Importantly, this constant term markedly changes the M dependence of $[\eta]$ for short chains, and its effect persists up to fairly long chains.

Though it seems to leave more to be improved, the work of Yoashizaki et al. is quite significant, because it has made clear that polymer hydrodynamics has to exactly account for the thickness of the chain when we are concerned with low–molecular–weight polymers in solution. It is clear that, even under the theta condition, the statistics of such polymers shows considerable deviations from the Gaussian behavior, and adds a further complication to the hydrodynamic formulation. More generally, the physical behavior of short polymer chains should reveal features reflecting the details of the chemical structure of the molecule and thus becomes system–dependent. The discussion on it is beyond the scope of this book, which focuses on global polymer properties controlled by the large–scale characteristics of the molecule, i.e., its length and flexibility.

References

1. P. J. Flory, "Principles of Polymer Chemistry," Cornell Univ. Press, Ithaca, N.Y., 1953.
2. H. Yamakawa, "Modern Theory of Polymer Solutions," Harper & Row, New York, 1971.
3. M. Muthukumar and B. G. Nickel, J. Chem. Phys. **80**, 5839 (1984).
4. E. Teramoto, Busseiron Kenkyu No. 39, 1 (1951).
5. M. Yamamoto, Busseiron Kenkyu No. 44, 36 (1951).
6. R. J. Rubin, J. Chem. Phys. **20**, 1940 (1952).
7. F. Bueche, J. Chem. Phys. **21**, 205 (1953).
8. T. B. Grimley, J. Chem. Phys. **21**, 185 (1953).
9. M. Fixman, J. Chem. Phys. **23**, 1656 (1955).
10. H. Yamakawa and G. Tanaka, J. Chem. Phys. **47**, 3991 (1967).
11. A. J. Barrett, Ph. D. Thesis, University of London, 1975; quoted in Ref. 30.
12. K. F. Freed, Adv. Chem. Phys. **22**, 1 (1972).
13. S. Aronowitz and B. E. Eichinger, Macromolecules **9**, 377 (1976); Y. Chikahisa, J. Chem. Phys. **52**, 206 (1978).
14. P. J. Flory, J. Chem. Phys. **17**, 303 (1949).
15. M. Kurata, W. H. Stockmayer, and A. Roig, J. Chem. Phys. **33**, 151 (1960).
16. W. H. Stockmayer, Makromol. Chem. **35**, 54 (1960).
17. H. Reiss, J. Chem. Phys. **47**, 186 (1967).
18. S. F. Edwards, Proc. Phys. Soc. (London) **85**, 613 (1965).
19. H. Yamakawa, J. Chem. Phys. **48**, 3845 (1968).
20. P.-G. de Gennes, Phys. Lett. A **38**, 339 (1972).
21. J. C. le Guillou and J. Zinn-Justin, Phy. Rev. Lett. **39**, 95 (1977).
22. J. F. Douglas and K. F. Freed, Macromolecules **17**, 1854, 2344 (1984).
23. F. T. Wall and J. J. Erpenbeck, J. Chem. Phys. **30**, 634, 637 (1959); F. T. Wall, L. A. Hiller, Jr., and D. J. Wheeler, J. Chem. Phys. **22**, 1036 (1954); F. T. Wall, L. A. Hiller, Jr., and W. F. Atchison, J. Chem. Phys. **23**, 913, 2314 (1955); **26**, 1742 (1957).
24. C. Domb, Adv. Chem. Phys. **15**, 229 (1969).
25. C. Domb, J. Chem. Phys. **38**, 2957 (1963).
26. C. Domb and F. T. Hioe, J. Chem. Phys. **51**, 1915 (1969).
27. C. Domb and G. S. Joyce, J. Phys. C: Solid State Phys. **5**, 956 (1972).
28. Z. Alexandrowicz, Phys. Rev. Lett. **50**, 736 (1983).
29. C. Domb and A. J. Barrett, Polymer **17**, 179 (1976).
30. M. Lax, A. J. Barrett, and C. Domb, J. Phys. A: Math. Gen. **11**, 361 (1978).
31. G. Tanaka, Macromolecules **13**, 1513 (1980).
32. W. H. Stockmayer, Brit. Polym. J. **9**, 89 (1977).

33. H. Fujita, unpublished.

34. H. Suzuki, Kaigai Kobunshi Kenkyu 31, 205 (1985); Brit. Polym. J. 14, 137 (1982).

35. B. H. Zimm and W. H. Stockmayer, J. Chem. Phys. 17, 1301 (1948).

36. N. C. Smith and R. J. Fleming, J. Phys. A: Math. Gen. 8, 938 (1975).

37. B. H. Zimm, W. H. Stockmayer, and M. Fixman, J. Chem. Phys. 21, 1716 (1953).

38. H. Yamakawa, A. Aoki, and G. Tanaka, J. Chem. Phys. 45, 1938 (1966).

39. F. T. Wall and F. T. Hioe, J. Phys. Chem. 74, 4416 (1970).

40. R. C. Jurs and J. E. Reissner, J. Chem. Phys. 55, 4948 (1971).

41. K. Suzuki, Bull. Chem. Soc. Jpn 41, 538 (1968).

42. A. K. Kron and O. B. Ptitsyn, Vysokomol. Soedin. A 9, 759 (1967).

43. Y. Miyaki, Y. Einaga, and H. Fujita, Macromolecules 11, 1180 (1978).

44. Y. Miyaki, Ph. D. Thesis, Osaka University, 1981.

45. Y. Miyaki, Y. Einaga, T. Hirosye, and H. Fujita, Macromolecules 10, 1356 (1977).

46. T. Hirosye, Y. Einaga, and H. Fujita, Polym. J. 11, 819 (1979).

47. T. Matsumoto, N. Nishioka, and H. Fujita, J. Polym. Sci. A-2 10, 23 (1972).

48. T. Kato, K. Miyaso, I. Noda, T. Fujimoto, and M. Nagasawa, Macromolecules 3, 777 (1970).

49. Y. Miyaki and H. Fujita, Macromolecules 14, 742 (1981).

50. J. Mazur, J. Res. Natl. Bur. Std. 69A, 335 (1965); J. Chem. Phys. 43, 4354 (1965).

51. C. Domb, J. Gillis, and G. Wilmers, Proc. Phys. Soc. (London) 85, 625 (1965).

52. M. E. Fisher, J. Chem. Phys. 44, 616 (1966).

53. J. des Cloizeaux, J. Phys. (Paris) 41, 223 (1980).

54. A. Baumgärtner, Z. Phys. B 42, 265 (1981).

55. S. Render, J. Phys. A: Math. Gen. 13, 3525 (1980).

56. A. Peterlin, J. Chem. Phys. 23, 2464 (1955).

57. A. J. Barrett, Macromolecules 17, 1561 (1984).

58. Ref. 4; E. Teramoto, M. Kurata, and H. Yamakawa, J. Chem. Phys. 28, 785 (1958).

59. B. H. Zimm, J. Chem. Phys. 14, 164 (1946).

60. A. C. Albrecht, J. Chem. Phys. 27, 1002 (1957).

61. Y. Tagami and E. Г. Casassa, J. Chem. Phys. 50, 2206 (1969).

62. M. Kurata and H.Yamakawa, J. Chem. Phys. 29, 311 (1958).

63. M. Kurata, M. Fukatsu, H. Sotobayashi, and H. Yamakawa, J. Chem. Phys.41, 139 (1964).

64. G. Tanaka and K. Šolc, Macromolecules 15, 791 (1982).

65. H. Yamakawa, J. Chem. Phys. 48, 2103 (1968).

66. A. J. Barrett, Macromolecules 18, 196 (1985).

67. T. Norisuye, K. Kawahara, A. Teramoto, and H. Fujita, J. Chem. Phys. 49, 4330 (1968).

68. G. C. Berry, J. Chem. Phys. 44, 4550 (1966).

69. K. Takashima, G. Tanaka, and H. Yamakawa, Polym. J. 2, 245 (1971).

70. M. Daoud and G. Jannink, J. Phys. (Paris) 37, 973 (1976).

71. M. Fukuda, M. Fukutomi, Y. Kato, and T. Hashimoto, J. Polym. Sci., Polym. Phys. Ed. 12, 871 (1974).

72. S. Akita, Y. Miyaki, and H. Fujita, Macromolecules 9, 774 (1976); T. Hirosye, Y. Einaga, and H. Fujita, Polym. J. 11, 819 (1979).

73. W. Gobush, K. Šolc, and W. H. Stockmayer, J. Chem. Phys. 60, 12 (1974).

74. J. F. Douglas and K. F. Freed, Macromolecules 17, 1854 (1984).

75. H. Fujita and T. Norisuye, Macromolecules 18, 1637 (1985).

76. H. Huber, S. Bantle, P. Lutz, and W. Burchard, Macromolecules 18, 1461 (1985).

77. K. Huber and W. H. Stockmayer, Macromolecules 20, 1400 (1987).

78. B. K. Varma, H. Fujita, M. Takahashi, and T. Nose, J. Polym. Sci., Polym. Phys. Ed. 22, 1718 (1984).

79. M. Ragnetti, D. Geiser, H. Höcker, and R. G. Oberthür, Makromol. Chem. 186, 1701 (1985).

80. S. Bantle, M. Schmidt, and W. Burchard, Macromolecules 15, 1604 (1982).

81. P. J. Flory and W. R. Krigbaum, J. Chem. Phys. 18, 1086 (1950).

82. H. Sotobayashi and J. Springer, Adv. Polym. Sci. 6, 473 (1969).

83. L. Zhang, D. Qui, and R. Quian, Polym. J. 17, 657 (1985).

84. Z. Tong, S. Ohashi, Y. Einaga, and H. Fujita, Polym. J. 15, 835 (1983).

85. N. Takano, Y. Einaga, and H. Fujita, Polym. J. 17, 1123 (1985).

86. R. Perzynski, M. Delsanti, and M. Adam, J. Phys. (Paris) 48, 115 (1987).

87. J. des Cloizeaux, J. Phys. (Paris) 42, 635 (1981); J. des Cloizeaux and I. Noda, Macromolecules 15, 1505 (1982).

88. J. F. Douglas and K. F. Freed, Macromolecules 18, 201 (1985).

89. H. Vink, Eur. Polym. J. 10, 149 (1974).

90. K. R. Kniewske and W.-H. Kulicke, Makromol. Chem. 184, 2173 (1983).

91. T. Sato, T. Norisuye, and H. Fujita, J. Polym. Sci., Polym. Phys. Ed. 25, 1 (1987).

92. P. J. Flory and H. Daoust, J. Polym. Sci. 25, 429 (1957).

93. J. G. Kirkwood and J. Riseman, J. Chem. Phys. 16, 565 (1948).

94. P. Debye and A. M. Bueche, J. Chem. Phys. 16, 573 (1948).

95. J. G. Kirkwood, J. Polym. Sci. 12, 1 (1953).

96.   P. E. Rouse, J. Chem. Phys. **21**, 1272 (1953).
97.   F. Bueche, J. Chem. Phys. **22**, 603 (1954).
98.   B. H. Zimm, J. Chem. Phys. **24**, 269 (1956).
99.   Y. Einaga, Y. Miyaki, and H. Fujita, J. Polym. Sci., Polym. Phys. Ed. **17**, 2103 (1979).
100. G. Meyerhoff and B. Appelt, Macromolecules **12**, 2103 (1979).
101. W. Kuhn, Kolloid Z. **68**, 2 (1934).
102. H. Yamakawa and M. Fujii, Macromolecules **7**, 128 (1974).
103. H. Fujita, N. Taki, T. Norisuye, H. Sotobayashi, and H. Fujita, J. Polym. Sci., Polym. Phys. Ed. **15**, 2255 (1977).
104. J. Shimada and H. Yamakawa, J. Polym. Sci., Polym. Phys. Ed. **16**, 1927 (1978).
105. W. H. Stockmayer and A. C. Albrecht, J. Polym. Sci. **32**, 215 (1958).
106. G. Weill and J. des Cloizeaux, J. Phys. (Paris) **40**, 99 (1979).
107. G. Tanaka, Macromolecules **15**, 1028 (1981).
108. A. J. Barrett, Macromolecules **17**, 1566 (1984).
109. P. J. Flory and T. G Fox, Jr., J. Am. Chem. Soc. **73**, 1904 (1951).
110. L. Mandelkern and P. J. Flory, J. Chem. Phys. **20**, 212 (1952).
111. Y. Miyaki, H. Fujita, and M. Fukuda, Macromolecules **13**, 588 (1980).
112. M. Schmidt and W. Burchard, Macromolecules **14**, 210 (1981).
113. T. Yoshizaki and H. Yamakawa, Macromolecules **10**, 539 (1977).
114. C. W. Pyun and M. Fixman, J. Chem. Phys. **42**, 3838 (1965); **44**, 2107 (1966).
115. M. Bixon and R. Zwanzig, J. Chem. Phys. **68**, 1890 (1978).
116. B. H. Zimm, Macromolecules **13**, 592 (1980).
117. J. G. de la Torre, A. Jimenez, and J. J. Freire, Macromolecules **15**, 148 (1982).
118. M. Fixman, Macromolecules **14**, 1710 (1981).
119. Y. Oono, Adv. Chem. Phys. **61**, 301 (1985).
120. Y. Oono and M. Kohmoto, J. Chem. Phys. **78**, 520 (1983).
121. S.-Q. Wang, J. F. Douglas, and K. F. Freed, J. Chem. Phys. **85**, 3674 (1986).
122. S.-Q. Wang, J. F. Douglas, and K. F. Freed, Macromolecules **18**, 2464 (1985).
123. T. Yoshizaki, I. Nitta, and H. Yamakawa, Macromolecules **21**, 165 (1988).

Chapter 3   Renormalization Group Theory

1.   Basic Ideas and Formulation

1.1   Introduction

The renormalization group (RG) theory has been developed to solve many-body problems in physics. de Gennes [1] was the first to point out its potential for approaching the molecular weight dependence of global properties of polymers with excluded–volume interactions. Since then much has been worked out to formulate a new theory of polymer solutions. This development has occurred after Yamakawa's book [2] appeared. Oono [3] wrote a good review article and Freed [4] published a comprehensive volume describing various applications of the RG theory to polymer systems. However, these works are quite sophisticated in context and hence considerable backgrounds in mathematics and physics are needed to get through them. This chapter outlines the basic ideas and techniques of the RG theory in elementary terms and illustrates their applications to some static properties of linear polymers. The treatment of dynamic behavior by the RG theory is more complicated and still at the development stage. Thus we will not concern ourselves with it in this introductory account.

1.2   Renormalization

In Chapter 1 we defined a global polymer property as one which does not reflect the microscopic details of the polymer molecule but is determined by its large–scale characteristics such as chain length and flexibility. However, the existence of such a property cannot be justified theoretically in advance, but has to be taken as a postulate. The RG theory is formulated and developed on this postulate.

Since global polymer behavior is not affected by local details of the polymer chain, its formulation may be made by use of a suitably coarse–grained polymer chain. We may choose the random flight chain, the spring–bead chain, the Edwards continuous chain, and others, depending on the global property concerned, the method of formulation, and the accuracy of the results to be derived. In what follows, we are concerned with the theories developed with the Edwards continuous chain.

As explained in Section 3 of Chapter 1, the Hamiltonian H of the Edwards continuous chain contains three parameters: chain length L, excluded–volume strength v, and cut–off length λ. Thus its equilibrium global property Q depends on $L, v,$ and λ, i.e., $Q = Q(L, v, \lambda)$. We allow λ to vary at fixed L and v. Then, H and hence Q may undergo changes. If the change in Q can be compensated, i.e., Q is made invariant, by appropriate variations of L and v, we say that Q is renormalizable and refer to this process as **renormalization**. The RG theory is the method dealing with renormalizable physical properties.

We denote any renormalizable quantity of the Edwards continuous chain by Q. Then there exist an infinite number of coarse–grained Edwards chains, each having different λ but the same value of Q. Such chains are called equivalent with respect to Q and their ensemble is called the renormalization group.

1.3 Universality Assumption

In Chapter 1, the Edwards chain in the limit of $\lambda \to 0$ was named the primary chain. We denote L and v for this chain by L_0 and v_0, and define $Q_B(L_0, v_0)$ by

$$Q_B(L_0, v_0) = \lim_{\lambda \to 0} Q(L, v, \lambda) \tag{1.1}$$

The condition that Q is renormalizable can then be expressed by saying that there exist transformations

$$L = f(L_0, v_0, \lambda) \tag{1.2}$$

$$v = g(L_0, v_0, \lambda) \tag{1.3}$$

which, for any value of $\lambda > 0$, satisfy the relation

$$Q(L, v, \lambda) = Q_B(L_0, v_0) \tag{1.4}$$

Since this relation should hold for any Q, the functions f and g must be independent of the kind of Q, i.e., universal. We take this expectation as a postulate and call it the **universality assumption.**

1.4 Renormalization Constants

The RG theory, unlike the two–parameter theory described in Chapter 2, deals with chains in d dimensions, where d is any positive number equal to or smaller than 4. This maneuver takes advantage of the fact that, as will be shown below, chains in 4 dimensions undergo no excluded–volume effect in any solvent (in the binary cluster approximation).

To start with, we note that, as mentioned in Section 3.2 of Chapter 1, the Hamiltonian of a d–dimensional Edwards continuous chain is given by eq 1–3.26 with $\mathbf{c}(s)$ and v defined as follows:

$$\mathbf{c}(s) = (d/b)^{1/2} \mathbf{r}(s) \tag{1.5}$$

$$v = (d)^{d/2} \beta_c b^{-(4+d)/2} \tag{1.6}$$

It can be readily checked that v has dimensions of (length)$^{-\epsilon/2}$. Here, ϵ is defined by

$$\epsilon = 4 - d \tag{1.7}$$

This parameter plays a very important role in RG methods.

When L_0 is doubled at fixed v, L for any λ is doubled. Hence, eq 1.2 may be rewritten $L = L_0 Z_L(v_0, \lambda)$. Since Z_L is dimensionless, it should depend on a single dimensionless variable made up of v_0 and λ. We may choose u_0 defined by

$$u_0 = v_0 \, \lambda^{\epsilon/2}/4\pi^2 \tag{1.8}$$

as such a variable. Correspondingly, we may define a dimensionless quantity u by

$$u = v \, \lambda^{\epsilon/2}/4\pi^2 \tag{1.9}$$

Since v_0 and v are supposed to be independent of chain length, eq 1.3 may be rewritten $v = v_0 Z_u(v_0, \lambda)$ or $u = u_0 Z_u(v_0, \lambda)$. Since Z_u is dimensionless, it should depend only on u_0 made up of v_0 and λ. In this way, we find that

$$L = L_0 Z_L(u_0) \tag{1.10}$$

$$u = u_0 Z_u(u_0) \tag{1.11}$$

which may be solved for L_0 and u_0 to give

$$L_0 = L Y_L(u) \tag{1.12}$$

$$u_0 = u Y_u(u) \tag{1.13}$$

The quantities Z_L, Z_u, Y_L, and Y_u are called renormalization constants. According to the above argument, they should be universal functions of the respective variables.

1.5   Renormalization Group Calculation

1.5.1   Practical Significance

Suppose we have been able to determine $Y_L(u)$ and $Y_u(u)$ as well as $Q_B(L_0, v_0)$ by some theoretical means. Since the latter can be written $Q_B(L Y_L(u), u Y_u(u)\lambda^{-\epsilon/2})$, we find out how Q_B depends on L, u, and λ. According to eq 1.4, this dependence gives us quantitative information about Q of the Edwards chain coarse-grained with cut-off length λ as a function of L and u. Then one may ask what practical value the function $Q(L, u, \lambda)$ so obtained has.

As will be explained in Section 1.7, the L dependence of $Q(L, u, \lambda)$ in a fixed solvent changes to that in a better solvent as λ is allowed to increase. In other words, by coarse–graining it is possible to look at what happens to the L dependence of Q when the solvent quality is changed from poor to good. In the two–parameter theory, the effect of solvent quality can be examined only by changing the parameter β or v. But, according to the RG theory, the solvent effect can be revealed by varying λ at fixed v. Thus if the asymptotic form of Q at large λ becomes available, it is possible to obtain information about how Q in very good solvents depends on L, i.e., the molecular weight dependence of Q of self-avoiding chains can be deduced. This possibility answers the question mentioned above. We should recall that by the two–parameter theory it is hopelessly difficult to explore analytically the behavior of self-avoiding chains.

1.5.2   Determination of Renormalization Constants

We must know $Y_L(u)$ and $Y_u(u)$ to convert $Q_B(L_0, v_0)$ into $Q(L, u, \lambda)$. In the primary chain, excluded–volume interactions can act between infinitesimally close points on the chain, since the chain is allowed to take on conformations which contain infinitesimally small loops. Owing to this property, unless v_0 is zero, Q_B diverges to infinity as $\epsilon \to 0$, as will be illustrated below. However, no such singularity appears in a coarse–grained chain with cut–off length λ because intrachain interactions in it do not act between contour points separated by less than λ. Therefore, $Q(L, u, \lambda)$ should remain finite in the limit $\epsilon \to 0$ if $\lambda \neq 0$. The renormalization constants Y_L and Y_u can be determined from this physical requirement, as exemplified below. Because of this operation, renormalization is sometimes misunderstood as a mere mathematical maneuver of removing singular behavior of the primary chain in the limit of $\epsilon \to 0$.

1.6   Renormalization Group Equation

Equation 1.4 may be rewritten $Q(L, u, \lambda) = Q_B(L_0, v_0)$. Differentiation of this with respect to λ at constant L_0 and v_0 gives

$$\lambda \frac{\partial}{\partial \lambda} Q(L, u, \lambda)|_{L_0, v_0} = 0 \qquad (1.14)$$

which is called the **renormalization group equation**. This equation can be rewritten

$$\left[\lambda \frac{\partial}{\partial \lambda} + A(u) \frac{\partial}{\partial u} + LB(u) \frac{\partial}{\partial L} \right] Q(L, u, \lambda)|_{L_0, v_0} = 0 \qquad (1.15)$$

73

where

$$A(u) = \lambda \left(\frac{\partial}{\partial \lambda} \right) u|_{v_0} = \lambda \left(\frac{\partial}{\partial \lambda} \right) Z_u(u_0) u_0|_{u_0 \to u} \qquad (1.16)$$

$$B(u) = \lambda \left(\frac{\partial}{\partial \lambda} \right) \ln Z_L(u_0)|_{u_0 \to u} \qquad (1.17)$$

where the notation $u_0 \to u$ signifies replacing u_0 in the derivatives by the right–hand side of eq 1.13.

Equation 1.15 is a partial linear differential equation of first order for $Q(L, u, \lambda)$ and can be integrated by the method of characteristics to give a general solution

$$Q = F \left(L \exp\left[-\int^u B(x)\, dx/A(x)\right], \quad \lambda \exp\left[-\int^u dx/A(x)\right] \right) \qquad (1.18)$$

where F denotes an arbitrary function.

1.7 Fixed Point and Solvent Effect

Because v_0 is proportional to the excluded–volume strength, u_0 increases from zero to infinity as the solvent is varied from a θ solvent to an extremely good one, unless λ is zero (see eq 1.8). Hence, the solvent effect on $Q(L, u, \lambda)$ can be seen by knowing how u varies with u_0 through eq 1.13. According to eq 1.8, u_0 can be varied by either changing v_0 at fixed λ or changing λ at fixed v_0. Thus, it is possible to calculate the solvent effect on Q by varying λ in a fixed non–θ solvent. This possibility is one of the great advantages of the RG theory. In particular, in the limit of $\lambda \to \infty$, we should be able to see asymptotic behavior of the chain in an extremely good solvent, i.e., the self–avoiding chain.

According to eq 1.13, u vanishes at $u_0 = 0$, i.e., under the θ condition. However, this equation does not imply that u diverges as u_0 increases indefinitely. Actually, u cannot increase more than a finite value u^* for which the integrands in eq 1.18 diverge, i.e., which satisfies the condition

$$A(u^\star) = 0 \qquad (1.19)$$

This limiting value of u corresponds to $u_0 = \infty$, and is called the fixed point of the system. Thus, the L dependence of Q of the self–avoiding chain is given by $Q(L, u^\star, \lambda)$.

As shown in Section 2, $u^\star$ for d–dimensional chains is given by

$$u^\star = \epsilon/8 + O(\epsilon^2) \qquad (1.20)$$

74

if d is close to but smaller than 4. Hence, u is zero for chains in 4 dimensions and remains in the order of ϵ for chains in space close to 4 dimensions whatever the solvent quality is. This fact suggests a method of perturbation calculation in which $Q(L, u, \lambda)$ for such chains is expanded in powers of the variable u. With the anticipation that $u \ll 1$ when v_0 is very small, this method may be carried out as follows. First, $Q_B(L_0, v_0)$ is calculated as a power series in v_0, then L_0 and v_0 in the series are converted into L and u by using the renormalization constants, and finally the result obtained is taken as the desired $Q(L, u, \lambda)$. Some examples valid up to first order in ϵ are shown in the next section.

2.   First–order Theory

2.1   Distribution Function of End Distance

From now on, for the simplicity of presentation, length is measured in units of the Kuhn segment length b. Thus $\mathbf{R}$, L, λ, etc. appearing below in this chapter are all dimensionless, i.e., real length divided by b. As $Q(L, u, \lambda)$ we first consider $G(\mathbf{R}; L, u, \lambda)$, the normalized distribution function of the end–to–end distance $\mathbf{R}$ of a d–dimensional Edwards chain. It can be shown that $G_B(\mathbf{R}; L_0, v_0)$ is given to first order in v_0 by

$$G_B = G_{B\theta} + v_0 \int_0^{L_0} \int_{s'}^{L_0} [G_{B\theta}(\mathbf{R}; L_0) G_{B\theta}(\mathbf{O}_{ss'})$$
$$G_{B\theta}(\mathbf{R}, \mathbf{O}_{ss'}; L_0)]\, ds\, ds' + O(v_0{}^2) \quad (2.1)$$

Here the subscript θ signifies the value for the primary chain under the θ condition ($v_0 = 0$), and

$$G_{B\theta}(\mathbf{R}; L_0) = (2\pi L_0)^{-d/2} \exp(-R^2/2L_0) \quad (2.2)$$

$$G_{B\theta}(\mathbf{O}_{ss'}) = [2\pi(s - s')]^{-d/2} \quad (2.3)$$

$$G_{B\theta}(\mathbf{R}, \mathbf{O}_{ss'}; L_0) = (2\pi)^{-d}[(s - s')(L_0 - s + s')]^{-d/2}$$
$$\times \exp[-R^2/(L_0 - s + s')] \quad (2.4)$$

After integration, eq 2.1 leads to

$$G_B = (2\pi L_0)^{-d/2} \exp(-R^2/2L_0)$$
$$\times \{1 - (v_0/4\pi^2)(2\pi L_0)^{\epsilon/2}[(2/\epsilon)(2 - R^2/2L_0)$$
$$+ (R^2/2L_0 - 1)(\gamma - 1 + \ln(R^2/2L_0)] + O(v_0{}^2)\} \quad (2.5)$$

where γ is the Euler constant. Equation 2.5 diverges as ϵ tends to zero. This is an example of the general fact that $Q_B(L_0, v_0)$ diverges in 4 dimensions.

75

With no loss of generality we may choose $Y_L(0) = Y_u(0) = 1$. Since $u \to 0$ as $v_0 \to 0$, it is relevant for the consideration of small v_0 to expand $Y_L(u)$ and $Y_u(u)$ as

$$Y_L(u) = 1 + a_1 u + a_2 u^2 + \ldots \tag{2.6}$$

$$Y_u(u) = 1 + b_1 u + b_2 u^2 + \ldots \tag{2.7}$$

In order that the distribution function under consideration is renormalizable, it is necessary that the function obtained by transforming L_0 and v_0 to L and u with the aid of eq 1.8, 1.12, and 1.13 as well as eq 2.6 and 2.7 gives the distribution function $G(\mathbf{R}; L, u, \lambda)$ of a coarse–grained chain with cut–off length λ. Furthermore, for the reason mentioned in Section 1.5.2, the resulting function has to remain finite at $\epsilon = 0$. It can be shown that if a_1 is chosen to be $-2/\epsilon$ this requirement is met to first order in u (in other words, terms of second and higher order in u may diverge as $\epsilon \to 0$). In this way, correct to first order in u, we obtain

$$G(\mathbf{R}; L, u, \lambda) = (2\pi L)^{-d/2} \exp(-R^2/2L)\{1 + u[(R^2/2L) \\ + (1 - R^2/2L)(\gamma + \ln(R^2/2L))(2 - (R^2/2L)\ln(2\pi L/\lambda))]\} \tag{2.8}$$

This result does not agree with that reported by Oono[3]. The difference is due to the fact that his G is not normalized.

To first order in u the value of b_1 cannot be determined. As shown below, it is evaluated from the first–order perturbation analysis of the second virial coefficient as $8/\epsilon$. Thus we find

$$Y_L(u) = 1 - (2/\epsilon)u + \ldots \tag{2.9}$$

$$Y_u(u) = 1 + (8/\epsilon)u + \ldots \tag{2.10}$$

Equation 1.16 can be rewritten

$$A(u) = (\epsilon u/2)/[1 + u(d\ln Y_u/du)] \tag{2.11}$$

which, on substitution of eq 2.10, gives

$$A(u) = (u/2)[\epsilon - 8u + O(u^2)] \tag{2.12}$$

Equation 1.20 immediately follows from this.

Equation 2.8 is supposed to hold if $u \ll 1$. It follows from eq 1.20 that $u^\star \ll 1$ if $\epsilon \ll 1$. Therefore we may conclude that eq 2.8 is valid regardless of solvent quality for chains whose dimensionality is close enough to 4. But, for

3–dimensional chains, this equation may be of use only in poor solvents near θ, since the condition $u^\star \ll 1$ may no longer hold for $\epsilon = 1$. In order to determine $G(\mathbf{R}; L, u, \lambda)$ for 3–dimensional chains in good solvents we must proceed to second–order (in u) perturbation calculations, though a considerable amount of work would be needed for carrying them through.

2.2   Mean–square End Distance

The mean–square end distance $\langle R^2 \rangle$ of a d–dimensional coarse–grained Edwards chain is given by

$$\langle R^2 \rangle = \frac{d\pi^{d/2}}{\Gamma(d/2 + 1)} \int_0^\infty R^{d+1} G(\mathbf{R}; L, u, \lambda)\, dR \tag{2.13}$$

where Γ denotes the gamma function. Substitution of eq 2.8 followed by integration leads to

$$\langle R^2 \rangle = Ld\{1 + u[\ln(2\pi L/\lambda) - 1] + O(u^2)\} \tag{2.14}$$

which, to first order in u, is equivalent to

$$\langle R^2 \rangle = Ld(2\pi L/\lambda)^u (1 - u) \tag{2.15}$$

We note that this transformation uses the relation

$$a^x = 1 + x \ln a + x^2 (\ln a)^2 + \dots \quad (x \sim 0) \tag{2.16}$$

In its mathematical developments the RG theory of polymer solutions often takes advantage of the approximation obtained by truncating this series at the term linear in x. This approximation is valid for any large a if the condition $x \ln a \ll 1$ is satisfied. Hence it is useful in formulating asymptotic behavior at $a \gg 1$ when x is sufficiently small.

Since $u^\star \sim \epsilon/8$ for chains close to 4 dimensions, $\langle R^2 \rangle$ of such a chain in the self–avoiding limit is given by replacing u in eq 2.15 with $\epsilon/8$, i.e.,

$$\langle R^2 \rangle \sim L^{1+(\epsilon/8)} \tag{2.17}$$

For $\epsilon = 1$ this gives

$$\langle R^2 \rangle \sim L^{1.125} \tag{2.18}$$

which must be inaccurate since eq 1.20 cannot be expected to hold up to $\epsilon = 1$. In fact, the most accurate asymptotic behavior of $\langle R^2 \rangle$ for 3–dimensional chains predicted by the RG theory is

$$\langle R^2 \rangle \sim L^{1.176} \tag{2.19}$$

77

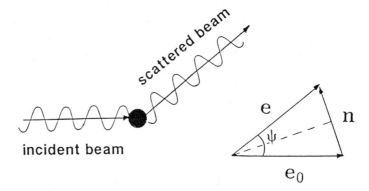

Fig. 3–1. Definition of the scattering vector **k**. e_0 and e are unit vectors in the directions of incident and scattering light, respectively. $\mathbf{n} = \mathbf{e} - \mathbf{e}_0$. $\mathbf{k} \equiv (2\pi/\Lambda)\mathbf{n}$. $|\mathbf{n}| = 2\sin(\psi/2)$.

which is due to le Guillou and Zinn–Justin[5]. Unfortunately, no success has as yet been achieved in the calculation of the prefactor of this proportionality.

In the binary cluster approximation considered here the θ condition corresponds to $v_0 = 0$, which gives $u = 0$. Thus we obtain from eq 2.15

$$\langle R^2 \rangle_\theta = Ld \tag{2.20}$$

which is exact for any d. The end distance expansion factor α_R of a d–dimensional chain is obtained from eq 2.15 and 2.20 to be

$$\alpha_R{}^2 = (2\pi L/\lambda)^u (1 - u) \tag{2.21}$$

correct to first order in u. It should be noted that the RG theory does not allow α_R to be expressed in terms of a single variable, contrary to the two-parameter theory.

2.3 Particle Scattering Function

The particle scattering function $P_B(k)$ for the primary chain of d–dimensions can be expressed by (see Yamakawa's book [2])

$$P_B(k) = \frac{2}{L_0{}^2} \int_0^{L_0} \int_{s'}^{L_0} ds\, ds' \int W_B(s' - s, \mathbf{r}_{ss'}) \exp(i\mathbf{k} \cdot \mathbf{r}_{ss'})\, d^d\mathbf{r}_{ss'} \tag{2.22}$$

Here, **k** is the scattering vector (see Figure 3–1 for the definition), $k = |\mathbf{k}| = (4\pi/\Lambda)\sin(\psi/2)$, with Λ the wavelength of incident light in the solution and ψ

78

the scattering angle, W_B the normalized distribution function of the two variables indicated, $\mathbf{r}_{ss'}$ the vector connecting contour points s and s' of the chain, and $d^d\mathbf{r}_{ss'}$ the volume element in space of dimension d. Actually, W_B contains L_0 and v_0 as parameters. Expanding W_B in powers of v_0 and substituting the result into eq 2.22, Ohta et al. [6] obtained correct to first order in u_0 after some complex calculations

$$P_B(k) = P_{B\theta} - 2u_0(2\pi L_0/\lambda)^{\epsilon/2}[(2/\epsilon)X(\sigma_0) + Y(\sigma_0)] \tag{2.23}$$

where

$$P_{B\theta} = 2[\sigma_0^{-1} - \sigma_0^{-2}(1 - e^{-\sigma_0})] \quad \text{(Debye function)} \tag{2.24}$$

$$X(y) = y^{-1}[1 + e^{-y} - 2y^{-1}(1 - e^{-y})] \tag{2.25}$$

$$\begin{aligned}
Y(y) = \int_0^1 &\{2y^{-2}e^{-y}f(y - yt + yt^2) \\
&- 2y^{-2}(1-y)f(-yt + yt^2) - 2(1 - e^{-yt+yt^2})[y^2t(1t)]^{-1} \\
&+ [y(1-t)]^{-1} - (1 - e^{-yt+yt^2})[y^2t(1-t)]^{-1}\} \, dt \\
&+ y^{-1}(1 + e^{-y}) - y^{-2}(1 - e^{-y}) - y^{-2}(1-y)e^{-y}f(y)
\end{aligned} \tag{2.26}$$

with

$$\sigma_0 = L_0 k^2/2 \tag{2.27}$$

$$f(y) = \int_0^y t^{-1}(e^t - 1) \, dt \tag{2.28}$$

According to the universality assumption, the renormalization constants should be the same for all renormalizable Q. Hence we may transform L_0 and u_0 in the above expression for $P_B(k)$ to L and u by eq 2.9 and 2.10. To first order in u, u_0 in eq 2.23 is simply replaced by u, and the term containing $1/\epsilon$ in this equation disappears by the transformation $L_0 \to L$. The resulting equation gives, correct to first order in u, the particle scattering function $P(k; L, u, \lambda)$ for a chain with cut–off length λ, i.e.,

$$P(k; L, u, \lambda) = 2[\sigma^{-1} - \sigma^{-2}(1 - e^{-\sigma})uY(\sigma)] \tag{2.29}$$

where

$$\sigma = (1/2)Lk^2(2\pi L/\lambda)^u \tag{2.30}$$

79

In a θ solvent where u vanishes, eq 2.29 reduces to the well-known Debye function. On the other hand, if u is replaced by its upper bound $u^\star = \epsilon/8$ valid for $\epsilon \to 0$, eq 2.29 gives the particle scattering function for self-avoiding chains of dimensions sufficiently close to 4. Ohta et al. [6] concluded from a numerical comparison of these extreme cases that the particle scattering function does not undergo a significant excluded-volume effect. This conclusion is consistent with what has long been noticed with actual polymers in various solvents, but the agreement should be accepted with reservation, since the theoretical result is concerned only with the case $\epsilon \sim 0$. We take note of a recent study by Tsunashima and Kurata [7], which showed that $P(k)$ for polystyrene in toluene deviated significantly from the Debye function in the region of high k.

2.4 Mean-square Radius of Gyration

If eq 2.29 is expanded in powers of σ we obtain

$$P(k; L, u, \lambda) = 1 - (\sigma/3)[1 - (13/12)u]$$
$$+ (\sigma^2/12)[1 - (79/30)u] + \ldots \quad (2.31)$$

On the other hand, in general, it can be shown that $P(k; L, u, \lambda)$ of a d-dimensional chain is expressed by a power series in k^2 as

$$P(k; L, u, \lambda) = 1 - (1/d)\langle S^2 \rangle k^2 + \ldots \quad (2.32)$$

where $\langle S^2 \rangle$ is the mean-square radius of gyration of the chain. This series can be derived from an expression that is obtained by removing the subscripts B and 0 from eq 2.22. Comparison of eq 2.31 and 2.32, with eq 2.30 taken into account, gives

$$\langle S^2 \rangle = (Ld/6)(2\pi L/\lambda)^u [1 - (13/12)u] \quad (2.33)$$

which is correct to first order in u. To the same approximation we obtain from eq 2.15 and 2.33

$$\langle S^2 \rangle / \langle R^2 \rangle = (1/6)[1 - (1/12)u] \quad (2.34)$$

For $\epsilon \sim 0$ this ratio in the self-avoiding limit is equal to $(1/6)[1 - (\epsilon/96)]$, which yields 0.165 for $\epsilon = 1$. As mentioned in Chapter 2, the corresponding value estimated from computer simulation experiments is 0.155. The difference cannot be neglected simply.

2.5 Second Virial Coefficient

The second virial coefficient $A_{2B}(L_0, v_0)$ for the primary chain of d dimensions is expressed, correct to first order in v_0, as follows:

$$A_{2B} = \frac{N_A v_0}{2VM^2}[\int \cdots \int \delta_{ss'}^{(2)} P(\{\mathbf{r}_1\}) P(\{\mathbf{r}_2\}) \, ds \, ds' \, d^d\{\mathbf{r}_1\} d^d\{\mathbf{r}_2\}$$
$$- \frac{v_0}{2} \int \cdots \int \delta_{ss'}^{(2)} \delta_{tt'}^{(2)} P(\{\mathbf{r}_1\}) P(\{\mathbf{r}_2\}) \, ds \, ds' \, dt \, dt' \, d^d\{\mathbf{r}_1\} \, d^d\{\mathbf{r}_2\}] \tag{2.35}$$

where N_A is the Avogadro constant, V the volume of the solution, $P(\{\mathbf{r}_p\})$ the normalized one–body distribution of chain p (hopefully, not confused with the particle scattering function $P(k)$), $\{\mathbf{r}_p\}$ the instantaneous conformation of chain p, and $\delta_{ss'}^{(2)}$ a symbol that is equal to unity when the contour point s of chain 1 and the contour point s' of chain 2 are in contact and to zero otherwise. Equation 2.35 can be obtained by transforming eq 2-2.2 into the continuous chain and then expanding the exponential term in powers of v_0. Correct to first order in v_0, $P(\{\mathbf{r}_p\})$ may be expressed by

$$P(\{\mathbf{r}_p\}) = \frac{P_\theta(\{\mathbf{r}_p\}) - (v_0/2) \iint \delta_{ss'}^{(1)} P_\theta(\{\mathbf{r}_p\}) \, ds \, ds'}{1 - (v_0/2) \int \cdots \int \delta_{ss'}^{(1)} P_\theta(\{\mathbf{r}_p\}) \, ds \, ds' \, d^d\{\mathbf{r}_p\}} \tag{2.36}$$

where $\delta_{ss'}^{(1)}$ is the single chain version of $\delta_{ss'}^{(2)}$.

After some complex calculations it follows from eq 2.35 with eq 2.36 that, correct to first order in u_0,

$$A_{2B} = (N_A/2)(2\pi L_0/M)^2 u_0 \lambda^{-\epsilon/2}$$
$$\times \{1 - u_0[(4/\epsilon) + 1 - 4\ln 2 + 2\ln(2\pi L_0/\lambda)]\} \tag{2.37}$$

Transforming L_0 and u_0 in this equation into L and u by use of eq 2.6 and 2.7, with $a_1 = -2/\epsilon$ taken into account, we find that the singularity due to $1/\epsilon$ disappears if b_1 is chosen to be $8/\epsilon$. Equation 2.10 is derived in this way. The resulting equation for A_{2B} yields $A_2(L, u, \lambda)$ if the second virial coefficient is renormalizable. Thus, to first order in u, we obtain

$$A_2 = 2N_A(\pi L/M)^2 u \lambda^{-\epsilon/2} (2\pi L/\lambda)^{-2u}[1 - u(1 - 4\ln 2)] \tag{2.38}$$

This formula was originally derived by Oono and Freed [8], but it was found to be in error by subsequent authors [9]. The derivation of the correct result needs Y_u valid to second order in u and is described in Section 3.7.

81

2.6 Remarks

Calculations of global polymer properties by the RG theory take advantage of the fact that u remains small in space close to 4 dimensions regardless of solvent. Thus, $Q(L, u, \lambda)$ is expanded in powers of u and the coefficients of the expansion are evaluated perturbatively. Since, as eq 1.20 indicates, the upper bound of u, i.e., $u^\star$, is of the order of ϵ, the resulting series in u can be converted into an expansion in ϵ. The question that always arises is how many terms of this ϵ-expansion must be determined to make reasonably accurate estimates of $Q(L, u, \lambda)$ for 3-dimensional chains ($\epsilon = 1$), especially in good solvents. No guiding principle is available, however. As shown above, the calculations to first order in ϵ yielded results that were quantitatively not very encouraging. Hence, it was natural that many authors attempted to calculate at least the terms of second-order in ϵ, expecting that more accurate predictions for 3-dimensional chains could be attained. Reviewing extensive literature on such attempts is beyond the author's ability. Here we refer to an example of second-order calculation due to Douglas and Freed [10].

3. Theory of Douglas and Freed

3.1 Basic Assumption and Relations

The theory of Douglas and Freed is applicable to global quantities Q which do not vanish under the θ condition. Hence, the second virial coeffiicient is excluded from its scope. Here we consider the case in which Q is $\langle R^2 \rangle$ for a d-dimensional Edwards chain. As noted in Section 1.4, the Hamiltonian H of this chain is given by eq 1-3.26 with $c(s)$ and v defined by eq 1.5 and 1.6, respectively. Hence, H is a functional of $c(s)$ and a function of L, v, and λ. We express this fact as

$$H = \mathcal{H}(c(s); L, u\lambda^{-\epsilon/2}, \lambda) \tag{3.1}$$

where eq 1.9 has been taken into account. By definition $\langle R^2 \rangle$ is given by

$$\langle R^2 \rangle = \int \mathbf{R}^2 G(\mathbf{R}) d^d \mathbf{R} / \int G(\mathbf{R}) d^d \mathbf{R} \tag{3.2}$$

where $G(\mathbf{R})$ is the distribution function for the chain end distance $\mathbf{R}$ related to the Hamiltonian by

$$G(\mathbf{R}) = \frac{\int D[\mathbf{r}(s)] \delta(\mathbf{R} - \mathbf{r}(L)\mathbf{r}(0)) \exp(-H/k_\mathrm{B}T)}{\int \int D[\mathbf{r}(s)] \delta(\mathbf{R} - \mathbf{r}(L) - \mathbf{r}(0)) \exp(-H/k_\mathrm{B}T) d^d \mathbf{R}} \tag{3.3}$$

With eq 3.1 substituted, this equation yields

$$G(\mathbf{R}) = \mathcal{G}(\mathbf{R}; L, u\lambda^{-\epsilon/2}, \lambda) \qquad (3.4)$$

Hence, it follows from eq 3.2 that

$$\langle R^2 \rangle = \mathcal{F}(L, u\lambda^{-\epsilon/2}, \lambda) \qquad (3.5)$$

Applying eq 1.18, the general solution to the renomalization group equation, to $\langle R^2 \rangle$, we see that $\mathcal{F}(L, u\lambda^{-\epsilon/2}, \lambda)$ is a function of two combined variables $L \exp\left[-\int^u B(x)\, dx/A(x)\right]$ and $\lambda \exp\left[-\int^u dx/A(x)\right]$. Thus we have the relation

$$\begin{aligned}
\mathcal{F}(L, & u\lambda^{-\epsilon/2}, \lambda) \\
&= F\left(L \exp\left[-\int^u B(x)\, dx/A(x)\right], \lambda \exp\left[-\int^u dx/A(x)\right]\right)
\end{aligned} \qquad (3.6)$$

We measure the contour length s of the chain by a new scale $t(>0)$, and introduce a conformation function $c'(s/t)$ defined by

$$c'(s/t) = t^{-1/2} c(s) \qquad (3.7)$$

Then H can be expressed as

$$H = \mathcal{H}'(c'(s); L/t, u(t/\lambda)^{\epsilon/2}, \lambda/t) \qquad (3.8)$$

where

$$\begin{aligned}
\mathcal{H}'(c'(s); L/t, & u(t/\lambda)^{\epsilon/2}, \lambda/t)/k_\mathrm{B}T = (1/2)\int_0^{L/t} [dc'(s)/ds]^2\, ds \\
&+ (u(t/\lambda)^{\epsilon/2}/2)\int_0^{L/t}\int_{\substack{0 \\ |s-s'|>\lambda/t}}^{L/t} \delta(c'(s) - c'(s'))\, ds\, ds'
\end{aligned} \qquad (3.9)$$

and the relation

$$\delta(c(s) - c(s')) = t^{-d/2}\delta(c'(s/t) - c'(s'/t)) \qquad (3.10)$$

has been used. It is important to note that since $c'(s)$ differs from $c(s)$, $\mathcal{H}'$ is not the same as $\mathcal{H}$ in functional form. With eq 3.8 substituted into 3.3 we find

$$G(\mathbf{R}) = t^{-d/2}\mathcal{G}'(\mathbf{R}/t^{1/2}; L/t, u(t/\lambda)^{\epsilon/2}, \lambda/t) \qquad (3.11)$$

83

which is substituted into eq 3.2 to yield

$$\langle R^2 \rangle = t\mathcal{F}'(L/t, u(t/\lambda)^{\epsilon/2}, \lambda/t) \tag{3.12}$$

Since $\mathcal{H}'$ differs from $\mathcal{H}$ it can be shown that except for $t = 1$, $\mathcal{F}'$ is different from $\mathcal{F}$ in eq 3.5. Nonetheless, Douglas and Freed equated $\mathcal{F}'$ to $\mathcal{F}$, which is an ad hoc assumption, finding no theoretical justification. It seems odd that Douglas and Freed did not clearly state this fact anywhere, in their original papers [9, 10] or in the recent book of Freed [4].

If we replace $\mathcal{F}'$ by $\mathcal{F}$, we can write eq 3.12 with eq 3.6 as follows:

$$\langle R^2 \rangle = tF\left((L/t)\exp\left[-\int^u B(x)\,dx/A(x)\right], (\lambda/t)\exp\left[-\int^u dx/A(x)\right]\right) \tag{3.13}$$

This equation is the starting point of the Douglas–Freed analysis, and shows that there holds a scaling law for $\langle R^2 \rangle$. Various interesting predictions follow this scaling property.

The Douglas–Freed analysis starts with introducing a function $u(t)$ defined by

$$\ln(2\pi t/\lambda) = \int_u^{u(t)} dx/A(x) \tag{3.14}$$

and, moreover, defining a function $L(t)$ by

$$L(t) = L\exp\left[\int_u^{u(t)} B(x)\,dx/A(x)\right] \tag{3.15}$$

Equation 3.13 can then be transformed into

$$\langle R^2 \rangle = tF\left((L(t)/t)\exp\left[-\int^{u(t)} B(x)\,dx/A(x)\right], \exp\left[-\int^{u(t)} dx/A(x)\right]\right) \tag{3.16}$$

Up to this point, t has been left as any positive number. Here we fix it to a value t' that satisfies $L(t')/t' = 1$. Then eq 3.16 is simplified to the form

$$\langle R^2 \rangle = L'f(u') \tag{3.17}$$

where L' and u' are the values of $L(t')$ and $u(t')$, respectively, and f stands for an unknown function. It follows from eq 3.14 and 3.15 that

$$\ln(2\pi L/\lambda) + \int_u^{u'} [B(x) - 1]\,dx/A(x) = 0 \tag{3.18}$$

84

and

$$L' = L \exp \left[\int_u^{u'} B(x)\, dx / A(x) \right] \tag{3.19}$$

These relations allow L' and u' in eq 3.17 to be determined as functions of basic parameters L, λ, and u when $A(x)$ and $B(x)$ are known. However, the form of f cannot be determined within the framework of the Douglas–Freed theory.

The θ condition corresponds to the limit $u \to 0$. In this limit, u' has to approach zero, since as seen from $A(x)$ shown in the next section, the integral on the left–hand side of eq 3.18 diverges as $u \to 0$. On the other hand, the integral on the right–hand side of eq 3.19 remains finite at $u = 0$. Hence it follows that $L' \to L$ as $u \to 0$. Thus, eq 3.17 gives

$$\langle R^2 \rangle_\theta = L f(0) \tag{3.20}$$

If this is compared with eq 2.20, $f(0)$ is found to be d. Division of eq 3.17 by eq 3.20 gives for the end distance expansion factor α_R

$$\alpha_R{}^2 = (L'/L) f_R(u') \tag{3.21}$$

where

$$f_R(x) \equiv f(x)/f(0) \tag{3.22}$$

Equation 3.22 indicates that when $f_R(x)$ is expanded in powers of x the leading term is unity. The factor L'/L in eq 3.21 can be determined as a function of L/λ and u by solving eq 3.18 for u' and then substituting the result into eq 3.19. Thus we can approach α_R, except for the factor $f_R(u')$.

3.2 Calculation of the End Distance Expansion Factor

The key to approaching α_R by the Douglas–Freed theory is to calculate $A(x)$ and $B(x)$ analytically. Kholodenko and Freed [11] carried through very tedious calculations to evaluate the coefficients a_2 and b_2 in eq 2.6 and 2.7, and obtained the following expressions:

$$A(u) = \frac{4u(u^\star - u)}{1 + (21/4)(u^\star + u)} \tag{3.23}$$

$$B(u) = u[1 - (5/2)u] \tag{3.24}$$

$$u^\star = (\epsilon/8)[1 + (21/32)\epsilon] \tag{3.25}$$

These are correct to the approximation that terms in $O(u^2)$, $O(u\epsilon)$, and $O(\epsilon^2)$ are neglected or inaccurate in comparison with unity. In what follows, we call this approximation the Douglas–Freed scheme and derive expressions rigorously consistent with it.

Substituting eq 3.23 and 3.24 into eq 3.14 and 3.15, imposing the condition $L(t')/t' = 1$, and using eq 3.25, we obtain

$$\left(\frac{2\pi L'}{\lambda}\right)^{\epsilon/2} = \frac{\bar{u}'}{\bar{u}}\left(\frac{1-\bar{u}}{1-\bar{u}'}\right)^{[1+(21/32)\epsilon]} \tag{3.26}$$

$$L' = L\left(\frac{1-\bar{u}}{1-\bar{u}'}\right)^{(1+\epsilon)/4}\exp\left[-(11/16)u^\star(\bar{u}'-\bar{u})\right] \tag{3.27}$$

where

$$\bar{u}' = u'/u^\star, \quad \bar{u} = u/u^\star \tag{3.28}$$

In principle, eq 3.26 and 3.27 can be simultaneously solved for $\bar{u}'$ and L' as functions of L, u, λ, and ϵ. More specifically, L'/L can be obtained as a function of two variables L/λ and $\bar{u}$ for a given ϵ.

After a complex analysis Douglas and Freed showed that the solutions to eq 3.26 and 3.27 correct in the Douglas–Freed scheme are given by

$$\bar{u}' = \omega \tag{3.29}$$

and

$$L' = L(2\pi L/\lambda)^{2\nu-1} \tag{3.30}$$

where

$$2\nu - 1 = (\epsilon/8)[1 + (21/32)\epsilon]\omega(2\pi L/\lambda)^{-E} - (3/128)\omega^2\epsilon^2 \tag{3.31}$$

with

$$E = (\epsilon/4)(1-\omega) \tag{3.32}$$

Here, ω is defined by

$$\omega = \eta/(1+\eta) \tag{3.33}$$

with

$$\eta = (2\pi L/\lambda)^{\epsilon/2}\bar{u}/(1-\bar{u}) \tag{3.34}$$

Substitution of eq 3.29 and 3.30 into eq 3.21 yields the final result correct in the Douglas–Freed scheme. It reads

$$\alpha_R{}^2 = (2\pi L/\lambda)^{2\nu-1}f_R(\omega) \tag{3.35}$$

Since this theory has no means to determine the function $f_R(\omega)$, Douglas and Freed proposed an empirical method in which $f_R(\omega)$ is represented by a polynomial in ω. Actually, they truncated it at ω^2 as

$$f_R(\omega) = 1 + a_R\omega + b_R\omega^2 \tag{3.36}$$

and evaluated the coefficients by the method described in Section 3.4. The fact that the leading term is unity can be verified by noting the condition that ω vanishes at $u = 0$ (under the θ condition).

86

3.3 Self-avoiding Chains

In a very good solvent, i.e., in the self-avoiding limit, u tends to u^*, so $\bar{u} \to 1$, which gives $\eta \to \infty$ and hence $\omega \to 1$. Thus we find with eq 3.31 and 3.32 that ν for 3-dimensional chains in the self-avoiding limit is

$$\nu = 0.5918 \tag{3.37}$$

which, on substitution into eq 3.35, leads to

$$\alpha_R{}^2 \sim L^{0.184} \tag{3.38}$$

or

$$\langle R^2 \rangle \sim L^{1.184} \tag{3.39}$$

The exponent 1.184 is between the familiar Flory value of 1.20 and the accurate estimate of 1.176 by a higher-order RG perturbation calculation [5]. This ought to be taken as a remarkable success of the Douglas-Freed theory.

The excluded-volume variable z for the 3-dimensional Edwards chain was defined by eq 2-1.11. If v in this equation is eliminated by use of eq 1.9 for $\epsilon = 1$ we obtain

$$z = (2\pi L/\lambda)^{1/2} u \tag{3.40}$$

This relation is very important because it serves to bridge the two-parameter theory and the RG theory. It will be shown below that under the condition $L/\lambda \gg 1$ these two theories are substantially the same. In the self-avoiding limit, u can be replaced by u^*, which equals 0.2070 for $\epsilon = 1$ when computed from eq 3.25, and ν is found to be 0.5918 for $\epsilon = 1$ as shown above. Thus, it follows from eq 3.35 and 3.40 that

$$\alpha_R{}^2 = 1.78\, z^{0.367} f_R(1) \quad (z \to \infty) \tag{3.41}$$

This result indicates that, in very good solvents, α_R essentially becomes a function of z only. In this connection, we note that the corresponding relation from computer simulation experiments by Domb and coworkers (see Chapter 2) is

$$\alpha_R{}^2 = 1.64\, z^{0.40} \quad (z \to \infty) \tag{3.42}$$

3.4 Poor Solvent Systems

By taking into account that eq 3.29 is correct in the Douglas-Freed scheme and that u^* is $O(\epsilon)$ we may rewrite eq 3.27 as

$$\frac{L'}{L} = \left(\frac{1-\bar{u}}{1-\omega}\right)^{(1+\epsilon)/4} \exp\left[-(11u^*/16)(\omega - \bar{u})\right] \tag{3.43}$$

We substitute for ω the expression obtained by putting eq 3.34 into eq 3.33 and impose the condition

$$(2\pi L/\lambda)^{\epsilon/2} \gg 1 \qquad (3.44)$$

Then eq 3.43 is transformed into

$$L'/L = [1 + \bar{u}(2\pi L/\epsilon)^{\epsilon/2}]^{(1+\epsilon)/4} \exp\left[-(11\bar{u}/16)u^{\star}(2\pi L/\lambda)^{\epsilon/2}\right] \qquad (3.45)$$

which is correct to first order in $\bar{u}$ and ϵ in comparison with unity. When applied to a 3–dimensional chain and eq 3.40 inserted, this yields

$$L'/L = (1 + 4.83z)^{1/2} \exp\left[-(11/16)z\right] \qquad (3.46)$$

in the region where $O(\bar{u}^2)$ may be neglected relative to unity, i.e., in poor solvent systems. Equation 3.46 differs from the corresponding one derived by Douglas and Freed [10]:

$$L'/L = (1 + 8z)^{1/4} \qquad (3.47)$$

As explained in Section 3.8, eq 3.47 is inconsistent with the Douglas–Freed scheme.

Various relations that appeared in deriving eq 3.46 allow us to obtain

$$\bar{u}' = 4.83z/(1 + 4.83z) \qquad (3.48)$$

for 3–dimensional chains subject to the condition

$$(2\pi L/\lambda)^{1/2} \gg 1 \quad \text{(eq 3.44 for three dimensions)} \qquad (3.49)$$

With eq 3.46 and 3.48, eq 3.21 gives

$$\alpha_R{}^2 = (1 + 4.83z)^{1/2} \exp\left(-0.688z\right) f_R \left(\frac{4.83z}{1 + 4.83z}\right) \qquad (3.50)$$

Thus, in the Douglas–Freed scheme, it follows that if eq 3.49 holds, the end distance expansion factor for 3–dimensional chains in poor solvents obeys the two–parameter theory. This consequence may be utilized to determine the function $f_R(\omega)$. With eq 3.36 substituted for this function, the right–hand side of eq 3.50 is expanded in powers of z and the result is compared with the familiar z expansion for $\alpha_R{}^2$ in the two-parameter theory, i.e., $\alpha_R{}^2 = 1 + (4/3)z - 2.075z^2 + \ldots$ (see Section 1.3 of Chapter 2). Then we find

$$a_R = -0.0817 \qquad (3.51)$$

To evaluate b_R correctly we need a perturbation calculation one order higher than that accomplished by Kholodenko and Freed [11]. Thus, Douglas and Freed [10] tentatively set b_R equal to zero. To this approximation eq 3.50 yields

$$\alpha_R{}^2 = (1 + 4.83z)^{1/2} \exp(-0.688z)[1 - 0.395z/(1 + 4.83z)] \qquad (3.52)$$

Expanding this in powers of z, we get –3.12 for the coefficient of z^2, which is considerably different from the correct value of –2.075.

3.5 Crossover Region

With $a_R = -0.0817$ and $b_R = 0$ we obtain $f_R(1) = 0.9183$. Hence, eq 3.41 reads

$$\alpha_R{}^2 = 1.63z^{0.367} \tag{3.53}$$

Thus the Douglas–Freed RG theory allows us to determine the self–avoiding limit of α_R as a function of z, regarding not only the exponent to z but also the prefactor. Probably the prefactor value 1.63 is approximate, but its proximity to the computer simulation value 1.64 (see Section 1.4 of Chapter 2) is worthy of note.

The above analysis has revealed that α_R of a 3–dimensional chain depends only on z in the regions $z \ll 1$ and $z \gg 1$. Since, as can be seen from eq 3.35, α_R is generally a function of two variables L/λ and ω or equivalently L/λ and z, there must be a crossover region of z in which the dependence of α_R on L/λ becomes apparent. But it is not easy to estimate the spread of this crossover region. The curves a and b in Figure 3-2 show $\alpha_R{}^2(z)$ calculated from eq 3.52 and 3.53, respectively. Both may be joined smoothly by the dashed curve as shown, suggesting that the crossover region should be quite narrow. Thus it may be concluded that α_R of 3–dimensional continuous chains depends to a good approximation on z only over the entire range from poor to good solvent, as the two–parameter theory predicts. However, this conclusion is based, among other things, on the condition $L/\lambda \gg 1$(eq 3.41). Since the two–parameter theory is a special case of the general theory in the limit $\lambda/L \to 0$ (see Section 1.1 of Chapter 2), it is not always surprising that the RG theory predicts a very narrow crossover region and essentially reduces to the two–parameter theory under the condition in which eq 3.49 holds.

3.6 Radius Expansion Factor

Exactly the same analysis as that described above can be made regarding the radius expansion factor α_S. Corresponding to eq 3.21, we can derive

$$\alpha_S{}^2 = (L'/L)f_S(\omega) \tag{3.54}$$

Assuming $f_S(\omega) = 1 + a_S\omega$ and using eq 2-1.26 for the z expansion of $\alpha_S{}^2$, we find $a_S = -0.0935$. The final results corresponding to eq 3.52 and 3.42 are, respectively,

$$\alpha_S{}^2 = (1 + 4.83z)^{1/2} \exp{(-0.688z)}[1 - 0.451z/(1 + 4.83z)] \tag{3.55}$$

and

$$\alpha_S{}^2 = 1.60z^{0.367} \tag{3.56}$$

The ratio $\alpha_S{}^2/\alpha_R{}^2$ at the self–avoiding limit is found to be 0.982 from eq 3.53 and 3.56, which is considerably different from the simulation experimental value ~ 0.93 (see Section 1.6 of Chapter 2).

89

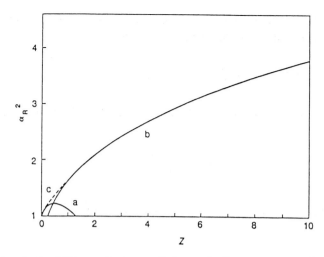

Fig. 3-2. RG predictions of the end distance expansion factor α_R. Curve a, in poor solvents; curve b, in very good solvents (self-avoiding chains); curve c, a smooth connection of a and b.

3.7 Second Virial Coefficient

Oono and Freed [8] used $Y_u(u) = 1 + (8/\epsilon)u$ (see eq 2.10) to transform u_0 in the expression for A_{2B} into u. However, this manipulation was in error because u is in the order of ϵ so that their Y_u was correct only to zeroth order in ϵ. The Y_u correct to first order in ϵ has to be used. It can be obtained in the process of deriving eq 3.23, yielding

$$Y_u = [1 - (u/u^\star)]^{-[1-(21/32)\epsilon]} \tag{3.57}$$

Omitting the analytical details, we show only the final formula for A_2 derived by Douglas and Freed [10]. It reads

$$A_2(L, u, \lambda) = (N_A/2M^2)\lambda^{d/2}(2\pi L/\lambda)^{\nu d}C(\omega) \tag{3.58}$$

where
$$C(\omega) = (\omega\epsilon/8)\{1 + (\epsilon/8)[(21/4) + (4\ln 2 - 1)\omega\} \tag{3.59}$$

These are exact in the Douglas–Freed scheme under the condition specified by eq 3.44.

It follows from eq 3.58 that A_2 for 3–dimensional chains varies linearly with $M^{-0.225}$ in the limit of very good solvent. This prediction is consistent with the experimental findings mentioned in Section 2 of Chapter 2.

90

Equation 2.33 for $\langle S^2 \rangle$ is correct to first order in ϵ in comparison with unity. To the same approximation this equation is equivalent to

$$\langle S^2 \rangle = (d/6)(2\pi L/\lambda)^{2\nu}(\lambda/2\pi)[1 - (13/96)\epsilon\omega] \qquad (3.60)$$

The penetration function Ψ for a d–dimensional chain is defined by

$$\Psi = \frac{2M^2 A_2 (d/3)^{d/2}}{(4\pi)^{d/2} N_A \langle S^2 \rangle^{d/2}} \qquad (3.61)$$

which reduces to eq 2–2.5 for $d = 3$. Substitution of eq 3.58 and 3.60 leads to

$$\Psi = (\epsilon/8)\omega\{1 + (\epsilon/8)[(21/4) + (4\ln 2 + (7/6))\omega]\} \qquad (3.62)$$

Since the left–hand side of eq 3.48 can be equated to ω by virtue of eq 3.29, eq 3.62 for $\epsilon = 1$ yields

$$\Psi = \frac{z}{1 + 4.83z} + \frac{1.45z^2}{(1 + 4.83z)^2} \qquad (3.63)$$

This formula indicates that Ψ for 3–dimensional chains becomes a universal function of z, as in the two–parameter theory, in the Douglas–Freed scheme and under the condition $(2\pi L/\lambda)^{1/2} \gg 1$. It predicts 0.269 for Ψ at the self–avoiding limit $(z = \infty)$. This limiting value is in accordance with 0.268 by Witten and Schäfer [12] and 0.269 by des Cloizeaux [13] from different RG calculations. Though slightly larger than the experimental estimates $0.22 - 0.25$, it represents a remarkable success of the RG theory.

Figure 3–3 shows $\Psi(z)$ calculated from eq 3.63, along with the Barrett curve B reproduced from Figure 2–6 of Chapter 2. It can be seen that the Douglas–Freed theory also predicts Ψ which increases monotonically to an asymptotic limit as z increases. Hence, this theory fails to explain the observed behavior mentioned in Chapter 2 that Ψ in a good solvent decreases to a limiting value with increasing α_S (or z). Finally, we note that when expanded in powers of z, eq 3.63 gives $\Psi = z(1 - 3.38z + \ldots)$, whereas the corresponding series obtained in the two–parameter theory is $\Psi = z(1 - 4.78z + \ldots)$.

As for field theory approaches to A_2, the reader is advised to consult Burch and Moore [14], Knoll et al. [15], Elderfield [16, 17], Stepanow [18], Stepanow and Staube [19], and Schäfer [20]. It is beyond the author's ability to give an account of them.

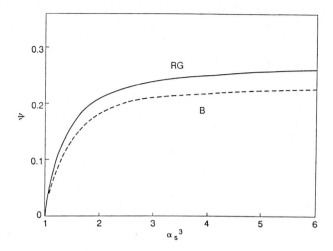

Fig. 3–3. Calculated penetration functions $\Psi(z)$. Solid line, Douglas–Freed theory (eq 3.63). Dashed line B, Barrett interpolation formula.

3.8 Remarks

1. In analyzing poor solvent systems Douglas and Freed used for L'/L not eq 3.46 but

$$L'/L = [(1 - \bar{u})/(1 - \omega)]^{1/4} \tag{3.64}$$

which is derived from eq 3.43 by neglecting ϵ in the exponent term $(1+\epsilon)/4$ and moreover by dropping the exponential function term. The argument of the latter term is in the order of ϵ. Thus, these operations amount to partially ignoring terms in $O(\epsilon)$ in comparison with unity. Hence, eq 3.64 is not completely consistent with the Douglas–Freed scheme which requires that all terms in $O(\epsilon)$ and $O(u)$ compared with unity should be rigorously evaluated and retained.

Substituting eq 3.33 for ω and then eq 3.34 for η, we find that eq 3.64 for 3–dimensional chains becomes

$$L'/L = [1 - \bar{u} + (z/u^\star)]^{1/4} \tag{3.65}$$

where eq 3.40 has been substituted. Douglas and Freed introduced the following restriction to $\bar{u}$:

$$\bar{u} \ll 1 + (z/u^\star) \tag{3.66}$$

For such $\bar{u}$ eq 3.65 may be replaced by $L'/L = [1+(z/u^\star)]^{1/4}$. It is reasonable to substitute for $u^\star$ the zeroth approximation value at $\epsilon = 1$, i.e., 1/8 (see eq 1.20). Then we arrive at eq 3.47.

92

Now, when eq 3.40 is inserted, eq 3.66 is written $1 \ll \bar{u}^{-1} + (2\pi L/\lambda)^{1/2}$, which indicates that, as long as the condition 3.49 is assumed, eq 3.66 is automatically satisfied regardless of solvent quality. In other words, this restriction need not be specially invoked when we are concerned with chains subject to the condition $(2\pi L/\lambda)^{1/2} \gg 1$.

2. In the Douglas–Freed theory, most results are obtained in closed form. Though not specially noted in the text, this is the consequence of the manipulations in which terms of second and higher order of ϵ or u in comparison with unity are dropped or artificially replaced by a relevant series of terms which, when added to the correctly determined zeroth and first terms, can be summed up in a closed form. Since there are a variety of choices for the series to be added, the closed form so obtained is not unique and may be grossly inaccurate for not–so–small values of ϵ or u. Hence, it will be legitimate not to put much credit on the numerical results predicted for 3–dimensional chains by the Douglas–Freed theory, even if they happen to come close to experimental ones.

3. The entire structure of the Douglas–Freed theory is founded on the assumption pointed out in Section 3, i.e., the approximation of $\mathcal{F}'$ in eq 3.12 by $\mathcal{F}$ in eq 3.5. At present we are unable to estimate how much the final results of the Douglas–Freed theory are affected by it. Thus it is relevant to keep in mind that despite its successful predictions the RG theory of Douglas and Freed leaves a serious problem to be resolved.

References

1. P.-G. de Gennes, Phys. Lett. A **38**, 339 (1972).
2. H. Yamakawa, "Modern Theory of Polymer Solutions," Harper & Row, New York, 1971.
3. Y. Oono, Adv. Chem. Phys. **61**, 301 (1985).
4. K. F. Freed, "Renormalization Group Theory of Macromolecules," Wiley-Interscience, New York, 1987.
5. J. C. le Guillou and J. Zinn-Justin, Phys. Rev. Lett. **39**, 95 (1977).
6. T. Ohta, Y. Oono, and K. F. Freed, Phys. Rev. A **25**, 2801 (1982).
7. Y. Tsunashima and M. Kurata, J. Chem. Phys. **84**, 6432 (1986).
8. Y. Oono and K. F. Freed, J. Phys. **15**, 1931 (1982).
9. J. F. Douglas and K. F. Freed, Macromolecules **17**, 1854 (1984).
10. J. F. Douglas and K. F. Freed, Macromolecules **17**, 2344 (1984); see also Ref. 9.
11. A. Kohlodensko and K. F. Freed, J. Chem. Phys. **78**, 7390 (1983).
12. T. Witten and L. Schäfer, J. Phys. A **11**, 1843 (1978).
13. J. des Cloizeaux, J. Phys. (Paris) **42**, 635 (1981).
14. D. J. Burch and M. A. Moore, J. Phys. **9**, 435 (1976).
15. A. Knoll, L. Schäfer, and T. A. Witten, J. Phys. (Paris) **42**, 767 (1981).
16. D. J. Elderfield, J. Phys. A **11**, 2483 (1978).
17. D. J. Elderfield, J. Phys. C **13**, 5883 (1980).
18. S. Stepanow, Ann. Phys. [7], **40**, 301 (1983).
19. S. Stepanow and E. Straube, J. Phys. Lett. **46**, L–1115 (1985).
20. L. Schäfer, Macromolecules **17**, 1357 (1984).

Chapter 4   Some Recent Topics

1.   Theory of Weill and des Cloizeaux

1.1   Basic Assumptions

As before, we consider a spring–bead chain consisting of $N + 1$ beads under the binary cluster approximation. In a θ solvent, the mean–square distance $\langle r_{ij}{}^2 \rangle$ between beads i and j is equal to $a^2 |i - j|$ regardless of the positions of the beads on the chain, where a is the mean length of one spring. In a non-θ solvent, if $\langle r_{ij}{}^2 \rangle$ is expressed as

$$\langle r_{ij}{}^2 \rangle = K |i - j|^{2\mu} \tag{1.1}$$

both K and μ are supposed to depend not only on $|i - j|$ but also on the individual values of i and j. This dependence is not yet fully elucidated. The work of Barrett expounded in Section 1.10 of Chapter 2 represents a great effort toward it.

According to the two–parameter theory and the RG theory as well, the state of a polymer chain in dilute solution crosses over from Gaussian to self–avoiding as the chain length increases in a fixed non-θ solvent, and the index α in the relation $\langle R^2 \rangle \sim N^{2\alpha}$ changes from 0.5 to 0.6 (more precisely, 0.588) accompanying this crossover. Here, as before, $\langle R^2 \rangle$ is the mean–square end distance of the chain, i.e., $\langle r_{ij}{}^2 \rangle$ for $i = 0$ and $j = N$. Thus, we may expect that μ also changes from 0.5 to 0.6 as the length of the i–j subchain becomes longer. Although the path of this increase is supposed to depend on the positions of beads i and j in the chain, we here consider what happens if this dependence is ignored. That is to say, we assume that μ depends only on $|i - j|$:

$$\mu = f(|i - j|) \tag{1.2}$$

where $f(x)$ is a function which tends to 0.5 as $x \to 0$ and to ν (0.6 or 0.588) as $x \to \infty$. Intuitively we may consider that $f(x)$ increases monotonically with x as sketched by the solid line in Figure 4-1.

As a rough approximation, Weill and des Cloizeaux [1] proposed replacing $f(x)$ by a step function (the dashed line in Figure 4-1) as

$$f(x) = 0.5 \quad (0 < x < N_\mathrm{c})$$

$$f(x) = \nu \quad (N_\mathrm{c} < x < \infty) \tag{1.3}$$

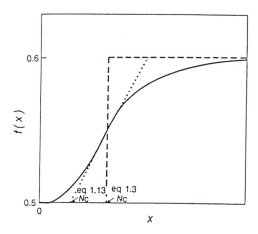

Fig. 4-1. Solid line, Expected change in $f(x)$ with $x = |i - j|$. Dashed line, Weill–des Cloizeaux approximation eq 1.3. Dotted line, François et al.'s approximation eq 1.13.

where N_c is an adjustable parameter. Furthermore, they set K equal to a^2 for $0 < x < N_c$ and determine it for $N_c < x < \infty$ in such a way that $\langle r_{ij}^2 \rangle$ is continuous at $x = N_c$. Thus we obtain

$$\langle r_{ij}^2 \rangle = a^2 |i - j| \quad (0 < x < N_c)$$
$$\langle r_{ij}^2 \rangle = a^2 N_c^{1-2\nu} |i - j|^{2\nu} \quad (N_c < x < \infty) \tag{1.4}$$

The theory based on this assumption is expected to give zeroth order predictions of global polymer properties, and is usually called the **blob theory**. However, this nomenclature does not seem relevant to the author (see Section 1.4). Hence, in this book, we call it the Weill–des Cloizeaux theory.

1.2 Expansion Factors

With eq 1.4 the end distance expansion factor α_R and the radius expansion factor α_S can be calculated analytically (in the continuous chain limit). The results are as follows:

$$\alpha_R^2 = y^{1-2\nu} \quad (0 < y < 1)$$
$$\alpha_R^2 = 1 \quad (1 < y) \tag{1.5}$$
$$\alpha_S^2 = y^2(3 - 2y) + 6y^{1-2\nu}[(1 - y^{1+2\nu})/(1 + 2\nu)$$
$$\quad - (1 - y^{2(1+\nu)})/2(1 + \nu)] \quad (y < 1) \tag{1.6}$$
$$\alpha_S^2 = 1 \quad (1 < y)$$

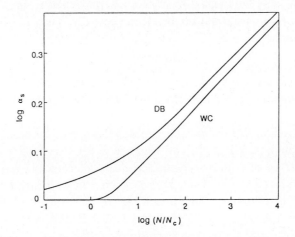

Fig. 4–2. Line WC, calculated from eq 1.6 with $\nu = 0.6$. Line DB, calculated from the Domb–Barrett interpolation formula combined with eq 1.9.

where

$$y = N_c/N \tag{1.7}$$

The asymptotic form of eq 1.6 at the limit of large N is

$$\alpha_S{}^2 = \frac{3}{(1 + 2\nu)(1 + \nu)} \left(\frac{N}{N_c}\right)^{2\nu - 1} \tag{1.8}$$

It follows from eq 1.5 and 1.8 with Flory's ν value of 0.6 that $\alpha_S{}^2/\alpha_R{}^2$ at the limit of large N is 0.856. But this value is too small compared to the value of 0.933 calculated from the Domb–Barrett interpolation formula (see Section 1.6 of Chapter 2).

Forcing eq 1.5 with $\nu = 0.6$ to agree with the asymptotic relation due to Domb et al. ($\alpha_R{}^2 = 1.64z^{0.4}$; see Section 1.4 of Chapter 2), we find the relation

$$1/y = 11.86z^2 \tag{1.9}$$

which allows the right–hand side of eq 1.6 to be expressed in terms of the excluded–volume variable z only. Thus we can conclude that the Weill–des Cloizeaux theory is a type of two–parameter theory.

Figure 4-2 compares the relation between $\log \alpha_S$ and $\log (N/N_c)$ calculated from eq 1.6 with $\nu = 0.6$ with that from the Domb–Barrett interpolation formula and eq 1.9. The line WC deviates significantly from the DB line, which, as has been shown in Chapter 2, fits experimental results very closely.

If $N \gg 1$, N may be equated to M/m, where M and m are the molecular weights of the entire chain and the bead, respectively. Then it follows from eq 1.9, 2–1.32, and 2–1.33 that

$$N_c \sim \beta^{-2} \qquad (1.10)$$

where β is the excluded–volume strength. Thus, as the solvent quality becomes better, N_c decreases. In the self–avoiding limit, this parameter diminishes to zero.

The Stokes radius R_H of a spring–bead chain can be calculated by the Kirkwood formula (see Ref. [2] of Chapter 2 for its details) if $\langle r_{ij}^{-1} \rangle$ is given as a function of i and j. Weill and des Cloizeaux [1] replaced this by $\langle r_{ij}^{-1} \rangle_\theta (\langle r_{ij}^2 \rangle_\theta / \langle r_{ij}^2 \rangle)^{1/2}$. Their approximation has already been discussed in Section 1.10 of Chapter 2, and, according to the estimate of Barrett shown there, the error introduced is as much as 10%. In this way, Weill and des Cloizeaux obtained for the friction expansion coefficient $\alpha_f = R_H/R_{H\theta}$ in the non–draining limit ($h \to \infty$)

$$\alpha_f = \frac{4y^{-1/2}}{2(3-y) + 3(y^{\nu-1}-1)/(1-\nu) + 3(y^{\nu-1}-y)/(2-\nu)} \quad (y < 1)$$
$$\alpha_f = 1 \quad (1 < y) \qquad (1.11)$$

Akcasu and Han [2] compared eq 1.11 with many experimental data but found only a semi–quantitative agreement.

The Weill–des Cloizeaux theory has acquired considerable popularity in recent years, because its basic assumption (eq 1.4) permits analytical calculation of various polymer properties in non–θ solvents. However, we should not expect it to be more than a semi–quantitative theory. For its improvement François et al. [3] proposed allowing μ in eq 1.2 to vary with $|i-j|$ according to either

$$f(x) = 0.60 - 0.10 \exp\{-c[(x/N_c) - 1]^2\} \quad (x > 0) \qquad (1.12)$$

or

$$\begin{aligned}
f(x) &= 0.50 \quad (N_c > x > 0) \\
f(x) &= 0.50 + 0.1(x - N_c)/N_c(k-1) \quad (kN_c > x > N_c) \\
f(x) &= 0.60 \quad (x > kN_c)
\end{aligned} \qquad (1.13)$$

where c and k are adjustable constants. The dotted line in Figure 4–1 shows $f(x)$ defined by eq 1.13. The slope ν_s of the $\log\langle S^2 \rangle$ vs. $\log(N/N_c)$ curves calculated by François et al. for $c = 0.02$ and $k = 15$ exhibited a maximum of

98

0.68 in the vicinity of $N/N_c = 15$. These authors used this fact to interpret their finding that ν_s for poly(acrylamide) in water reached a maximum above 0.6 at some finite molecular weight. However, as far as the author is aware, none of the previous data on flexible linear polymers have displayed such a maximum of ν_s. Further work on aqueous poly(acrylamide) is required.

1.3 Computer Simulation Data

We denote by $\langle r_n{}^2 \rangle$ the average of $\langle r_{ij}{}^2 \rangle$ obtained when $n = j - i (j > i)$ is fixed and i is varied from 0 to $N - n$. Then we have

$$\langle r_n{}^2 \rangle = (N - n + 1)^{-1} \sum_{i=0}^{N-n} \langle r_{i(i+n)}{}^2 \rangle \tag{1.14}$$

The average end distance expansion factor $\alpha_R(n)$ for subchains containing n beads and located at different positions is defined by

$$\alpha_R(n)^2 = \langle r_n{}^2 \rangle / \langle r_n{}^2 \rangle_\theta \tag{1.15}$$

The Weill–des Cloizeaux approximation, eq 1.4, can be rewritten in terms of $\alpha_R(n)$ as

$$\alpha_R(n) = 1 \quad (n < N_c), \quad \alpha_R(n) = (n/N_c)^{0.1} \quad (N_c < n) \tag{1.16}$$

where ν has been taken to be 0.6. To check the validity of this approximation Mattice [4] and Curro and Schaefer [5] undertook computer simulations and obtained very interesting results.

Figure 4–3 shows Mattice's data obtained for a rotational isomer polymethylene chain with $N = 1000$ in solvents of different qualities. The system of curves reported by Curro and Schaefer on a chain with $N = 299$ is similar in overall shape to that in Figure 4–3. The following features may be worth noting:

(i)   As n decreases $\alpha_R(n)$ approaches unity regardless of solvent quality;

(ii)   The slope $\delta(n)$ for $\log \alpha_R{}^2$ against $\log n$ varies with n as well as solvent quality. It is larger the better the solvent;

(iii)   In good solvents, $\delta(n)$ exceeds 0.2 before apparently tending to a constant as n increases. In the result of Curro and Schaefer, $\delta(n)$ becomes negative after a certain n.

Features (ii) and (iii) greatly differ from that predicted by eq 1.16, indicating the complex nature of the swelling of subchains in a polymer molecule. This complexity is a reflection of the interactions of the subchain with its flanking

99

end chains. It does not seem easy to interpret these simulation data theoretically, especially the finding of negative $\delta(n)$.

If the average end distance expansion factor for a subchain with fixed n and i is denoted by $\alpha_R(n|i)$, we have

$$\alpha_R(n|i)^2 = \langle r_{i(i+n)}{}^2 \rangle / \langle r_{i(i+n)}{}^2 \rangle_\theta \qquad (1.17)$$

Figure 4–4 illustrates the i dependence (i.e., position dependence) of this expansion factor with n as a parameter, calculated by Mattice [4] for a chain with $N = 100$. It can be seen that the subchain with fixed n undergoes a larger expansion as it is located nearer the center of the entire chain. This finding implies that excluded–volume repulsion tends to prevent the inner part of the polymer coil from becoming crowded with segments, as should be expected intuitively.

Matsushita and coworkers [6] prepared a homopolymer of deuterated (d) styrene, a diblock copolymer of ordinary (h) and d styrenes, and a triblock copolymer of h, d, and h styrenes and measured the radii of gyration of the d portions in carbon disulfide by small–angle neutron scattering. The d styrene content was 6.3% in the diblock sample and 10.7% in the triblock one, while the molecular weights of the d parts were nearly equal for both homopolymer and copolymers $(2.5–2.9\times10^4)$. The expansion factors for the d portions were found to be $1.0_6 \pm 0.03$, $1.1_8 \pm 0.03$, and $1.2_6 \pm 0.05$ for the homo, diblock, and triblock samples, respectively. These values indicate that the labeled part in the middle of the chain undergoes a larger expansion than that at the end of the chain. But, as Matsushita et al. mentioned, the difference is comparable to the limit of experimental accuracy. It is highly desirable that further experiments of this kind are undertaken, because the results ought to deepen our understanding of chain statistics and to assist a more accurate formulation of global polymer properties.

1.4 Remarks

The Weill–des Cloizeaux theory assumes that no expansion takes place within subchains shorter than aN_c. On this assumption we may consider a chain model which consists of Gaussian subcoils, each being made of N_c beads and interacting with one another. With such a subcoil viewed as a blob, this model is often referred to as the blob model. However, it is important to recognize that the Weill–des Cloizeaux theory concerns interactions between beads but not those between blobs. Therefore, contrary to many other authors, the author does not consider it relevant to call it the blob theory.

When N_c is set equal to zero, the Weill–des Cloizeaux theory appears to reduce to the earlier ϵ theory of Peterlin [7] and Ptitsyn and Eizner [8] (see

100

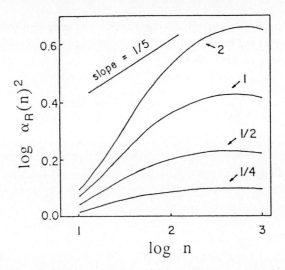

Fig. 4–3. Subchain expansion factor as a function of sub-chain length n in a polymethylene chain of 1000 bonds. Larger figures attached to the curves are for better solvents.

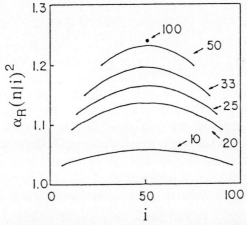

Fig. 4–4. Subchain expansion factors as a function of sub-chain location i at different subchain lengths (indicated by the number attached to each curve).

Ref. [9]). However, the following difference has to be noted. In the former, ν is an essentially fixed parameter (usually set equal to 0.6 for non-θ solvents) and N_c is related to the excluded–volume strength β, whereas, in the latter, ϵ (which corresponds to $2\nu - 1$ in the former) is treated as an adjustable parameter associated with solvent quality. Anyway it is clear that these theories are too

101

naive to describe the precise picture of global behavior of polymer molecules with excluded–volume interactions and hence their application should be limited to semi–quantitative considerations.

2.   Coil–Globule Transition

2.1   Change to the Globular State

In the binary cluster approximation the excluded–volume strength β becomes negative in a poor solvent below the θ temperature, and the mean force acting between the segments of a polymer chain becomes attractive in such a solvent. The attraction leads to a shrinkage of the polymer coil below its size under the θ condition, and the degree of shrinkage should increase as the temperature is lowered. As the coil shrinks, its segment density increases and the chain is collapsed (folded together). Thus, the coil becomes more like a compact globule. This change in the state of a polymer chain from the open coil to the globular particle is called the coil-globule transition. Stockmayer [10] in 1960 was the first to call attention to it as an interesting object of research, but progress in experimental and theoretical work on it has taken place in relatively recent years. Williams et al. [11] summarized the major contributions made by the end of the 1970s. This section presents a brief overview of more recent investigations.

2.2   Theoretical Considerations

In this book we have proceeded so far assuming, either explicitly or implicitly, that the binary cluster approximation holds. This assumption is reasonable for good solvents and even for marginal ones, but it will no longer be valid in poor solvents near or below the θ temperature and at least the residual ternary interaction of chain segments will have to be taken into account in order to formulate polymer behavior in such solvents [12].

The coil–globule transition of a linear flexible polymer in dilute solution has been formulated by Orofino and Flory [13], de Gennes [14], Moore [15], Sanchez [16], and others. The results obtained give substantially similar predictions except for one point which will be explained below. Here we refer to an elementary theory due to de Gennes [14], though it is virtually identical to the earlier theory of Orofino and Flory.

Using the mean–field approximation and taking binary and ternary cluster interactions into account, we can derive

$$G/k_{\mathrm{B}}T = (3/2)\alpha_{\mathrm{S}}^{2} - \ln \alpha_{\mathrm{S}}^{3} + (N/2)[\rho W_2(T) + \rho^2 W_3(T)] \qquad (2.1)$$

for the free energy G of a spring–bead chain consisting of N beads ($N \gg 1$).

Here, $k_B T$ has the usual meaning, ρ is the mean number density of beads in the polymer domain, i.e.,

$$\rho \sim N/\langle S^2 \rangle^{3/2} \sim N^{-1/2} \alpha_S^{-3/2} \tag{2.2}$$

and W_2 and W_3 are the binary and ternary interaction energies per cluster, respectively. We note that

$$W_2 \sim \beta \tag{2.3}$$

It follows from eq 2.1 that, on minimizing G, α_S is given by

$$\alpha_S{}^5 - \alpha_S{}^3 - y \alpha_S{}^{-3} = 2.60z \tag{2.4}$$

where y is a dimensionless parameter defined by

$$y = AW_3 \tag{2.5}$$

with A being inversely proportional to the sixth power of the mean spring length a, and z is the excluded–volume variable. It is reasonable to assume that, in the region not far from θ, W_2 varies with T as

$$W_2 \sim \tau \tag{2.6}$$

where

$$\tau = 1 - (\Theta/T) \tag{2.7}$$

The reduced temperature τ and hence z are negative in the region below θ where coil–globule transitions take place. Lowering temperature can be described by increasing $|z|$.

de Gennes assumed W_3 and hence y to be independent of temperature, though no justification for this assumption seems to be available at present. Figure 4–5 displays the results calculated from eq 2.4. Each curve eventually approaches the abscissa axis as $|z|$ increases, implying that the chain completely collapses to a geometric point at the limit of low temperature. The curves for $y > 0.038$ show a smooth coil–globule transition, whereas those for $y < 0.038$ become two–valued in a certain range of z. The latter suggests that a discontinuous shrinkage of the polymer coil takes place at some temperature which depends on y. According to statistical mechanics, however, such a first–order–like transition cannot occur in a chain of finite length. Accordingly, the behavior of the curves for $y < 0.038$ must be considered an artifact brought about by the use of the mean–field approximation to the intrachain interaction energy. It can be shown from eq 2.4 that in the limit of large N, the chain changes abruptly from a coil to a completely collapsed globule at the θ temperature. In other words, de

103

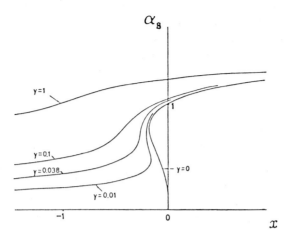

Fig. 4–5. Radius expansion factor α_S as a function of z calculated by eq 2.4 for several fixed values of y [14]. x is equal to $2.60z$. Note that the curves for $y > 0$ give α_S larger than unity at $x = 0$.

Gennes' theory predicts a first–order coil–globule transition at this temperature. However, in the theories of Moore [15] and Sanchez [16], α_S in the limit of large N undergoes a transition at θ which was named second order by Lifschitz et al. [17] (i.e., the slope of α_S plotted against z changes discontinuously at $z = 0$.) Experimental checking of these different predictions offers experimentalists a very challenging problem.

Equation 2.4 also predicts that if y is independent of M as well as T, α_S is scaled by a combined variable $|\tau|M^{1/2}$ and that, in the limit of large values of this variable, α_S asymptotically behaves as

$$\alpha_S \sim |\tau|^{-1/3} M^{-1/6} \tag{2.8}$$

provided $y \neq 0$. Vidakovic and Rondelez [18] modified eq 1.11, for a description of α_f in the coil–globule transition region by assigning $1/3$ to ν. But this modification seems too ad hoc, since the Weill–des Cloizeaux theory ignores the ternary cluster interaction energy.

According to eq 2.4, α_S does not become unity at $z = 0$ (i.e., $\beta = 0$) unless $y = 0$, which appears to contradict the definition of α_S, which says that $\alpha_S = 1$ at the θ temperature. However, this does not matter. In the binary cluster approximation, the θ condition is equivalent to $z = 0$. But when the ternary cluster interaction energy is taken into account, this equivalence no longer holds so that α_S becomes unity at a nonzero z at which the effects of binary and ternary cluster interactions on A_2 just cancel to give $A_2 = 0$.

Effects of ternary cluster interactions on α_f and α_η are not yet formulated. Hence, experimental data for these hydrodynamic expansion factors in the coil–globule transition region are usually compared with the theory of α_S.

2.3 Experimental Information

2.3.1 *Experimental Problems*

Usually, a polymer solution in a poor solvent begins to separate into two phases when it is cooled to a temperature T_p below θ.[1] The locus of T_p plotted against the mass concentration c (or concentrations of other units such as weight fraction and volume fraction) is called the cloud–point curve. In general, this curve varies with the molecular weight M of the dissolved polymer, as illustrated in Figure 9-1. Physical measurements on dilute polymer solutions have to be conducted in the narrow tapered region between the ordinate axis and the left branch of the cloud–point curve for the polymer under study. As can be seen from the figure, this region becomes narrower at higher M and lower T. Thus, Swislow et al. [19] had to lower c below 10^{-5} g/cm^3 when they studied an ultrahigh polystyrene sample with $M = 2.7 \times 10^7$ in cyclohexane at 32°C by quasi–elastic light scattering.

Below θ, attractive forces act between the segments of different chains as well as those of each individual chain, tending to aggregate or associate polymer molecules. This effect makes it difficult to prepare molecularly dispersed solutions. Solutions containing aggregates are usually in a non–equilibrium state or in a metastable state at best. Experimental data taken on them are not simple to analyze.

Polymer samples used for coil–globule transition studies must be as close to monodispersity as possible. When a polydisperse solution is cooled to a certain temperature below θ its part containing high–molecular–weight fractions enters two–phase regions while the rest still maintains homogeneity. Behavior of such a system cannot be analyzed by the theory of uniform solutions.

In order to approach the entire picture of a coil–globule transition we must measure expansion factors by lowering T and increasing M as much as possible so that a maximal range of the scaling variable $|\tau| M^{1/2}$ may be covered. However, as can be understood from the above account, it is not easy to meet these two requirements at the same time. Thus, the values of $|\tau| M^{1/2}$ attained for the system polystyrene + cyclohexane still remain less than 100, which is not large enough to estimate the asymptotic behavior of the transition. We note that Chu

---

[1] For some combinations of polymer and poor solvent, phase separation occurs upon rising temperature. In this book, we will not consider this type of phase separation.

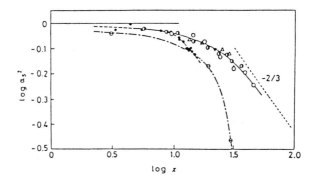

Fig. 4–6. Early data on PS in CH, showing $\log \alpha_S^3$ as a function of $\log x$, where $x = -\tau M_w^{1/2}$. $\ominus$, Slagowski et al. (M fixed) [21]; •, Nierlich et al. (M fixed) [22]; other symbols, Miyaki and Fujita (M varied) [23]. The dashed straight line of slope $-2/3$ indicates the asymptotic behavior predicted by eq 2.8.

et al. [20] have very recently reached 167 for a polystyrene with $M = 8.6 \times 10^6$ in methyl acetate at $25°C$ ($\theta = 43°C$).

Experimental work on coil–globule transition by direct measurement of chain dimensions was not reported for more than 15 years after Stockmayer had called attention to this phenomenon. Probably, experimentalists were not confident enough of being able to circumvent the technical difficulties mentioned above. Many coil–globule transition data have since been published, but, interestingly, they are not always consistent with one another. Thus, the reported transitions are roughly divided into two types, very sharp in one and rather gradual in the other. This divergence reflects the difficulties inherent in the measurement of chain dimensions below the θ temperature.

2.3.2   Early Data

Slagowski et al. [21] in 1976 were the first to report a light scattering study of coil–globule transition. Working on narrow–distribution polystyrene with $M = 4.44 \times 10^7$ in cyclohexane at 35.4, 35, 34.5, and 34°C, they found that α_S^2 decreased from 1 to 0.342 in a temperature range as narrow as 1.40°C below θ. Thus it appeared that the polystyrene coil shrinks very sharply with decreasing temperature below θ. In a small–angle neutron scattering study reported in 1978, Nierlich et al. [22] examined polystyrene with $M = 2.9 \times 10^4$ in cyclohexane down to 11.2°C and showed that as the temperature was lowered

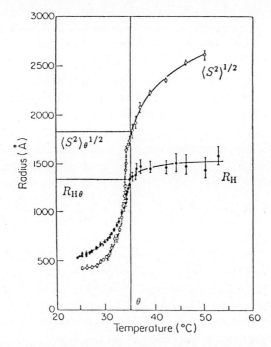

Fig. 4-7. Temperature dependence of statistical radius (o) and Stokes radius (•) for PS ($M = 2.6 \times 10^7$) in CH [26].

below θ, α_S first stayed almost constant and then decreased sharply following eq 2.8. Light scattering experiments on polystyrene in cyclohexane reported in 1981 by Miyaki and Fujita [23] and also by Oyama et al. [24], however, revealed much more gradual transitions. This can be seen in Figure 4-6, which summarizes the early data in the form of $\log \alpha_S^2$ plotted against $\log |\tau| M^{1/2}$.

Pritchard and Carolin [25] in 1980, studying polystyrene with $M = 3.7 \times 10^4$ in cyclohexane by quasi-elastic light scattering, observed a gradual decrease in α_f upon cooling below θ. However, the quasi-elastic light scattering experiment on the same system (but $M = 2.6 \times 10^7$) reported in the same year by Sun et al. [26] showed extremely sharp drops in both α_S and α_f, as illustrated in Figure 4-7.

This dramatic result was received as suggesting that the coil–globule transition of a polymer become first order at the limit of infinite chain length. In Figure 4-7 we should take note of the fact that the chain tends to collapse not to a geometric point but to a finite size ($\alpha_S \sim 0.25$; $\alpha_f \sim 0.37$), in contrast to the prediction of de Gennes' theory. Another point to notice is that α_f decreases more gradually than α_S with lowering temperature. A similar difference was

107

found by Bauer and Ullman [27] for polystyrene in cyclohexane, though they did not observe sharp transitions as were done by the Tanaka group [19, 26].

2.3.3 Recent Data

The coil–globule transition in the system polystyrene + cyclohexane was further pursued by Perzynski et al. [28] in 1982 and by Vidakovic and Rondelez [18] in 1984 in terms of α_f and by Perzynski et al. [29] in 1984 in terms of the α_η. Gradual transitions were observed in all these studies and, as in those by previous workers, they confirmed that the expansion factors were scaled by the combined variable $|\tau|M^{1/2}$. However, we must add that Stepanek et al. [30] in 1982 observed very sharp transitions for both radius of gyration and Stokes radius when they studied polystryrene ($M = 2 \times 10^6$) in dioctyl phthalate.

Thus, in the middle of the 1980s, we had two sets of contradictory data for polystyrene, one giving a very sharp transition and the other a gradual one. In order to look for the origin of such a discrepancy Chu and coworkers launched careful static and dynamic light scattering studies. Figure 4–8 depicts the data of Park et al. [31] for polystyrene in cyclohexane as a plot of $Y \equiv \alpha_S{}^3|\tau|M^{1/2}$ against $X \equiv |\tau|M^{1/2}$. Here the filled symbols refer to metastable solutions, and all other symbols to stable ones. The latter are seen to fall on a master curve which rises monotonically and tends to level off at $Y \sim 20$. This indicates that Y is scaled by X as repeatedly confirmed by previous workers. The master curve gives a gradual coil–globule transition. On the other hand, the data points for metastable solutions exhibit a maximum and deviate appreciably from those for stable solutions at $X > 40$.

In Figure 4–9, the solid curve shows the master curve fitting the stable solution data in Figure 4–8, while the points indicate the above–cited data of Slagowski et al., Nierlich et al., and Tanaka's group, which all gave very sharp transitions. Interestingly, the data of the first and third groups (and those of Stepanek et al. as well) display behavior similar to the metastable solution data in Figure 4–8. This finding suggests that the polystyrene solutions examined by these groups were not thermodynamically stable. Park et al. state that the ultrahigh polystyrene sample used by Tanaka's group was not narrow enough in molecular weight distribution so that the test solutions would have phase–separated if they had been left standing long. But Tanaka might not readily accept such a criticism.

Park et al. [31] obtained essentially similar results for α_f, and concluded that a metastable solution gives a sharper change in α_f with X than does a stable one. Thus, according to Chu's group, the previously observed very sharp coil–globule transitions in polystyrene solutions can be ascribed to the metastable state of the test solutions used. Finally, we note that Selser [32] observed for poly(α–methylstyrene) in cyclohexane only very slight shrinkages in the sub-θ region. At present, we do not know how to interpret this finding.

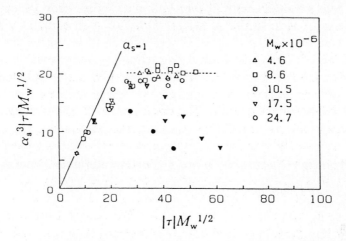

Fig. 4–8. Variation of α_S with T and M for PS in CH in the coil–globule transition region [31]. Filled symbols refer to metastable solutions and all others to stable solutions.

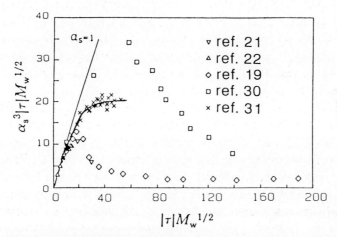

Fig. 4–9. Comparison of the master curve with some literature data indicated by reference numbers.

2.3.4 Plateau Values

For X at which Y levels off the coil–globule transition enters the region governed by the asymptotic relation 2.8. Figure 4–9 shows that Y reaches a plateau value of about 20 ($g^{1/2}$ $mol^{-1/2}$) at $X > 40$. These values indicate that the transition begins to show asymptotic behavior already at $\alpha_S \sim 0.80$. Park et al. [31] found that $Y' \equiv \alpha_f^3 |\tau| M^{1/2}$ for polystyrene in cyclohexane tended to level off much more slowly. In a more recent study which extended measurements of α_f to higher X (up to 95) by using an ultrahigh polystyrene sample and an improved instrument, Chu [33] found Y' to reach a plateau value of 43 (± 2) at $X > 80$. It is worth noting that α_f begins to exhibit asymptotic behavior more slowly than does α_S. Chu et al. [20] performed similar work with coil–globule transition in polystyrene in methyl acetate (this system has upper and lower θ temperatures). The results were essentially similar to their findings on cyclohexane solutions, except that the plateau value of Y was 50.

2.4 Problems Concerning Ternary Cluster Interaction Energy

2.4.1 Basic Relations

As repeatedly noted thus far, the residual ternary cluster interaction energy is expected to play a role in poor solvents near θ. Thus, since the pioneering work of Orofino and Flory [13] many theoretical studies have been made to clarify its effects on chain dimensions and the second virial coefficient. However, they have turned out to reveal some serious problems, which still remain unsolved.

Orofino and Flory used the smooth–density sphere model to include the (residual) ternary cluster interaction as in de Gennes' theory expounded in Section 2.2. However, it is relevant for a more rigorous theory to start with adding this energy to the Edwards Hamiltonian (see Section 3.2 of Chapter 1). To do this we first consider the spring–bead chain model. We redenote the excluded–volume strength β by β_2, rename it the binary excluded–volume strength, and introduce the ternary excluded–volume strength β_3 defined by

$$\Delta u_{ijk} = \beta_3 \delta_{ij} \delta_{jk} \qquad (2.9)$$

Here Δu_{ijk} is the residual potential energy of a ternary cluster formed by beads i, j, and k. In proceeding to a continuous chain, both β_2 and β_3 are allowed to approach zero in such a way that β_{2c} and β_{3c} defined by

$$\beta_{2c} = \lim_{\Delta s \to 0} \beta_2 (b'/\Delta s)^2 \qquad (2.10)$$

$$\beta_{3c} = \lim_{\Delta s \to 0} \beta_3 (b'/\Delta s)^3 \qquad (2.11)$$

remain finite (Section 3.2 of Chapter 1 should be consulted for b' and Δs). We note that b' converges to the Kuhn segment length b in the continuous chain limit.

It can be shown that the Edwards Hamiltonian extended to include the ternary

110

cluster interaction energy is represented by

$$H/k_\mathrm{B}T = (1/2) \int_0^1 [d\mathbf{f}(x)/dx]^2 dx$$

$$+ (2\pi)^{3/2}(z_2/2) \int_0^1 \int_{|x-x'|>\lambda/L}^1 \delta(\mathbf{f}(x) - \mathbf{f}(x'))dx dx'$$

$$+ (2\pi)^3(z_3/6) \int_0^1 \int_1^1 \int_{\substack{|x-x'|>\lambda/L \\ |x'-x''|>\lambda/L \\ |x''-x|>\lambda/L}}^1 \delta(\mathbf{f}(x) - \mathbf{f}(x'))$$

$$\times \delta(\mathbf{f}(x') - \mathbf{f}(x''))dx dx' dx'' \qquad (2.12)$$

where

$$\mathbf{f}(x) = (3/Lb)^{1/2}\mathbf{r}(s) \quad (x = s/L) \qquad (2.13)$$

with $\mathbf{f}(x)$ representing the instantaneous conformation of the chain with a reduced contour length x as a parameter and λ a cut–off length. Non–dimensional
variables z_2 and z_3 are defined by

$$z_2 = (2\pi)^{-3/2}v_2 L^{1/2} \qquad (2.14)$$
$$z_3 = (2\pi)^{-3}v_3 \qquad (2.15)$$

where

$$v_2 = (3)^{3/2}\beta_{2c}b^{-7/2} \qquad (2.16)$$
$$v_3 = (3)^3\beta_{3c}b^{-6} \qquad (2.17)$$

The quantities z_2 and v_2 are the same as z and v, respectively, that were defined
in Chapter 2. It is important to note that z_2 is proportional to $L^{1/2}$, while z_3
is independent of L. It can be shown that z_j defined similarly for a j–body
$(j = 4, 5, \dots)$ cluster is inversely proportional to the $-(j-3)/2$-th power of
L. Hence, z_j for $j \geq 4$ asymptotically vanishes as $L \to \infty$. However, this
fact does not mean that only the binary and ternary cluster interactions may
be taken into account in formulating global behavior of long chains in dilute
solution. Inclusion of the higher–order cluster interactions in the analysis offers
an interesting problem to theorists.

111

2.4.2 Calculated Results

Yamakawa [34] was the first to calculate by perturbation the end distance expansion factor α_R on the basis of eq 2.12, and, recently, Cherayil et al. [35] made a similar calculation with the result which may be written

$$\alpha_R{}^2 = 1 + z_2[4/3 + O((\lambda/L)^{1/2})] \\ + z_3[O((L/\lambda)^{1/2}) - 4\pi - 32/\pi] \qquad (2.18)$$

The term multiplied by z_2 converges to the well–known value of 4/3 as $\lambda/L \to 0$. This fact implies that, in the binary cluster approximation, perturbation calculations lead to meaningful formulas at the limit of vanishing cut–off. On the other hand, the term multiplied by z_3 diverges as $\lambda/L \to 0$. This consequence is inconsistent with the theory of Orofino and Flory in which the coefficient of z_3 is independent of L (the latter does not incorporate the cut–off length). Yamakawa's earlier formula contains the same inconsistency.

If we do not let λ/L go to zero in order to prevent the z_3 term from divergence, the α_R of a continuous chain near the θ condition turns out to depend on the cut–off length. This consequence contradicts the postulate that, being a global property, α_R should be independent of the extent of chain coarse–graining. We do not know how to cope with this dilemma.

Cherayil et al. [35] also calculated the second virial coefficient A_2 for continuous chains to first order in z_2 and z_3, obtaining

$$A_2 = [(2\pi/3)^{3/2} N_A (Lb)^{3/2}/2M^2][z_2 + 4z_3(\sigma - 2)] \qquad (2.19)$$

where σ is a parameter essentially proportional to $(L/\lambda)^{1/2}$ (theoretically $\sigma > 4$). We find that the z_3 term in this expression diverges at the limit $\lambda/L \to 0$, as does the corresponding term for α_R and that A_2 becomes cut–off dependent when the ternary cluster interaction energy is incorporated in its calculation. The latter also contradicts the postulate that A_2 is a global polymer property.

The θ temperature is defined as a temperature at which A_2 vanishes. Hence, according to eq 2.19, θ should satisfy the relation:

$$z_2(\theta) + 4z_3(\theta)(\sigma - 2) = 0 \qquad (2.20)$$

which indicates that θ depends on L (hence M) and λ. However, as far as the author is aware, few data reveal a definite molecular weight dependence of the

θ temperature except in the region of oligomers (see Ref. [36] for example). To circumvent this discrepancy we may consider that both binary and ternary excluded-volume strengths, β_{2c} and β_{3c}, simultaneously vanish at θ.

According to the calculation of Cherayil et al. [35], the third virial coefficient at a temperature θ' at which z_2 vanishes is expressed to first order in z_3 by

$$A_3 = [(2\pi Lb/3)^3 N_A{}^2/3M^3]z_3 \qquad (2.21)$$

Therefore, if β_{2c} and β_{3c} vanish at θ, not only A_2 but also A_3 must become zero at this temperature. However, this conclusion does not agree with the experimental results mentioned in Section 2.9 of Chapter 2. Hence it is legitimate to consider that in general β_{3c} remains nonzero at the θ temperature. Then we must conclude that β_{2c} does not vanish at the same temperature since otherwise eq 2.20 is not satisfied. However, if $\beta_{2c}(\theta) \neq 0$ and $\beta_{3c}(\theta) \neq 0$, θ becomes M-dependent, inconsistent with experiment. Furthermore, when these conditions of β's hold, the condition $A_2(\theta) = 0$ may no longer mean the vanishing of excluded-volume effect on chain conformations in θ solvents. In other words, the chain may behave as non-Gaussian at θ. This prediction has been substantiated by computer simulations [37].

At present, we do not foresee any plausible ideas for resolving these serious problems which arise when the binary cluster approximation is forced to be abandoned in poor solvents in the vicinity of the θ condition. For this reason some current theorists consider that polymer behavior in such solvents is far more difficult to treat than that in good solvents (where the binary cluster approximation is supposed to be good), in contrast to the concept that prevailed in early days.

3.   Characteristic Frequency

3.1   Autocorrelation Functions

In recent years, the quasi-elastic light scattering (QELS) method has become increasingly routine in experimental studies of polymer dynamics in both dilute and concentrated solutions. It allows us to obtain information about the possible modes of Brownian motion of a polymer molecule either isolated or entangled with others. Leaving its technical details and various practical applications to reference books [38], we here give a brief account of its important aspects which are closely related to the current theory of dilute polymer solutions.

We consider an experiment in which a monochromatic laser beam (with a frequency of the order of $10^{14}\,\text{s}^{-1}$) is incident into a small volume in a dilute polymer solution. Light scattered from that volume is received by a detector including a highly sensitive photomultiplier (PM). The electric current $i(t)$ generated in the PM at time t is proportional to $|E(t)|^2$, the intensity of the electric field $E(t)$ of the light received by the detector at that time. We may write $|E(t)|^2$ as $E^\star(t)E(t)$, where $E^\star(t)$ denotes the complex conjugate of $E(t)$, i.e., if $E = E_+ + iE_-$, $E^\star = E_+ - iE_-$, with $i = \sqrt{-1}$.

Now, $E(t)$ is the sum of the electric fields $E^{(n)}(t)$ of the beams impinging on the detector from scattering points $1, 2, \ldots, N$ in the small volume considered. Since these points are moving thermally, the $E^{(n)}(t)$ and hence $E(t)$ fluctuate from time to time. Techniques using electronic devices to measure the resulting fluctuation of $i(t)$ have been developed in recent years. The QELS method aims at extracting information about the motion of scattering points from the $i(t)$ data so obtained. To this end we introduce two functions $C_1(\tau)$ and $C_2(\tau)$ defined by

$$C_1(\tau) = \lim_{T \to \infty} T^{-1} \int_0^T E^\star(t)E(t + \tau)\, dt \tag{3.1}$$

$$C_2(\tau) = \lim_{T \to \infty} T^{-1} \int_0^T |E(t)|^2 |E(t + \tau)|^2\, dt$$

$$= B^2 \lim_{T \to \infty} T^{-1} \int_0^T i(t)i(t + \tau)\, dt \tag{3.2}$$

where B is a proportionality constant determined by equipment. Usually, $C_1(\tau)$ and $C_2(\tau)$ are called the **field autocorrelation function** and the **current autocorrelation function**, respectively. The former cannot be measured but is relatively easy to treat theoretically, whereas the reverse is the case for the latter. Thus it is desirable to find a relation between these two functions of τ.

For this purpose we make the following assumption. The bivariant probability $P(E(t), E(t + \tau))$ that the electric field of scattered light at the detector has a given value of $E(t)$ at time t and a given value of $E(t + \tau)$ at time $t + \tau$ is Gaussian. This means that

$$P(E_+(t), E_+(t + \tau)) = \frac{1}{2\pi\sigma^2(1 - r^2)^{1/2}}$$

$$\times \exp\left[-\frac{E_+(t)^2 - 2rE_+(t)E_+(t + \tau) + E_+(t + \tau)^2}{2\sigma^2(1 - r^2)}\right] \tag{3.3}$$

where

$$\sigma^2 = \langle E_+(t)^2 \rangle \tag{3.4}$$

$$r = \langle E_+(t)E_+(t + \tau)\rangle / \langle E_+(t)^2 \rangle \tag{3.5}$$

and a similar expression for $P(E_-(t), E_-(t+\tau))$ obtained by replacing the subscript + in eq 3.3 by $-$. Here, the angle brackets signify the time average defined as

$$\langle X \rangle = \lim_{T \to \infty} T^{-1} \int_0^T X \, dt \tag{3.6}$$

As shown in the Appendix, if the above assumption holds, we have

$$C_2(\tau) = |C_1(\tau)|^2 + |C_1(0)|^2 \tag{3.7}$$

when $E(t)$ is unpolarized, and

$$C_2(\tau) = 2|C_1(\tau)|^2 + |C_1(0)|^2 \tag{3.8}$$

when $E(t)$ is plane–polarized. Equation 3.7 is usually called the Siegert relation [38, 39]. In actual QELS experiments, plane–polarized light is incident to the solution, so that ideally $E(t)$ should be plane–polarized and eq 3.8 can be applied. However, since the experimental setup and conditions cannot realize this ideal situation, QELS data are usually analyzed with an empirical relation

$$C_2(\tau) = \gamma|C_1(\tau)|^2 + |C_1(0)|^2 \tag{3.9}$$

where γ is a parameter which absorbs all non–idealities.

We divide the measuring time interval T into J sampling times chosen to be sufficiently small (of the order of 10^{-3} to 10^{-6} s), and denote the number of photons out of the PM in each sampling time Δt by n_k for the k-th Δt. Since n_k is proportional to the electric current at $t = k\Delta t$, we have for $J \gg 1$

$$C_2(\tau) \sim J^{-1} \sum_{k=1}^{J} n_k n_{k+j} \tag{3.10}$$

where $\tau = j\Delta t$. The sum in this expression can be evaluated by sending the PM outputs into an autocorrelator. Recent advances in electronics and computers have dramatically increased the span and accuracy of $C_2(\tau)$ determinations.

3.2 Dynamic Structure Factor

In Figure 4–10, plane–polarized (in the z direction) light of frequency ω travels in the y direction and impinges on N non–absorbing scatterers contained in a small volume around the origin O. A detector is placed in the $x-y$ plane at a large distance R from O. If all the scatterers are optically isotropic, light reaching the detector is plane–polarized in the z direction. Hence its electric field $E(t)$ is written $e_z E(t)$, with e_z the unit vector pointing in the z direction. According

115

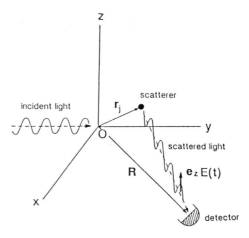

Fig. 4–10. Geometry of incident and scattered light.

to electromagnetic theory (see, for example, Ref. [9]), $E(t)$ for $|\mathbf{r}_j| \ll |\mathbf{R}|$ is represented by

$$\mathbf{E}(t) = \sum_{j=1}^{N} \mathbf{A}_j \exp\left[i\mathbf{k} \cdot \mathbf{r}_j(t)\right] \exp\left(i\omega t\right) \exp\left(-i\mathbf{k} \cdot \mathbf{R}\right) \qquad (3.11)$$

where $\mathbf{r}_j(t)$ is the position vector of scatterer j at time t, $\mathbf{k}$ the scattering vector defined in Section 2.3 of Chapter 3, and $\mathbf{A}_j$ a complex quantity which depends on t. Fluctuation of $\mathbf{E}(t)$ around a sinusoidal change occurs owing to the variation of the $\mathbf{r}_j(t)$ with time. Inserting eq 3.11 into eq 3.1 and assuming $\mathbf{A}_j$ to be constant (which is valid for spherical scatterers), we obtain

$$C_1(\tau) = \sum_{j=1}^{N}\sum_{\ell=1}^{N} \mathbf{A}_j^{\star}\mathbf{A}_\ell \langle \exp\left[-i\mathbf{k} \cdot \mathbf{r}_j(t) + i\mathbf{k} \cdot \mathbf{r}_\ell(t+\tau)\right]\rangle \, \exp\left(-i\omega\tau\right) \quad (3.12)$$

We treat the case in which all scatterers are identical (which means that all $\mathbf{A}_j$ are equal to, say, $\mathbf{A}$) and the time average may be replaced by an ensemble average. Then we may set t equal to zero and obtain

$$C_1(\tau) = \exp\left(-i\omega\tau\right)|\mathbf{A}|^2 S(k,\tau) \qquad (3.13)$$

where $S(k,\tau)$ is the **dynamic structure factor** defined by

$$S(k,\tau) = \sum_{j=1}^{N}\sum_{\ell=1}^{N} \langle \exp\left[i\mathbf{k} \cdot (\mathbf{r}_\ell(\tau) - \mathbf{r}_j(0))\right]\rangle \qquad (3.14)$$

with the angle brackets signifying an ensemble average. If we define $\rho_\tau(k)$ by

$$\rho_\tau(k) = \sum_{j=1}^{N} \exp\left(i\mathbf{k} \cdot \mathbf{r}_j(\tau)\right) \tag{3.15}$$

we may express $S(k,\tau)$ as

$$S(k,\tau) = \langle \rho_0(0)^* \rho_\tau(k) \rangle \tag{3.16}$$

Equation 3.15 can be transformed into

$$\rho_\tau(k) = \int \rho(\mathbf{r}(\tau)) \exp\left(i\mathbf{k} \cdot \mathbf{r}(\tau)\right) d\mathbf{r}(\tau) \tag{3.17}$$

where $\rho(\mathbf{r}(\tau))$ denotes the local number density of N scatterers at t. Thus, $\rho_\tau(k)$ is the Fourier transform of the local number scatterer density at time τ.

Substitution of eq 3.13 into eq 3.9 yields

$$C_2(\tau) = C_2(\infty)\{1 + \gamma[S(k,\tau)/S(k,0)]^2\} \tag{3.18}$$

The parameters $C_2(\infty)$ and γ may be estimated from the limiting values of $C_2(\tau)$ at $\tau \to \infty$ and zero. In this way, eq 3.18 allows $S(k,\tau)/S(k,0)$ to be determined from $C_2(\tau)$ data. We note that $C_2(\tau)$ depends on the sampling time Δt so that more correctly it should be denoted $C_2(\tau, \Delta t)$.

3.3 Characteristic Frequency for Gaussian Chains

Now we join the N scatterers by identical Gaussian springs to form a spring-bead chain. Then our object becomes a flexible linear polymer in dilute solution. We introduce a quantity $\Omega(k)$ defined by

$$\Omega(k) = -\lim_{\tau \to 0} \frac{d\ln[S(k,\tau)/S(k,0)]}{d\tau} \tag{3.19}$$

which is sometimes called the **characteristic frequency**. In terms of it we define $D(0)$ having the dimensions of a diffusion coefficient by

$$D(0) = \lim_{k \to 0} \Omega(k)/k^2 \tag{3.20}$$

This quantity is related to the average speed of unsteady center–of–mass translation immediately after the chain is set in motion by application of an external force. Hence, it is sometimes referred to as the initial diffusion coefficient. At

117

the moment when the chain starts moving the external force is equally distributed overall beads of the chain. As the chain travels, the differing translational speeds of various conformations are successively averaged out owing to the relaxation of internal chain motions, and the applied force is unequally distributed among the beads. Finally, the translation of the chain reaches a steady state characterized by a diffusion coefficient denoted by D_0 in Section 3.1 of Chapter 2. Schmidt et al. [40] formulated the drift from $D(0)$ to D_0 for a Gaussian chain by using the Zimm theory of polymer dynamics, and showed that the latter is smaller than the former only by 1.7%. In Section 3.5 we explain that what can actually be evaluated by QELS is not $D(0)$ but D_0.

To calculate $\Omega(k)$ theoretically we must determine $\{r_j(\tau)\}$ by solving the equations of motion for the chain. Akcasu and Gurol [41] in 1976 attacked this on the basis of the Kirkwood diffusion equation [42], and Akcasu et al. [43] presented a more general theory in a review article of 1980. In what follows, without going to mathematical details, we summarize some important results on Ω for Gaussian chains. Benmouna and Akcasu [44] and Akcasu et al. [43] extended the calculation of Ω to non-Gaussian chains by invoking the Weill-des Cloizeaux approximation, eq 1.4. However, as mentioned in Section 1, this approximation seems too crude to explore excluded-volume effects on Ω quantitatively.

According to Burchard et al. [45] who slightly modified the Akcasu–Gurol theory, we have for non-draining Gaussian chains

$$\frac{\Omega(k)}{k^2 D(0)} = \frac{(9u/32)[u + E(u) + 2(u^2 - 1)G(u)]}{u^2 - 1 + \exp(-u^2)} \tag{3.21}$$

where

$$u^2 = a^2 k^2 N/6 = k^2 \langle S^2 \rangle_\theta \tag{3.22}$$

$$E(u) = \exp(-u^2) \int_0^u \exp(t^2)\, dt \tag{3.23}$$

$$G(u) = \int_0^u t^{-1} E(t)\, dt \tag{3.24}$$

By the theory of Schmidt et al. [40], $D(0)$ in eq 3.21 may be replaced by D_0 with a negligible error. Remarkably, eq 3.21 was derived without invoking the preaveraging approximation to the Oseen interaction tensor. Akcasu and Gurol [43] showed that the non-preaveraged $D(0)$ is given by

$$D(0) = 16 k_{\mathrm{B}} T / [(6\pi)^{3/2} \eta_0 a N^{1/2}] = 0.196 k_{\mathrm{B}} T / (\eta_0 a N^{1/2}) \tag{3.25}$$

where η_0 is the solvent viscosity and a the mean spring length. Interestingly, this formula happens to agree with D_0 derived by the Kirkwood approximation (see Ref. [9]) in the non-draining limit.

When the preaveraging approximation is used, the following formula is derived in place of eq 3.21 [45]:

$$\frac{\Omega(k)_{\mathrm{pre}}}{k^2 D(0)} = \frac{(3\pi^{1/2}u/16)[(2u^2 - 1)\Phi(u) + 2\pi^{1/2}u\exp(-u^2)]}{u^2 - 1 + \exp(-u^2)} \tag{3.26}$$

where $\Phi(u)$ is the error function defined by

$$\Phi(u) = \frac{2}{\pi^{1/2}} \int_0^u \exp(-t^2)\, dt \tag{3.27}$$

It follows from eq 3.21 and 3.26 [45] that

$$\lim_{k\to\infty} \Omega(k)/k^3 = k_{\mathrm{B}}T/(16\eta_0) \quad \text{(without preaveraging)} \tag{3.28}$$

$$\lim_{k\to\infty} \Omega(k)_{\mathrm{pre}}/k^3 = k_{\mathrm{B}}T/(6\pi\eta_0) \quad \text{(with preaveraging)} \tag{3.29}$$

The latter asymptotic relation was first obtained by de Gennes [46] in 1967, and the factor 1/16 (= 0.0625) in eq 3.28 was previously estimated to be 0.055 by Akcasu and Gruol [41]. The factor $1/6\pi$ in eq 3.29 is about 15% smaller than 1/16, indicating that the error due to the preaveraging is rather serious. But, more importantly, these equations contain no parameter connected with local polymer structure. In order to obtain by QELS information about the local dynamics of polymer molecules we have to formulate the dynamic structure factor on a more realistic chain model.

3.4   Comparison with Experiment

Probably Han and Akcasu [47] in 1981 were the first to perform a systematic QELS study in which the k dependence of Ω was determined as a function of the molecular weight M of the polymer in both good and θ solvents. Using the $D(0)$ value given by eq 3.25, they calculated $\Omega(k)/(k^2 D(0))$ for polystyrene in cyclohexane at the θ temperature as a function of $k\langle S^2\rangle_\theta^{1/2}$ (= u). The result is shown in Figure 4–11, with different marks for different M. Though not corrected for concentration, i.e., not referred to infinite dilution, these data suffice to substantiate the theoretical prediction of eq 3.21 that $\Omega(k)/(k^2 D(0))$ (under the θ condition) depends only on a single variable u.

It is important to observe in Figure 4–11 that, contrary to the prediction of eq 3.21, the plotted points converge not to unity but a value considerably smaller

119

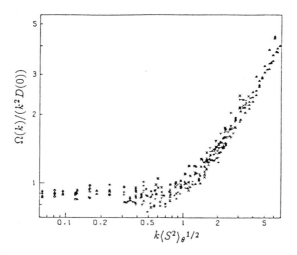

Fig. 4-11. Values of $\Omega(k)/(k^2 D(0))$ for PS in CH at the θ temperature plotted against $k\langle S^2\rangle_\theta^{1/2}$ [47]. Different symbols are for different values of molecular weight M. These data are for non-zero polymer concentrations, but presumably not very different from infinite-dilution values.

than unity. As explained below, the experimentally determined $\Omega(k)/k^2$ in the limit $k \rightarrow 0$ is not $D(0)$ but D_0. Hence, Figure 4-11 shows that the steady diffusion coefficient evaluated by QELS is distinctly smaller than $D(0)$. However, according to the theory of Schmidt et al. [40] mentioned above, the D_0 value predicted by the Kirkwood diffusion equation (or the Kirkwood–Riseman theory) differs only slightly from $D(0)$. Thus, we find that the Kirkwood–Riseman theory considerably overestimates true values of D_0. This fact is responsible for the discrepancy, noted in Section 3.5 of Chapter 2, between the experimental values of the hydrodynamic factor ρ_θ and the prediction by the Kirkwood–Riseman theory.

Equation 3.21 combined with eq 3.25 and the relation $\langle S^2\rangle_\theta = a^2 N$ also predicts that $\Omega(k)/(k^3 k_B T \eta_0^{-1})$ under the θ condition should become a universal function of u. The data of Han and Akcasu [47, 48] for polystyrene in cyclohexane shown in Figure 4-12 confirm this prediction. Curves A and B in the figure have been calculated from eq 3.21 and 3.26, respectively. Their overall trend is very similar to that of the data points, but the theoretical values appear consistently above the measured ones (vertical shifts bring them to good agreement with experiment, but such an adhoc procedure is not preferable). In particular, the asymptotic value of $\Omega/(k^3 k_B T \eta_0^{-1})$ at large u is estimated to be

120

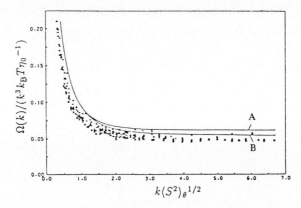

Fig. 4–12. Values of $\Omega/(k^3 k_B T \eta_0^{-1})$ as a function of $k\langle S^2 \rangle_\theta^{1/2}$. Various symbols: data for PS in CH at θ (from Ref. [48]). Curve A, calculated with no preaveraging approximation. Curve B, calculated with preaveraging approximation. Puzzlingly, curve B is closer to experimental data than is curve A.

about 0.050, while it is 0.0625 for curve A and 0.053 for curve B. Interestingly, curve B based on the preaveraging approximation comes closer to the experimental points than does curve A which does not invoke this approximation. Though puzzling, this finding should be counted one of the most significant in recent studies of dilute polymer solutions. A similar puzzling result was also obtained by Tsunashima et al. [49] in their recent work on polystyrene. Han and Akcasu [47] write "Considering the uncertainties in the extraction of $\Omega(q)$ from $S(q,t)$ (their q and t are our k and τ, respectively), the agreement (between theory and experiment as shown in Figure 4–12) may be regarded as satisfactory." However, it seems to the author that their statement is a bit too optimistic. Though they may look small, the above–mentioned numerical gaps between theory and experiment have to be narrowed by further theoretical work.

Lee et al. [50] calculated the characteristic frequency by renormalization group theory to first order in ϵ ($= 4 - d$). However, the agreement of their formula with experiment was not very encouraging.

3.5 Remarks

1. Theoretical formulation of the function $S(k, \tau)$ offers a challenge, which is not easy to overcome even for Gaussian chains. Probably, Akcasu and coworkers have gone most deeply into it as far as flexible chains are concerned. For example, they [43] derived the following at the limit of sufficiently long Gaussian

121

spring–bead chains with the preaveraged Oseen interaction tensor:

$$\frac{S(k,\tau)}{S(k,0)} = \frac{\exp\left[-\alpha\phi_0(\tau)\right] + 2\sum_{s=1}^{\infty}\exp\left[-\alpha\phi_s(\tau)\right]}{1 + 2\sum_{s=1}^{\infty}\exp\left(-\alpha s\right)} \tag{3.30}$$

Here

$$\alpha = k^2 a^2/6 \tag{3.31}$$

$$\phi_s(\tau) = s + \frac{1}{2\pi}\int_{-\pi}^{\pi}\frac{1 - \exp\left(-\gamma_p\tau\right)}{1 - \cos p}\cos ps\,dp \tag{3.32}$$

where

$$\gamma_p = 2W(1 - \cos p)[1 + 2BZ(p)] \tag{3.33}$$

with

$$Z(p) = \sum_{n=1}^{\infty}\frac{\cos pn}{n^{1/2}} \tag{3.34}$$

$$W = (3/a^2)(k_{\mathrm{B}}T/\zeta) \tag{3.35}$$

$$B = \zeta/(\eta_0 a\pi(6\pi)^{1/2}) \tag{3.36}$$

a being the mean length of a spring, ζ the friction coefficient of a bead, and η_0 the solvent viscosity. Unfortunately, eq 3.30 is too complicated to be useful for routine analysis of QELS data.

2. For sufficiently small k (correctly, $k\langle S^2\rangle^{1/2} \ll 1$) $S(k,\tau)$ can be expressed as [40]

$$\ln \mathcal{S}(k,\tau) = -k^2\tau D(\tau) + O(k^4) \tag{3.37}$$

where $\mathcal{S}(k,\tau) \equiv S(k,\tau)/S(k,0)$ and $D(\tau)$ is the diffusion coefficient at time τ after the chain starts translation. In principle, eq 3.37 allows $D(\tau)$ to be determined as a function of τ. However, according to the theory of Schmidt et al. [40], the τ dependence of $D(\tau)$ is so weak that we should in effect find an essentially linear relation between $\ln \mathcal{S}(k,\tau)$ for small k and $k^2\tau$. This prediction has been borne out by numerous experiments. Since the time resolution power of any photon–autocorrelator is limited, $\ln \mathcal{S}(k,\tau)$ can actually be measured only at τ larger than a sampling time for photon–correlation. The measurement has to be made over a range of τ because otherwise the slope of $\ln \mathcal{S}(k,\tau)$ vs. $k^2\tau$ cannot be evaluated accurately. As can be seen from eq 3.37, the necessary range of τ becomes wider as k is lowered. Hence, in the limit $k \to 0$, the plot for $\ln \mathcal{S}(k,\tau)$ vs. $k^2\tau$ substantially consists of data points for τ much larger than the sampling time, and its slope yields a $D(\tau)$ value close to $D(\infty)$. In other words, what can actually be determined by QELS at the limit $k \to 0$ is the steady–state diffusion coefficient D_0 rather than $D(0)$. In fact, the latter is not measurable, because we cannot decrease sampling times for photon–correlation measurements below a certain limit. We again note that what Akcasu and Gluorl [43] were able to calculate without invoking the preaveraging approximation is not $D(\infty)$ but $D(0)$ given by eq 3.25.

3. The QELS method can be utilized to evaluate $D(\infty)$ of a single polymer in concentrated solutions. To this end, chemically labeled chains are dispersed at a very low concentration in a given concentrated solution of unlabeled chains of the same or different species and $S(k,\tau)$ data for the labeled species are analyzed by the method established for dilute solutions. QELS measurements on concentrated solutions in which polymer chains entangle with one another usually reveal a decay of $S(k,\tau)$ much faster than that observed in dilute solutions. This decay is associated with the relaxation of local concentration fluctuations, and defines a dynamic quantity called the cooperative diffusion coefficient and discussed in Chapter 7.

4. Macrorings

4.1 Introduction

Ring or cyclic polymers in dilute solution were already given attention in the 1940s, but they remained a subject of theoretical interest until DNA macrorings were discovered in living cells [51–53] in the early 1960s. However, we still had to wait more than a decade until synthetic ring polymers became available to polymer physical chemists. Thus, ring poly(dimethylsiloxane) was prepared [54] by a ring–linear chain equilibration reaction and subsequent separation of the linear species by fractionation. Ring polystyrene was obtained [55–59] by a ring–closure reaction of bifunctional anionic polystyrene with dibromo (or dichloro)–p–styrene followed by fractional separation of the linear species. With these successes in chemistry it has become possible to examine the molecular weight dependence of global properties of ring polymers in dilute solution. In the following, we describe some typical dimensional and hydrodynamic data available from recent studies and related theories.

4.2 θ Temperature

In cyclohexane, the θ temperature for ring polystyrene was found to be 28.5°C, which is 6°C lower than that for linear polystyrene [59, 60]. In deuterocyclohexane (d–cyclohexane) the former was 34°C and the latter was 40°C [61]. Furthermore, in decalin, the corresponding temperatures were 15 and 19°C [62]. Thus, in these poor solvents, the second virial coefficient $A_2(r)$ for ring polystyrene was positive at the θ temperature $\theta(l)$ for linear polystyrene, where

123

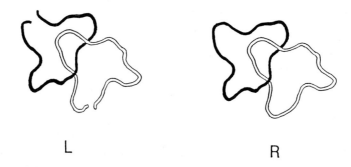

L   R

Fig. 4–13. L, entangled linear chains. R, intercrossed ring
chains (the state obtained by joining each linear chain in L
at its ends).

l and r denote linear and ring species, respectively. These findings revealed a
significant fact that repulsion can act between ring polymers coming close in a
solvent in which segment–segment interactions vanish. However, the existence
of such repulsion had already been predicted theoretically by Frank–Kamenetskii
et al. [63]

When two linear chains approach in a solvent in which the excluded–volume
strength β vanishes (we consider this in the binary cluster approximation), the
conformational state of one chain is not perturbed by that of the other. Thus,
the two chains can realize any entangled state. For example, they can be in a
state as indicated by L in Figure 4–13. We join the ends of each of them to
obtain a pair of intercrossed ring chains as indicated by R in the figure. Obviously, rings cannot realize the state R unless knotting takes place during their
polyzmerization process. Hence, the total number of configurations possible for
a pair of unknotted ring chains is smaller than that for a pair of linear chains,
provided that all other conditions are identical. Thus, the entropy of the former decreases below that of the latter. This is equivalent to the phenomenon
appearing when linear chains repel each other in a non-θ solvent. Hence, it follows that even when no volume exclusion between segments is present and thus
individual chains are in the unperturbed state, $A_2(r)$ becomes positive. Iwata
[64] formulated the repulsion due to the topological feature of ring polymers to
derive an expression for $A_2(r)$. He showed that it explains the data of Roovers
and Toporowski [59] for ring polystyrene in cyclohexane. In this book, we do
not go into an account of his theory, because it is highly sophisticated.

From the above considerations it is clear that measurements of chain dimensions and hydrodynamic properties of an isolated, unperturbed ring polymer

have to be made in a θ solvent for its linear counterpart, not in a solvent in which $A_2(\mathbf{r})$ vanishes. This precaution is taken into account in the reported experiments on ring polymers, except one.

4.3 Unperturbed Macrorings

4.3.1 Chain Dimensions

As early as 1946 Kramers [65] derived the theoretical relation

$$\langle S^2(\mathbf{r})\rangle_\theta / \langle S^2(l)\rangle_\theta = 1/2 \tag{4.1}$$

and this formula was confirmed by many subsequent authors. Here the subscript θ signifies the θ condition for the linear species. Experimentally, eq 4.1 was verified for the first time by Roovers [60]. In Figure 4–14, his light scattering data on $\langle S^2(\mathbf{r})\rangle_\theta$ for ring polystyrene in cyclohexane at 35°C ($\theta(l)$) are shown as a function of molecular weight M. This figure also shows very recent data by small–angle neutron scattering (SANS) of Hadziioannou et al. [66] on ring polystyrene in d–cyclohexane at 33°C ($\theta(\mathbf{r})$). The two sets of data are consistent and fitted by the solid line which is obtained by combining eq 4.1 with $\langle S^2(l)\rangle_\theta = 8.3 \times 10^{-4} M$ (nm^2) established by Miyaki et al. [67] for polystyrene in cyclohexane. In passing, Hadziioannou et al. [66] showed that the particle scattering function $P(k)$ for ring polystyrene at $\theta(\mathbf{r})$ agrees with Casassa's function [68]:

$$P(k) = (1/h)\exp\left(-h^2\right) \int_0^h \exp\left(t^2\right) dt \tag{4.2}$$

where $h = 2^{-1/2} k \langle S^2(\mathbf{r})\rangle_\theta^{1/2}$, with k being the magnitude of the scattering vector.

4.3.2 Transport Coefficients

The intrinsic viscosity $[\eta(\mathbf{r})]_\theta$ and the friction coefficient $f(\mathbf{r})_\theta$ of an unperturbed ring polymer were first calculated by Bloomfield and Zimm [69] and by Fukatsu and Kurata [70] using the Kirkwood–Riseman theory with the preaveraged Oseen interaction tensor. In the non–draining limit their calculations yield

$$[\eta(\mathbf{r})]_\theta / [\eta(l)]_\theta = 0.662 \tag{4.3}$$

$$f(\mathbf{r})_\theta / f(l)_\theta = 0.848 \tag{4.4}$$

Equation 4.4 was also derived by Ptitsyn [71].

Figure 4-15 depicts experimental data of $[\eta(\mathbf{r})]_\theta$ for polystyrene in cyclohexane as a function of M [57, 60, 61, 72, 73]. We precluded a few data for the

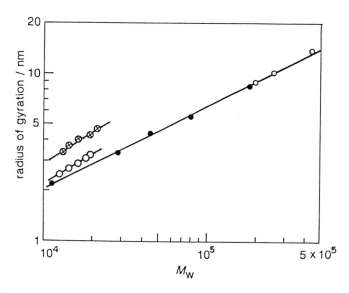

Fig. 4–14. Statistical radii of ring PS in various solvents. $\circ$, in CH at $35°C\,(\theta(l))$ [60]; $\bullet$, in d–CH at $33°C\,(\theta(r))$ [66]; $\bigcirc$, in d–toluene at $22°C$ [81]. $\otimes$, linear PS in d–toluene at $22°C$ [81].

samples which the authors considered contaminated with the linear precursor or its polycondensates. The data from different laboratories are consistent with one another and can be fitted by the indicated thick line with slope 0.5. The thin line has been derived by combining eq 4.3 with $[\eta(l)]_\theta = 8.8 \times 10^{-4}M^{1/2}$ (100 cm^3/g) which was established by Einaga et al. [74] It gives somewhat higher $[\eta(r)]_\theta$ values than the thick line, but this degree of agreement would suffice to warrant the validity of eq 4.3. However, looking at it more closely, we find that Roovers' data are better fitted by a straight line with a slope slightly smaller than 0.5. In fact, Roovers [60] reported 0.465.

Dodgson and Semlyen [75] found $[\eta(r)]_\theta/[\eta(l)]_\theta$ to be 0.67, in excellent agreement with eq 4.3, for poly(dimethylsiloxane) in butanone at $20°C$ over the range $2500 < M_w < 16000$. However, the index of the Houwink–Mark–Sakurada relation was not 0.50 but 0.58. This is probably due to the fact that their rings were not long enough.

Figure 4–16 shows two sets of data for the Stokes radius $R_H(r)_\theta$ of polystyrene, one of Hadziioannou et al. [66] by quasi–elastic light scattering in deuterated cyclohexane at $33°C$ and the other of Roovers [60] by velocity sedimentation in cyclohexane at $35°C$. The solid line fitting the plotted points has been obtained by combining eq 4.4 with $R_H(l)_\theta = 0.0229M^{1/2}$ (nm) established by Schmidt and Burchard [76]. We may conclude that eq 4.4 has been

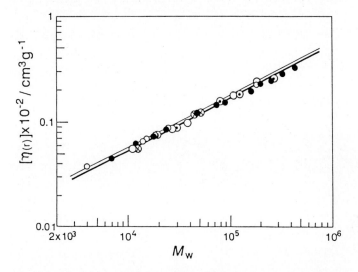

Fig. 4–15. $[\eta(\mathbf{r})]$ data for PS in CH at $\theta(l)$. o, [57]; •, [60]; ◯, [61]; ⊗, [72]; ⊙, [73]. Thick line, fitting plotted points. Thin line, calculated from eq 4.3 and the equation of Einaga et al. [74] for $[\eta(l)]_\theta$.

substantiated experimentally.

4.4 Perturbed Macrorings

4.4.1 Chain Dimensions

The radius expansion factor $\alpha_S(\mathbf{r})$ for perturbed spring–bead ring chains was calculated to first order in the excluded–volume variable z by Casassa [77], who obtained

$$\alpha_S(\mathbf{r})^2 = 1 + (\pi/2)z + \cdots \qquad (4.5)$$

The coefficient of z is larger than the corresponding value 134/105 for the linear counterpart (see eq 2–1.26), indicating that when compared at the same chain length, the ring swells more by intrachain repulsion than the linear chain. This consequence is reasonable since chain segments in the ring are confined to a smaller domain and hence have a higher probability of collision than those in the linear counterpart.

Computer simulations of self-avoiding ring chains have given information on the ratio $\langle S^2(\mathbf{r})\rangle/\langle S^2(l)\rangle$ at the limit of large chain length. By suitable extrapolation of data from Monte Carlo simulations of off–lattice chains of N (the number of bonds) up to 99, Bruns and Naghizadeh [78] estimated this ratio to be 0.559. Chen [79] found 0.568 on the basis of Monte Carlo calculations on

127

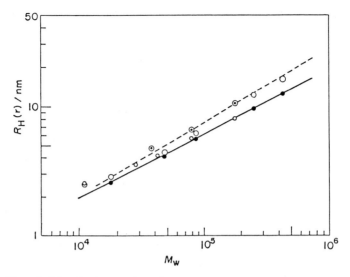

Fig. 4–16. Stokes radii of ring PS. ○, in d–CH at 33°C
[66]; •, in CH at 35°C [60]; ◯, in toluene at 35°C [60];
⊙, in THF at 25°C [88]. Solid line, calculated from eq 4.4
and the equation of Schmidt and Burchard [76] for $R_H(l)_\theta$.
Dashed line, Roovers' data multiplied by 0.899.

random–flight chains of N from 16 to 256. Interestingly, Prentis [80] obtained
Chen's value by first–order renormalization group calculation. When these find-
ings are compared with eq 4.1, the ratio of $\langle S^2(\mathbf{r})\rangle$ to $\langle S^2(l)\rangle$ increases only
from 0.50 to about 0.56–0.57 when the excluded–volume strength varies from
zero to infinity.

Radii of gyration of ring and linear polystyrenes in d–cyclohexane at 22°C
were measured by Ragnetti et al. [81] by SANS. Their results are included
in Figure 4–14. The vertical displacement between the two lines fitting the
plotted points (the slope of each line is close to 0.60) gives 0.56 for the ratio
$\langle S^2(\mathbf{r})\rangle/\langle S^2(l)\rangle$. Higgins et al. [82, 83] found this ratio to be 0.53 ± 0.05 from
SANS measurements on ring and linear poly(dimethylsiloxane) in d–benzene
at 25°C. These results are in good agreement with the simulation values mentioned above. However, such agreement will have to be accepted with some
reservation, since the data were limited to M below 2×10^4 so that they were
not likely to manifest asymptotic behavior at the self–avoiding limit.

4.4.2 Transport Coefficients

First–order calculations of $[\eta(\mathbf{r})]$ and $f(\mathbf{r})$ at the non–draining limit were
made by Norisuye and Fujita [84] and then by Shimada and Yamakawa [85],

using the Kirkwood–Riseman theory with the preaveraged Oseen interaction tensor. The results are

$$\alpha_\eta(\mathbf{r})^3 = 1 + 1.180z + \cdots \qquad (4.6)$$

$$\alpha_f(\mathbf{r}) = 1 + 0.776z + \cdots \qquad (4.7)$$

Referring to the corresponding series for the linear chain (see Section 3.4.2 of Chapter 2), we obtain

$$\alpha_\eta(\mathbf{r})^3 / \alpha_\eta(l)^3 = 1 + 0.04z + \cdots \qquad (4.8)$$

$$\alpha_f(\mathbf{r}) / \alpha_f(l) = 1 + 0.17z + \cdots \qquad (4.9)$$

These indicate that the excluded–volume effects on $[\eta]$ and f are stronger for ring chains than for linear ones as in the case for chain dimensions.

Attempts to calculate $[\eta(\mathbf{r})]$ and $f(\mathbf{r})$ valid for large z were made by Bloomfield and Zimm [69] and by Fujita and Yu [86] using the ϵ theory of Peterlin and Ptitsyn touched upon in Section 1.4, but their results were found to be in error. Chen [79] applied the Kirkwood formula (see Ref. [9]) to Monte Carlo data for self–avoiding random flight ring and linear chains and obtained

$$f(\mathbf{r}) / f(l) = 0.899 \qquad (4.10)$$

In Figure 4–17, the experimental data of $[\eta(\mathbf{r})]$ for polystyrene in two good solvents, toluene at 25 or 35°C [57, 60] and tetrahydrofuran (THF) at 25°C [61, 72], are plotted against M_w. The dashed line, which is derived by multiplying the data of Einaga et al. [74] for polystyrene in benzene by 0.662 taken from eq 4.3, is indicated for comparison. Incidentally, the data of Lutz et al. [61] and Mckenna et al. [72] come close to it, while those of Roovers [60] and Geiser and Höcker [57] appear below it. Dodgson and Semlyen [75] reported $[\eta(\mathbf{r})]/[\eta(l)]$ to be about 0.60 for poly(dimethylsiloxane) in cyclohexane and toluene (good solvents) at 25°C.

According to the data of Hild et al. [62] and Qian and Cao [87] for polystyrene in various solvents, it seems certain that $[\eta(\mathbf{r})]/[\eta(l)]$ decreases upon improving solvent quality. For example, this ratio was 0.66 and 0.55 in cyclohexane and benzene at 25°C, respectively, at $M = 55000$ [62]. However, the available data are still inadequate for drawing a quantitative relation between $[\eta(\mathbf{r})]/[\eta(l)]$ and solvent quality.

Figure 4–16 includes the $R_H(\mathbf{r})$ data for polystyrene in toluene at 35°C by Roovers [60](velocity sedimentation) and those in tetrahydrofuran at 25°C by Duval et al. [88] (QELS) The dashed line has been derived from the best–fit line to Roovers' data combined with eq 4.9. Edwards et al. [89] and Higgins et al. [90] found $R_H(\mathbf{r})/R_H(l)$ $(= f(\mathbf{r})/f(l))$ to be 0.84 ± 0.016 for poly

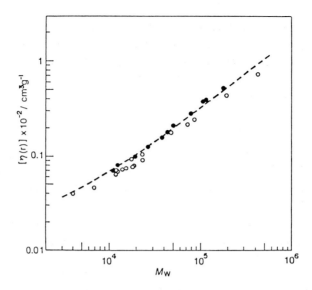

Fig. 4–17. $[\eta(\mathbf{r})]$ data for PS in good solvents. $\circ$, in toluene at 25 or 35°C [57, 60]; $\bullet$, in THF at 25°C [61, 72]. Dashed line, data of Einaga et al. [74] multiplied by 0.662.

(dimethylsiloxane) of M_w below 2×10^4 in toluene at 25°C from classic gradient diffusion and dynamic light and neutron scattering measurements, respectively. This value is consistent with eq 4.4 for unperturbed rings. Probably, the macrorings studied by these authors were too low in molecular weight to be adequate for a test of eq 4.10, which is expected to be valid at high molecular weights.

4.5 Second Virial Coefficient

Roovers and Toporowski [59] found that $A_2(\mathbf{r})$ for polystyrene in cyclohexane has a definite positive value at $\theta(l)$. This was a very important finding, because it demonstrated that repulsion arises between a pair of unperturbed macrorings. Casassa [68] calculated $A_2(\mathbf{r})$ to first order in z. In terms of the reduced second virial coefficient h defined in Section 2.2 of Chapter 2 his result is expressed as

$$h(z) = 1 - 4.457z + \cdots \tag{4.11}$$

The coefficient of z for linear chains is -2.865 (eq 2-2.7). Equation 4.11 is incomplete because it does not take the repulsion found by Roovers and Toporowski into account.

Experimental values of $A_2(\mathbf{r})$ for polystyrene [81, 88] and poly(dimethylsiloxane) [91, 92] in various good solvents were smaller than those of $A_2(l)$

130

in poor solvents. For example, for the latter polymer of $M_w \sim 2 \times 10^4$, $A_2(\mathbf{r})$ was about 20% lower than $A_2(l)$. However, these data do not permit a definite conclusion on the molecular weight dependence of $A_2(\mathbf{r})$, since they are not only badly scattered but limited to relatively low molecular weights. More experimental work is urgently required.

4.6 Effect of Self–knotting

Macrorings can take up a variety of topological states, depending on the number and type of self–knots that are formed during the course of their chemical synthesis. Figure 4–18 (a) illustrates some simple examples of self–knotted rings. Monte Carlo simulations [63, 93] showed that the probability of knot formation increases with the number of random walk steps and that 30–40% of the generated rings are knotted at 150 steps. However, the fraction of knotted rings was very small for self–avoiding walks [94]. Average dimensions of a knotted ring should be smaller than those of the corresponding unknotted one. ten Brinke and Hadziioannou [93] investigated this prediction by computer simulations of random walks on a body–centered cubic lattice and showed that the exponent ν in the relation $\langle S^2(\mathbf{r}) \rangle \sim M^{2\nu}$ was 0.503 for unknotted rings, 0.542 for knotted rings, and 0.492 for all rings including both knotted and unknotted ones. The absolute values of $\langle S^2(\mathbf{r}) \rangle$ depended on the knotted structure of the ring as expected.

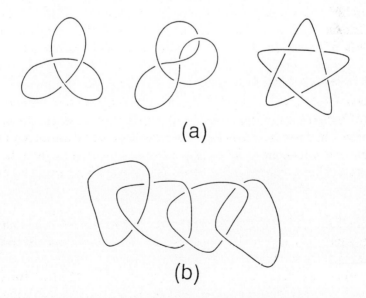

(a)

(b)

Fig. 4–18. (a) self–knotted rings. (b) catena–rings.

There is a possibility that catena-rings as illustrated in Figure 4-18 (b) are formed depending on polymerization conditions. Little work has as yet been done on the theory and experiment of such macrorings in dilute solution.

Appendix

Derivation of eq 3.7 and 3.8

Case of Unpolarized Light

In general, we have

$$C_2(\tau) = \langle [E_+(t)^2 + E_-(t)^2][E_+(t+\tau)^2 + E_-(t+\tau)^2] \rangle$$
$$= \langle E_+(t)^2 E_+(t+\tau)^2 \rangle + \langle E_-(t)^2 E_-(t+\tau)^2 \rangle$$
$$+ 2\langle E_+(t)^2 \rangle \langle E_-(t)^2 \rangle \qquad (4A\text{-}1)$$

because the E_+ and E_- components of E are independent and $\langle E_+(t+\tau)^2 \rangle = \langle E_+(t)^2 \rangle$ and $\langle E_-(t+\tau)^2 \rangle = \langle E_-(t)^2 \rangle$. If the bivariant distribution of $E_+(t)$ and $E_+(t+\tau)$ is given by eq 3.3, replacement of time average by ensemble average leads to

$$\langle E_+(t)^2 E_+(t+\tau)^2 \rangle = \int_{-\infty}^{\infty} \int_{-\infty}^{\infty} P(x,y)\, x^2\, y^2 dx\, dy$$
$$= \sigma^4 (1 + 2r^2) = \langle E_+(t)^2 \rangle^2 + 2\langle E_+(t)E_+(t+\tau) \rangle^2 \qquad (4A\text{-}2)$$

In a similar way, we get

$$\langle E_-(t)^2 E_-(t+\tau)^2 \rangle = \langle E_-(t)^2 \rangle^2 + 2\langle E_-(t)E_-(t+\tau) \rangle^2 \qquad (4A\text{-}3)$$

Substitution of eq 4A-2 and 4A-3 into eq 4A-1 yields for unpolarized light

$$C_2(\tau) = 4\langle E_+(t)^2 \rangle^2 + 4\langle E_+(t)E_+(t+\tau) \rangle^2 \qquad (4A\text{-}4)$$

because unpolarized light satisfies the conditions:

$$\langle E_+(t)^2 \rangle = \langle E_-(t)^2 \rangle \qquad (4A\text{-}5)$$

$$\langle E_+(t)E_+(t+\tau) \rangle = \langle E_-(t)E_-(t+\tau) \rangle \qquad (4A\text{-}6)$$

For $C_1(\tau)$ we obtain, again using the independence of E_+ and E_- as well as eq 4A-6,

$$|C_1(\tau)| = |\langle [E_+(t) - iE_-(t)][E_+(t+\tau) + iE_-(t+\tau)] \rangle|$$
$$= 2\langle E_+(t)E_+(t+\tau) \rangle \qquad (4A\text{-}7)$$

With eq 4A-7 it follows from eq 4A-4 that

$$C_2(\tau) = |C_1(0)|^2 + |C_1(\tau)|^2 \qquad (4A\text{-}8)$$

which is eq 3.7, i.e., the Siegert relation.

Case of Plane-polarized Light

In this case, $E_-(t) = E_-(t + \tau) = 0$. Hence, eq 4A-1 reduces to

$$C_2(\tau) = \langle E_+(t)^2 E_+(t + \tau)^2 \rangle \qquad (4A-9)$$

and eq 4A-7 to

$$|C_1(\tau)| = \langle E_+(t) E_+(t + \tau) \rangle \qquad (4A-10)$$

Equation 3.8 follows by introducing eq 4A-9 and 4A-10 into eq 4A-2.

References

1. G. Weill and J. des Cloizeaux, J. Phys. (Paris) **40**, 99 (1979).
2. A. Z. Akcasu and C. C. Han, Macromolecules **12**, 276 (1979).
3. J. François, T. Schwartz, and G. Weill, Macromolecules **13**, 564 (1980).
4. W. L. Mattice, Macromolecules **14**, 1491 (1981); Macromolecules **14**, 1485 (1981).
5. J. G. Curro and D. W. Schaefer, Macromolecules **13**, 1199 (1980).
6. Y. Matsushita, I. Noda, M. Nagasawa, T. P. Lodge, E. J. Amis, and C. C. Han, Macromolecules **17**, 1785 (1984).
7. A. Peterlin, J. Chem. Phys. **23**, 2464 (1955).
8. O. B. Ptitsyn and Yu. E. Eizner, Zh. Tekh. Fiz. **32**, 2464 (1958).
9. H. Yamakawa, "Modern Theory of Polymer Solutions," Harper & Row, New York, 1971.
10. W. H. Stockmayer, Makromol. Chem. **35**, 54 (1960).
11. C. Williams, F. Brochard, and H. L. Frisch, Ann. Rev. Phys. Chem. **32**, 433 (1981).
12. B. H. Zimm, J. Chem. Phys. **14**, 164 (1946); this was the first reference to the ternary cluster interaction energy. See also Y. Oono, J. Phys. Soc. Jpn. **41**, 228 (1976) and T. Oyama and Y. Oono, J. Phys. Soc. Jpn. **42**, 1348 (1948).
13. T. A. Orofino and P. J. Flory, J. Chem. Phys. **26**, 1067 (1957).
14. P.-G. de Gennes, J. Phys. Lett. **36**, L55(1975); see also C. B. Post and B. H. Zimm, Biopolymers **18**, 1487 (1979).
15. M. A. Moore, J. Phys. A **10**, 305 (1977).
16. I. C. Sanchez, Macromolecules **12** 980 (1979).
17. I. N. Lifshitz, A. Y. Grosberg, and A. R. Khokhlov, Rev. Mod. Phys. **50**, 683 (1978).
18. P. Vidakovic and F. Rondelez, Macromolecules **17**, 418 (1984).
19. G. Swislow, S. T. Sun, I. Nishio, and T. Tanaka, Phys. Rev. Lett. **44**, 796 (1980).
20. B. Chu, I. H. Park, Q.-W. Wang, and C. Wu, Macromolecules **20**, 2833 (1987).
21. E. Slagowski, G. Tsai, and D. McIntyre, Macromolecules **9**, 687 (1976).
22. M. Nierlich, J. P. Cotton, and B. Farnoux, J. Chem. Phys. **69**, 1379 (1978).
23. Y. Miyaki and H. Fujita, Polym. J.**13**, 749 (1981).
24. T. Oyama, K. Shiokawa, and K. Baba, Polym. J. **13**, 167 (1981).
25. M. J. Pritchard and D. Caroline, Macromolecules **13**, 957 (1980).
26. S. T. Sun, I. Nishio, G. Swislow, and T. Tanaka, J. Chem. Phys. **73**, 5971 (1980).
27. D. R. Bauer and R. Ullman, Macromolecules **13**, 392 (1980).
28. R. Perzynski, M. Adam, and M. Deslanti, J. Phys. (Paris) **43**, 129 (1982).

29. R. Perzynski, M. Delsanti, and M. Adam, J. Phys. (Paris) **45**, 1765 (1984).

30. P. Stepanek, C. Konak, and B. Sedlacek, Macromolecules **15**, 1214 (1982).

31. I. H. Park, Q.-W. Wang, and B. Chu, Macromolecules **20**, 1965 (1987).

32. J. C. Selser, Macromolecules **18**, 585 (1985).

33. B. Chu, lecture notes (Critical phenomena and polymer coil–to–globule transition), private communication.

34. H. Yamakawa, J. Chem. Phys. **45**, 2606 (1966).

35. B. J. Cherayil, J. F. Douglas, and K. F. Freed, J. Chem. Phys. **83**, 5293 (1985).

36. K. Huber and W. H. Stockmayer, Macromolecules **20**, 1400 (1987). Figure 3 of this paper indicates that A_2 for polystyrene in cyclohexane at 35°C (the θ temperature for $M > 10^4$) increases sharply from zero with decreasing M below 10^4. This fact implies that θ for oligostyrene in cyclohexane becomes lower as M decreases.

37. W. Burns, Macromolecules **17**, 2826 (1984).

38. B. Chu, "Laser Light Scattering," Academic Press, New York, 1974; Ed. B. J. Bern and R. Pecora, "Dynamic Light Scattering," Wiley–Interscience, New York, 1976.

39. According to E. Jakeman and E. R. Pike, J. Phys. A2, 411 (1969), the original reference to this relation is A. J. F. Siegert, M. I. T. Rad. Lab. Rep. No. 465 (1943), which is not available to the author.

40. M. Schmidt, W. H. Stockmayer, and M. L. Mansfield, Macromolecules **15**, 1609 (1982).

41. A. Z. Akcasu and H. Gurol, J. Polym. Sci., Polym. Phys. Ed. **14**, 1 (1976).

42. J. G. Kirkwood, J. Polym. Sci. **12**, 1 (1954).

43. A. Z. Akcasu, M. Benmouna, and C. C. Han, Polymer **21**, 866 (1980).

44. M. Benmouna and A. Z. Akcasu, Macromolecules **11**, 1187 (1978).

45. W. Burchard, M. Schmidt, and W. H. Stockmayer, Macromolecules **13**, 580 (1980).

46. P.-G. de Gennes, Physics (USA) **3**, 37 (1967).

47. C. C. Han and A. Z. Akcasu, Macromolecules **14**, 1080 (1980).

48. D. W. Schaefer and C. C. Han, in "Dynamic Light Scattering," Ed. R. Pecora, Wiley–Interscience, New York, 1985, Chap. 5.

49. Y. Tsunashima, N. Nemoto, and M. Kurata, Macromolecules **17**, 425 (1984).

50. A. Lee, W. Baldwin, and Y. Oono, Phys. Rev. A **30**, 968 (1984).

51. W. Fries and R. L. Sinsheimer, J. Mol. Biol. **5**, 424 (1962).

52. R. Weil and J. Vinograd, Proc. Natl. Acad. Sci. US **73**, 3852 (1963).

53. J. C. Wang, in "Cyclic Polymers," Ed. J. A. Semlyen, Elsevier, London, 1986, Chap. 7.

54. J. C. Semlyen, Pure & Appl. Chem. 53, 1797 (1981).

55. G. Hild, A. Kohler, and R. Rempp, Eur. Polym. J. 16, 525 (1980).

56. D. Geiser and H. Höcker, Polym. Bull. 2, 591 (1980).

57. D. Geiser and H. Höcker, Macromolecules 13, 653 (1980).

58. B. Vollmert and J. X. Hunag, Makromol. Chem., Rapid Commun. 1, 333 (1980).

59. J. Roovers and P. M. Toporowski, Macromolecules 16, 843 (1983).

60. J. Roovers, J. Polym. Sci. Polym. Phys. Ed. 23, 1117 (1985).

61. P. Lutz, G. M. Mckenna, P. Rempp, and C. Strazielle, Makromol. Chem., Rapid Commun. 7, 599 (1986).

62. G. Hild, C. Strazielle, and P. Rempp, Eur. Polym. J. 19, 721 (1983).

63. M. D. Frank-Kamenetskii, A. V. Lukashin, and A. V. Vologodskii, Nature 258, 398 (1975).

64. K. Iwata, Macromolecules 18, 115 (1985).

65. H. A. Kramers, J. Chem. Phys. 14, 415 (1946).

66. G. Hadziioannou, P. M. Cotts, G. ten Brinke, C. C. Han, P. Lutz, C. Strazielle, P. Rempp, and A. J. Kovacs, Macromolecules 20, 493 (1987).

67. Y. Miyaki, Y. Einaga, and H. Fujita, Macromolecules 11, 1180 (1980).

68. E. F. Casassa, J. Polym. Sci. Part A 3, 605 (1965).

69. V. A. Bloomfield and B. H. Zimm, J. Chem. Phys. 44, 315 (1966).

70. M. Fukatsu and M. Kurata, J. Chem. Phys. 44, 4538 (1966).

71. O. B. Ptitsyn, Zh. Tekhn. Fiz. 29, 75 (1959).

72. G. M. Mckenna, G. Hadziioannou, P. Lutz, G. Hild, C. Strazielle, C. Straupe, P. Rempp, and A. J. Kovacs, Macromolecules 20, 498 (1987).

73. Z. He, M. Yuan, X. Zhang, X. Wang, J. Haung, C. Li, and L. Wang, Eur. Polym. J. 22, 597 (1986).

74. Y. Einaga, Y. Miyaki, and H. Fujita, J. Polym. Sci., Polym. Phys. Ed. 17, 2103 (1979).

75. K. Dodgson and J. A. Semlyen, Polymer 18, 1265 (1977).

76. M. Schmidt and W. Burchard, Macromolecules 14, 210 (1981).

77. E. F. Casassa, cited in Ref. 9.

78. W. Bruns and J. Naghizadeh, J. Chem. Phys. 65, 747 (1976).

79. Y. Chen, J. Chem. Phys. 78, 5192 (1983).

80. J. J. Prentis, J. Chem. Phys. 76, 1574 (1982).

81. M. Ragnetti, D. Geiser, H. Höcker, and R. C. Oberthür, Makromol. Chem. 186, 1701 (1985).

82. J. S. Higgins, K. Dodgson, and J. A. Semlyen, Polymer 20, 553 (1979).

83. K. Dodgson and J. S. Higgins, in "Cyclic Polymers," Ed. J. A. Semlyen, Elsevier, London, 1986, Chap. 5.

137

84.   T. Norisuye and H. Fujita, J. Polym. Sci., Polym. Phys. Ed. **16**, 999 (1978).
85.   J. Shimada and H. Yamakawa, J. Polym. Sci., Polym. Phys. Ed. **16**, 1927 (1978).
86.   H. Fujita and H. Yu, cited in Ref. 9.
87.   R. Qian and T. Cao, Makromol. Chem. **188**, 1757 (1987).
88.   M. Duval, P. Lutz, and C. Strazielle, Makromol. Chem., Rapid Commun. **6**, 71 (1985).
89.   C. J. C. Edwards, R. F. T. Stepto, and J. A. Semlyen, Polymer **21**, 781 (1980).
90.   J. S. Higgins, K. Ma, L. K. Nicholson, J. B. Hayter, K. Dodgson, and J. A. Semlyen, Polymer **24**, 793 (1983).
91.   C. J. C. Edwards, R. F. T. Stepto, and J. A. Semlyen, Polymer **23**, 869 (1982).
92.   C. J. C. Edwards and R. F. T. Stepto, in "Cyclic Polymers," Ed. J. A. Semlyen, Elsevier, London, 1986, Chap. 4.
93.   G. ten Brinke and G. Hadziioannou, Macromolecules **20**, 480 (1987).
94.   Y. Chen, J. Chem. Phys. **75**, 2447 (1981).

Chapter 5   Stiff Chains

1.   Introduction

Up to this point we have confined ourselves to ideally flexible chains. Thus, the theories developed on the models of such chains (for example, the spring-bead chain) should no longer be adequate for polymers whose chemical structure suggests considerable stiffness of the chain backbone. Many chain models may be used to formulate a theory of stiff or semi-flexible polymers in solution, but the most frequently adopted is the **wormlike chain** mentioned in Section 1.3 of Chapter 1; it is sometimes called the **KP chain**. This physical model was introduced long ago by Kratky and Porod [1] to represent cellulosic polymers. However, significant progress in the study of its dilute solution properties, static and dynamic, has occurred in the last two decades.

The present chapter aims to describe some typical contributions from recent studies on stiff polymers in dilute solution. We will be mainly interested in (i) applicability of the wormlike chain model to actual polymers, (ii) validity of the hydrodynamic theories [2–4] recently developed for this model, and (iii) the onset of the excluded-volume effect on the dimensions of semi-flexible polymers. Yamakawa [5, 6] has generalized the wormlike chain model to one that he named the helical wormlike chain. In a series of papers he and his collaborators have made a great many efforts to formulate its static and dynamic properties in dilute solution. In fact, the theoretical information obtained is now comparable in both breadth and depth to that of the wormlike chain (see Ref. [6] for an overview). Unfortunately, however, most of the derived expressions are too complex to be of use for quantitative analysis and interpretation of experimental data. Thus, we only have a few to be considered with reference to the practical aspects of the helical wormlike chain, and have to be content with mentioning the definition and some basic features of this novel model.

2.   Theories for Unperturbed Wormlike Chains

2.1   Dimensional Properties

2.2.1   Persistence Length

We consider a freely-rotating chain consisting of N bonds (segments) of length a linked linearly at a supplementary angle ψ. We assume that no bead-bead interaction is operating, i.e., the chain is not perturbed by volume exclusion. Until Section 4 below our considerations are limited to the case in which this assumption is obeyed. Thus, for simplicity, the term unperturbed is omitted when we refer to wormlike chains. The wormlike chain is a continuous version

139

of this freely–rotating chain derived by letting $N \to \infty$, $a \to 0$, and $\psi \to 0$ in such a way that Na (= chain length) and $a/(1 - \cos\psi)$ remain finite at L and q, respectively:

$$L = \lim_{\substack{N \to \infty \\ a \to 0}} Na \tag{2.1}$$

$$q = \lim_{\substack{\psi \to 0 \\ a \to 0}} a/(1 - \cos\psi) \tag{2.2}$$

Here L is called the contour length and q the **persistence length** of the wormlike chain. These are model parameters.

The mean–square end–to–end distance $\langle R^2 \rangle$ and the mean–square radius of gyration $\langle S^2 \rangle$ of a wormlike chain are represented by

$$\langle R^2 \rangle = 2qL - 2q^2(1 - e^{-L/q}) \tag{2.3}$$

$$\langle S^2 \rangle = (qL/3) - q^2 + (2q^3/L)[1 - (q/L)(1 - e^{-L/q})] \tag{2.4}$$

which can be derived from the corresponding expressions for the freely–rotating chain by applying eq 2.1 and 2.2. It is important to note that both $\langle R^2 \rangle$ and $\langle S^2 \rangle$ are expressed in terms of L and q only.

It can be shown that q is equal to the average projection of **R**, the end distance vector, of an infinitely long wormlike chain onto the tangent vector at the chain end. Hence, on average, an infinite wormlike chain with a larger q is more extended in the direction of its end tangent vector. This suggests that q can be taken as a measure of chain stiffness.

It follows from eq 2.3 and 2.4 that

$$\langle R^2 \rangle = 6\langle S^2 \rangle = 2Lq \quad (L/q \to \infty) \tag{2.5}$$

$$\langle R^2 \rangle = 12\langle S^2 \rangle = L^2 \quad (L/q \to 0) \tag{2.6}$$

The former is valid for Gaussian chains (Chapter 2) and the latter for straight rods. Hence the wormlike chain takes on a variety of conformations intermediate between Gaussian coils and rods depending on the value of a dimensionless parameter L/q. It is due to this property that the wormlike chain is used to model polymer molecules with stiffness. However, what can be obtained with wormlike chains is only a fraction of the infinitely numerous conformations realized by actual polymer molecules.

Comparing eq 2.5 with eq 1–3.15, we find

$$2q = b \tag{2.7}$$

where b is the Kuhn segment length for continuous chains (see Section 3.2 of Chapter 1). Thus, the ratio $L/2q$ is referred to as the number of Kuhn's segments in the wormlike chain. It is informative to note that b represents a measure of stiffness of a continuous chain.

2.1.2   Wire Model

We represent a very thin elastic wire of length L by a differentiable space curve and denote the unit tangent vector at a contour point s by $\mathbf{u}(s)$, where s is the length measured along the curve from one of its ends. The bending energy $U(s)$ stored per unit contour length at the point s when the wire undergoes a small deformation from the straight configuration is given by

$$U(s) = (\alpha_0/2)|\partial\mathbf{u}(s)/\partial s|^2 \tag{2.8}$$

where α_0 stands for the bending force constant of the wire. We can ignore the torsional energy because the wire is very thin. According to the path integral method of Saito et al. [7] or the method of Landau and Lifshitz [8] based on elasticity theory, it can be shown that

$$\langle \mathbf{u}(s) \cdot \mathbf{u}(s') \rangle = \exp\left(-2\lambda|s - s'|\right) \tag{2.9}$$

where the angle brackets signify the average over all possible conformations of the wire and λ is defined by

$$(2\lambda)^{-1} = \alpha_0(k_B T)^{-1} \tag{2.10}$$

with $k_B T$ having the same meaning as in previous chapters. The end distance vector $\mathbf{R}$ of the wire is expressed by

$$\mathbf{R} = \int_0^L \mathbf{u}(s)\, ds \tag{2.11}$$

Hence

$$\langle R^2 \rangle = \int_0^L \int_0^L \langle \mathbf{u}(s) \cdot \mathbf{u}(s') \rangle\, ds\, ds' \tag{2.12}$$

Substitution of eq 2.9 yields

$$\langle R^2 \rangle = (L/\lambda) - (2\lambda^2)^{-1}(1 - e^{-2\lambda L}) \tag{2.13}$$

which, upon comparison with eq eq 2.3, gives the relation

$$(2\lambda)^{-1} = q \tag{2.14}$$

141

Thus we see that a wormlike chain with persistence length q is equivalent to a thin elastic wire whose bending force constant α_0 is given by

$$\alpha_0 = k_B T q \tag{2.15}$$

Equation 2.15 may be considered to be another definition of the persistence length (note that its original definition is eq 2.2). It states that the persistence length is a measure of chain stiffness associated with bending. Some authors prefer the notation $(2\lambda)^{-1}$ to q, but we use the latter throughout this book.

Equation 2.9 indicates that the orientation correlation of the tangents at contour points s and s' diminishes exponentially with the contour length $\Delta s = |s - s'|$ between the two points, and the decay is more rapid for smaller q, i.e., for more flexible chains. Thus, in a Gaussian chain ($q = 0$), there exists no correlation between the orientations of any paired segments. On the other hand, in a rod chain, the reverse is the case as can be intuitively understood.

2.2 Distribution Functions and Moments

Most basic to the theoretical formulation of wormlike chains is the distribution function $G(\mathbf{r}, \mathbf{u} | \mathbf{u}_0; s)$, which represents the probability density that the tangent $\mathbf{u}(s)$ and the position $\mathbf{r}(s)$ of a contour point s have given values $\mathbf{u}$ and $\mathbf{r}$, respectively, when the chain end $s = 0$ is at a laboratory–fixed coordinate origin and the tangent $\mathbf{u}(0)$ at that time has a given value $\mathbf{u}_0$. Since wormlike chains are Markoffian, this function satisfies a Fokker–Planck equation, which reads

$$\left(\frac{\partial}{\partial s} - \frac{1}{2q}\Delta_u{}^2 + \mathbf{u} \cdot \Delta_r\right)G(\mathbf{r}, \mathbf{u} | \mathbf{u}_0; s) = \delta(s)\delta(\mathbf{r})\delta(\mathbf{u} - \mathbf{u}_0) \tag{2.16}$$

where

$$\Delta_u{}^2 = \frac{1}{\sin\theta}\frac{\partial}{\partial\theta}(\sin\theta)\frac{\partial}{\partial\theta} + \frac{1}{(\sin\theta)^2}\frac{\partial^2}{\partial\phi^2} \tag{2.17}$$

with θ and ϕ being the polar angles of the unit vector $\mathbf{u}$, and δ the three-dimensional delta function. The right–hand side of eq 2.16 indicates the given boundary conditions at the chain end.

Integration of eq 2.16 over the entire range of $\mathbf{r}$ gives another Fokker–Planck equation

$$\left(\frac{\partial}{\partial s} - \frac{1}{2q}\Delta_u{}^2\right)G(\mathbf{u} | \mathbf{u}_0; s) = \delta(s)\delta(\mathbf{u} - \mathbf{u}_0) \tag{2.18}$$

where

$$G(\mathbf{u} | \mathbf{u}_0; s) = \int G(\mathbf{r}, \mathbf{u} | \mathbf{u}_0; s)\,d\mathbf{r} \tag{2.19}$$

Equation 2.16 cannot be solved exactly. Attempts have thus been made to obtain approximate expressions for $G(\mathbf{r}, \mathbf{u}|\mathbf{u}_0; s)$ as expansions from the Gaussian coil and rod limits in powers of q/s and s/q, respectively. The expansion from the coil is called Daniels' approximation. In this, all terms containing $\exp(-\text{constant} \times s/q)$ are neglected in comparison with the terms varying as integral powers of q/s. The second Daniels approximation, which means being valid to second order in q/s, and the deviation from the rod limit valid to first order in s/q were calculated by Gobush et al. [9] and Yamakawa and Fujii [10], respectively. However, the results are of limited practical use, converging only for $s/2q$ much larger or much smaller than unity. Despite its fundamental importance, work on $G(\mathbf{r}, \mathbf{u}|\mathbf{u}_0; s)$ has made no significant progress since these pioneering studies.

More attention has been directed toward the characteristic function $I(\mathbf{k}; s)$, which is defined by $G(\mathbf{r}; s)$ and related to even moments $\langle \mathbf{r}^{2n} \rangle$ as follows:

$$I(\mathbf{k}; s) = \int G(\mathbf{r}; s) \exp(i\mathbf{k} \cdot \mathbf{r}) \, d\mathbf{r}$$

$$= \sum_{n=0}^{\infty} (-1)^n \mathbf{k}^{2n} \langle \mathbf{r}^{2n} \rangle / (2n+1)! \tag{2.20}$$

where $G(\mathbf{r}; s)$ is the probability density that the contour point s is found at a specified position $\mathbf{r}$. As seen in the next section, $I(\mathbf{k}; s)$ is essential for calculating the particle scattering function $P(k)$. Equation 2.20 shows that if the $\langle \mathbf{r}^{2n} \rangle$ are obtained, $I(\mathbf{k}; s)$ and hence $G(\mathbf{r}; s)$ can be calculated. In this case, the recurrence formula for $\langle \mathbf{r}^{2n} \rangle$ due to Hermans and Ullman [11] (including Nagai's modification [12]) or Yamakawa's operational method [13] is useful. However, in practice, it is impossible to obtain analytically $\langle \mathbf{r}^{2n} \rangle$ valid for any values of n and s/q. Thus, Yamakawa et al. [14–16] invoked the following approximation. Starting from the rod limit, they first calculated $\langle \mathbf{r}^{2n} \rangle$ valid for any n up to fifth order in $s/2q$. Then they computed $\langle \mathbf{r}^{2n} \rangle$ or directly $I(\mathbf{k}; s)$ for $s/2q$ comparable to or larger than unity by applying a Laguerre polynomial expansion (which is an extension of the Hermite polynomial method for $G(\mathbf{r}; s)$ or $I(\mathbf{k}; s)$ originally developed by Nagai [17] and by Jernigan and Flory [18]) and the weighting function method [15]. From these results they were able to interpolate $I(\mathbf{k}; s)$ which applies over almost the entire ranges of $s/2q$ and k $(= |\mathbf{k}|)$.

2.3 Isotropic Scattering Functions

Because of their structural features stiff polymers are often optically anisotropic, and hence their light scattering data have to be analyzed with care.

143

However, in this section, we consider optically isotropic wormlike chains. Their $P(k)$ is expressed by

$$P(k) = L^{-2} \int_0^L (L - s)\, ds \int G(\mathbf{r}; s) \exp(i\mathbf{k} \cdot \mathbf{r})\, d\mathbf{r} \qquad (2.21)$$

where k denotes the scattering vector. With the first relation in eq 2.20, this equation can be rewritten

$$P(k) = L^{-2} \int_0^L (L - s) I(\mathbf{k}; s)\, ds \qquad (2.22)$$

The well–known expressions for $P(k)$ in the limits of Gaussian coil and rod are as follows (see Yamakawa's book frequently referred to in the preceding chapters):

$$P(k) = 2x^{-2}(x - 1 + e^{-x}) \quad \text{(Gaussian coils)} \qquad (2.23)$$

where $x = k^2 \langle S^2 \rangle$, and

$$P(k) = 2x^{-2}[x \int_0^x \frac{\sin t}{t}\, dt + \cos x - 1] \quad \text{(rods)} \qquad (2.24)$$

where $x = kL$. Analytical expressions for $P(k)$ valid near the coil limit [19] (the first–order deviation from eq 2.23) and the rod limit [20] (the deviations up to the fifth order from eq 2.24) were obtained, but their value for actual data analysis is limited.

Yoshizaki and Yamakawa [21] used the interpolated $I(\mathbf{k}; s)$ of Yamakawa et al. to calculate $P(k)$ which applies to k that covers wavelengths used in light scattering, small–angle X–ray, and neutron scattering experiments. The solid line in Figure 5–1 illustrates their $k^2 P(k)$ vs. k curve computed for a wormlike chain with $L = 100$ nm and $q = 10$ nm. The dotted and dashed lines are included for comparison. The former shows $k^2 P(k)$ calculated from eq 2.23 for a Gaussian coil whose $\langle S^2 \rangle$ is that of the wormlike chain, and the latter that calculated from eq 2.24 for a rod with $L = 100$ nm. The dotted line levels off to $2/\langle S^2 \rangle$, while the dashed line approaches an asymptote of slope π/L. With increasing k, the solid line first follows the line for the coil, deviating upward, and tends to merge with the asymptote for the rod. This limiting behavior of the solid line, along with the general relation $P(k)^{-1} = 1 + k^2 \langle S^2 \rangle / 3 + \cdots$ for small k, implies that light, small–angle X–ray, and neutron scattering from a linear polymer in dilute solution sees the overall dimensions of the chain at small k and its local stiff conformations at large k.

144

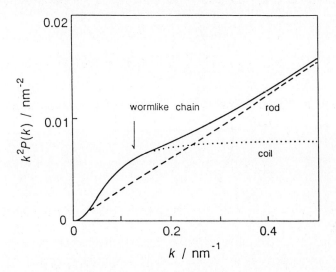

Fig. 5-1. Theoretical scattering curves as $k^2 P(k)$ vs. k for a wormlike chain with $L = 100$ nm and $q = 10$ nm [21] (solid line), a Gaussian coil with the same $\langle S^2 \rangle$ as the wormlike chain (dotted line), and a thin rod with $L = 100$ nm (dashed line).

On the basis of this general scattering behavior Kratky and Porod [1] predicted long ago that a plot of $k^2 P(k)$ against k should exhibit a "transition" from the asymptotic plateau for a Gaussian coil to the asymptote for a rod. The transition point may be defined as the point of intersection of the two asymptotes. Then, k^*, the value of k at this point, is equal to $6/\pi q$. Thus, q can be estimated if the two asymptotes are experimentally observable so as to be able to determine k^* definitely. This method has been applied to several flexible polymers. However, actual scattering curves do not always exhibit a distinct plateau region.

2.4 Transport Coefficients

2.4.1 The Yamakawa–Fujii Theory

The most important of recent theoretical studies on semi–flexible polymers is probably the formulation of Yamakawa and Fujii [2,3] for the steady transport coefficients of the wormlike cylinder. This hydrodynamic model, depicted in Figure 5-2, is a smooth cylinder whose centroid obeys the statistics of wormlike chains. In the figure, r* denotes the normal radius vector drawn from a contour

145

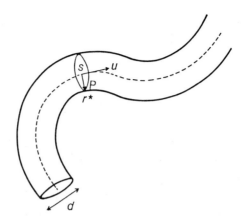

Fig. 5-2. Wormlike cylinder. Dashed line, cyclinder axis
whose shape obeys the statistics of wormlike chains.

point s on the centroid to a point P on the cylinder surface. Thus

$$\mathbf{r}^\star \cdot \mathbf{u} = 0 \quad \text{and} \quad |\mathbf{r}^\star| = d/2 \tag{2.25}$$

where d is the diameter of the cyclinder.

The Yamakawa–Fujii theory [2, 3] was developed by using the Kirkwood–
Riseman formalism with the effect of chain thickness approximately taken
into account. The following remarks may be in order. The Oseen interaction
tensor was preaveraged. Force points were distributed along the centroid of
the wormlike cylinder (not over the entire domain occupied by the cylinder).
The no–slip hydrodynamic condition was approximated by equating the mean
solvent velocity over each cross–section of the cylinder to the velocity of the
cylinder at that cross–section (Burgers' approximate boundary condition).

The basic equations for f and $[\eta]$ thus derived by Yamakawa and Fujii read

$$3\pi\eta_0 L/f = (1/8) \int_{-1}^{1}\int_{-1}^{1} K(x,y)\,dx dy \tag{2.26}$$

$$[\eta] = (N_A L^3/M) \int_{-1}^{1} \psi(x,x)\,dx \tag{2.27}$$

Here, η_0, N_A, and M have the same meaning as before, and $\psi(x,y)$ is the
solution to the following integral equation:

$$\int_{-1}^{1} K(x,\xi)\psi(\xi,y)\,d\xi = g(x,y) \tag{2.28}$$

146

where

$$K(x, y) = L\langle |\mathbf{r} - \mathbf{r}^\star|^{-1} \rangle \tag{2.29}$$

$$g(x, y) = (\pi/L^2)\langle \mathbf{S}(s) \cdot \mathbf{S}(s') \rangle \tag{2.30}$$

with

$$x = (2s/L) - 1, \quad y = (2s'/L) - 1 \tag{2.31}$$

$\mathbf{r}$ being the vector between contour points s and s', and $\mathbf{S}(s)$ the vector from the center of mass of the wormlike cylinder to the contour point s. The angle brackets indicate the double average over the conformation of the cylinder centroid and the direction of $\mathbf{r}^\star$.

The non-draining approximation is not invoked in deriving eq 2.26 and 2.27. When Burgers' boundary condition is used, no room is left for the introduction of this approximation. However, if d is set equal to zero, eq 2.26 and 2.27 reduce to the original Kirkwood-Riseman theory at the non-draining limit. The contribution of d tends to vanish as q/L approaches zero. The Yamakawa-Fujii theory for perfectly flexible wormlike cylinders is thus equivalent to the original Kirkwood-Riseman theory for non-draining Gaussian coils. This menas that the use of the wormlike cylinder model cannot resolve the serious discrepancies between theoretical and measured hydrodynamic factors discussed in Section 3 of Chapter 2.

2.4.2 Friction Coefficient

We first consider the rod limit ($q/L \to \infty$), for which we obtain

$$\langle |\mathbf{r} - \mathbf{r}^\star| \rangle = (t^2 + d^2/4)^{-1/2} \tag{2.32}$$

with $t = |s - s'|$. Substituting this into eq 2.26 and integrating, we arrive at

$$3\pi\eta_0 L/f = \ln(L/d) + 2\ln 2 - 1 + O(d/L) \tag{2.33}$$

The same expression can be derived without the use of the preaveraged Oseen tensor and the Kirkwood approximation [22]. Furthermore, if applied to a rigid sphere and an ellipsoid of rotation (either prolate or oblate), eq 2.26 happens to give the known exact results, the Stokes formula for the former and the Perrin formular for the latter [23]. Thus we may consider that eq 2.33 is exact as long as d/L is much smaller than unity.

Yamakawa and Fujii obtained $\langle |\mathbf{r} - \mathbf{r}^\star|^{-1} \rangle$ for $t/2q > \sigma$ by use of the second Daniels approximation to $G(\mathbf{r}, \mathbf{u}|\mathbf{u}_0; s)$ and assumed for $t/2q < \sigma$ a cubic equation as

$$\langle |\mathbf{r} - \mathbf{r}^\star|^{-1} \rangle = (t^2 + d^2/4)^{-1/2}[1 + f_1(t/2q) + f_2(t/2q)^2 + f_3(t/2q)^3] \tag{2.34}$$

where σ is treated as a parameter independent of $d/2q$ and the f_i are unknown functions of $(d/2q)^2$, though $f_1(0)$ is known to be 1/3 [10, 24]. First, $\sigma, f_2(0)$, and $f_3(0)$ were determined in such a way that $\langle|r|^{-1}\rangle$ for $t/2q$ above and below σ have the same first and second derivatives with respect to $t/2q$ at their intersection $t/2q = \sigma$. The results were $\sigma = 2.278, f_2(0) = 0.1130$, and $f_3(0) = -0.02447$. Next, the f_i for a series of $d/2q$ were calculated by forcing the corresponding $\langle|r-r^\star|^{-1}\rangle$ above and below $t/2q = \sigma$ to become continuous up to the second derivatives at $t/2q = \sigma$. The results were then fitted to

$$f_i = a_{i0} + a_{i1}(d/2q)^2 + a_{i2}(d/2q)^4 \quad (i = 1, 2, 3) \qquad (2.35)$$

Yamakawa and Fujii verified the accuracy of the resulting eq 2.34 by Monte Carlo calculations of $\langle|r - r^\star|^{-1}\rangle$ for freely–rotating chains with very small complementary bond angles. Their finding was confirmed later by Norisuye et al. [23], who evaluated $(t^2 + d^2/4)^{1/2}\langle|r - r^\star|^{-1}\rangle$ near the rod limit up to the fifth power in $t/2q$ by using Nagai's modification [12] of the Hermans–Ullman recurrence formula [11].

The final expression of f for $L/2q > 2.278$ is

$$3\pi\eta_0 L/f = (4/3)(6/\pi)^{1/2}(L/2q)^{1/2} + b_1 + b_2(L/2q)^{-1/2}$$
$$+ b_3(L/2q)^{-1} + b_4(L/2q)^{-3/2} \qquad (2.36)$$

where b_i $(i = 1, 2, 3, 4)$ is a tabulated function of $d/2q$. As anticipated, the leading term is identical to the formula for non–draining Gaussian chains obtained by the Kirkwood approximation. The second term b_1 vanishes at $d/2q = 0.322$. Hence the Yamakawa–Fujii theory predicts that if a wormlike cylinder happens to have a $d/2q$ value of 0.322, its diffusion coefficient D_0 in the unperturbed state varies linearly with $M^{-1/2}$ over a wide range of M. But this prediction does not seem definitive for the interpretation of the established experimental fact $D_{0\theta} \sim M^{-1/2}$ (see Section 3 of Chapter 2).

The Yamakawa–Fujii theory neglects the friction on the end surfaces of the wormlike cylinder. Norisuye et al. [23] and Yoshizaki and Yamakawa [25] estimated them by capping each end of a wormlike cylinder with a hemisphere and a straight spheroid, respectively. The results showed negligible effects on f unless the axial ratio p of the cylinder is smaller than 10, implying that the Yamakawa–Fujii theory of f is applicable down to very low M.

2.4.3 Intrinsic Viscosity

Equation 2.28 can be solved analytically at the limit of long straight rods by a polynomial expansion method [3], yielding

$$[\eta] = (\pi N_A L^3/24M)[\ln(L/d) + 2\ln 2 - (7/3) + O(d/L)]^{-1} \qquad (2.37)$$

Yamakawa [26] showed that if the Oseen tensor is not preaveraged, eq 2.37 is replaced by

$$[\eta] = (2\pi N_A L^3 / 45M)[\ln(L/d) + 2\ln 2 - (25/12) + O(d/L)]^{-1} \quad (2.38)$$

Although the numerical factor $2\pi/45$ in eq 2.38 is about 7% larger than the corresponding one $\pi/24$ in eq 2.37, the two equations give $[\eta]$ which agree within 1.2% in the range $60 < L/d < 250$.

Yamakawa and Fujii solved eq 2.28 numerically by using the same approximation for $\langle |\mathbf{r} - \mathbf{r}^\star|^{-1} \rangle$ as that employed for f. The resulting values of $[\eta]$ are available in tabular form as a function of L, q, and d.

Importantly, Yoshizaki and Yamakawa [25] found that, in contrast to f, $[\eta]$ of a wormlike cylinder undergoes significant end surface effects until the axial ratio p reaches about 50, on the basis of numerical solutions to the Navier–Stokes equation with the no–slip boundary condition for spheroid cylinders, spheres, and prolate and oblate ellipsoids of rotation. They constructed an empirical interpolation formula for $[\eta]$ of a spheroid cylinder which reduces to eq 2.37 for $p \gg 1$ and to the Einstein value at $p = 1$. Then, with its aid, Yamakawa and Yoshizaki [4] formulated a modified theory of $[\eta]$ for wormlike cylinders which agrees with the Yamakawa–Fujii theory [3] for $L/2q > 2.278$ and with the Einstein value at $L/d = 1$, regardless of $d/2q$ smaller than 0.1. However, no formulation has as yet been made for $L/2q < 2.278$ and $d/2q > 0.1$, i.e., for short flexible cylinders. In what follows, the Yamakawa–Yoshizaki modification is referred to as the Yamakawa–Fujii–Yoshizaki theory.

3. Experimental Data and Comparison with Theory

3.1 Equilibrium Properties

As can be seen from the above discussions, the chain dimensions and the particle scattering function of an unperturbed wormlike chain at infinite dilution are expressed as functions of two independent parameters L and q. Since L is proportional to the molecular weight M, it turns out that these properties of a series of homologous polymers in the unperturbed state are determined by M, q, and M_L. Here, M_L is defined by

$$M_L = M/L \quad (3.1)$$

when the polymers are modeled by wormlike chains and called the shift factor by Yamakawa and coworkers.

When $\langle S^2 \rangle$ data for an unperturbed polymer are given as a function of M, curve-fitting of eq 2.4 to them allows the parameters q and M_L to be evaluated. However, this method is not always effective, especially for relatively flexible polymers. Murakami et al. [27] showed that eq 2.4 can be approximated by

$$(M/\langle S^2 \rangle)^{1/2} = (3M_L/q)^{1/2}[1 + (3qM_L)/2M] \tag{3.2}$$

The difference between $\langle S^2 \rangle/(2q)^2$ values computed from this equation and the exact one (eq 2.4) is no more than 1% for $M/(2qM_L) > 2$. Thus unless the polymer is too short or too stiff, a linear relation should be found when the quantity on the left-hand side of eq 3.2 is plotted against M^{-1}, and q and M_L can be estimated from its slope and intercept. Recently, Zhang et al. [28] showed that for $M/(2qM_L) < 2$ eq 3.2 may be replaced by

$$(M^2/12\langle S^2 \rangle)^{2/3} = M_L{}^{4/3} + (2/15)(M_L{}^{1/3}/q)M \tag{3.3}$$

The method of Murakami et al. [27] was applied to sharp fractions of poly (hexyl isocyanate)(PHIC), a typical stiff polymer, in hexane, with the result that $q = 42 \pm 1$ nm and $M_L = 715 \pm 15$ nm^{-1}. In Figure 5–3, the data of Murakami et al. [27] are compared with the theoretical curve calculated from eq 2.4 with $q = 42$ nm and $M_L = 715$ nm^{-1}. Except for the four highest molecular weight fractions, the agreement between theory and experiment is very good, indicating that the wormlike chain is an excellent model for PHIC. This chain behaves as a flexible one in the region where the curve of $\ln(\langle S^2 \rangle/M)$ vs. $\ln M$ becomes horizontal. Hexane is judged to be a good solvent for PHIC from measured values of the second virial coefficient. Hence, in Figure 5–3, the upward deviations of the data points for M above 3×10^6 from the calculated curve are quite likely to reflect excluded-volume effects on $\langle S^2 \rangle$.

Figure 5–4 compares particle scattering function data for PHIC in hexane [27] with the theory corresponding to the indicated values of q and M_L, though the comparison is limited to the samples exhibiting unperturbed wormlike chain behavior. It is seen that a close agreement between theory and experiment can be obtained with the parameter values consistent with those estimated above from $\langle S^2 \rangle$ data.

The M_L value of 715 nm^{-1} yields 0.18 nm for h defined by

$$h = m/M_L \tag{3.4}$$

where m is the molar mass of one hexyl isocyanate residue. The quantity h represents the contour length of the wormlike chain per monomer unit of the corresponding polymer molecule. Its value of 0.18 nm happens to agree with the

150

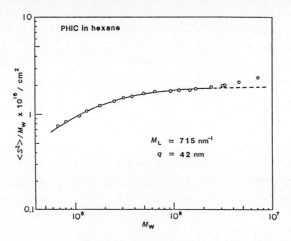

Fig. 5–3. Comparison of experimental and theoretical radii of gyration. o, PHIC in hexane [27]. Curve, unperturbed wormlike chain (eq 2.4) with $q = 42$ nm and $M_L = 715$ nm^{-1}.

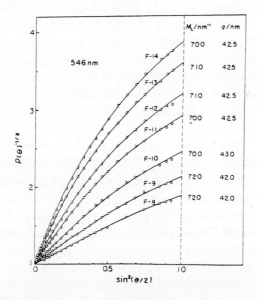

Fig. 5–4. Comparison of experimental and theoretical particle scattering functions. o, PHIC for different M in hexane [27]. Curves, unperturbed wormlike chains with indicated values of q and M_L.

pitch per monomer of the 8_3 helix deduced theoretically by Troxell and Scheraga

151

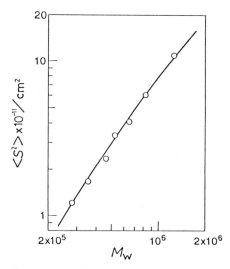

Fig. 5–5. Comparison between experimental and theoretical radii of gyration. o, double–stranded DNA in 0.2 M aqueous NaCl containing 2mM EDTA and 2 mM NaPO$_4$ (pH 7) [31]. Curve, unperturbed wormlike chain with $q = 68$ nm and $M_L = 1970$ nm^{-1}.

[29] for poly(methyl isocyanate). Shmueli et al. [30] proposed 0.19 nm for the pitch per monomer of the same helix in crystalline poly(butyl isocyanate). We may thus consider that the local conformation of PHIC in hexane is very similar to the 8$_3$ helix.

Figure 5–5 shows that radius of gyration data [31] for double–stranded DNA, another typical stiff polymer, can be described accurately by eq 2.4 with $q = 68$ nm and $M_L = 1970$ nm^{-1}. These parameter values have been estimated by the method of Murakami et al., and the M_L value is in close agreement with 1950 nm^{-1} that can be derived from the well–established geometry of the DNA double helix. Kirste and Oberthür [32] showed that the k dependence of $k^2 P(k)$ for a DNA sample measured by light and small–angle X–ray scattering can be represented by the theory of unperturbed wormlike chains.

Many other semi–flexible polymers are known which can be reasonably well modeled on wormlike chains. The M_L values obtained for most of such polymers are close to those calculated from crystallographic data, suggesting that stiff polymers in dilute solution maintain their crystalline conformations at least locally.

Values of $\langle S^2 \rangle / M$ for polymers not as stiff as PHIC and DNA level off to a coil limit at relatively low M. Hence, scattering measurements have to be extended to low M in order to determine the parameters q and M_L for such polymers. For this purpose small–angle X–ray scattering (SAXS) is more advantageous because it is concerned with k values much larger than those in the case of light scattering (LS).

The filled and unfilled circles in Figure 5–6 illustrate, respectively, SAXS and LS values of $\langle S^2 \rangle / M$ for polystyrene in cyclohexane at the θ temperature [33–35]. Pronounced deviations from the coil limit begin to occur when M is decreased below 10^4. Curve–fitting of eq 2.4 to these data gives $q = 1.0$ nm and $M_L = 390 \text{nm}^{-1}$. The solid line has been computed with these parameter values substituted into eq 2.4, and its close agreement with the data points shows that the wormlike chain mimics well unperturbed polystyrene from low to high molecular weights. The M_L value of 390 nm^{-1} is fairly close to 408 nm^{-1} which would be expected if the polystyrene molecule were to assume the all–trans conformation.

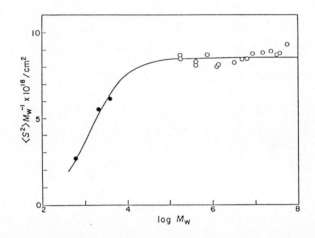

Fig. 5–6. SAXS (•) and LS (○) radius of gyration data for PS in CH at θ [33–35]. Curve, calculated from eq 2.4 with $q = 1.0$ nm and $M_L = 390$ nm^{-1}.

3.2 Transport Coefficients

3.2.1 Experimental Determination of Model Parameters

The hydrodynamic theories of Yamakawa et al. [2–4] contain the cylinder diameter d as well as q and M_L as parameters. These parameters can be determined from M dependent data of f or $[\eta]$, provided that the data exhibit behavior characteristic of rigid rods at low M and that of semi–flexible chains at high M. A simple procedure is then as follows. First, we plot $M^2 / [\eta]$ (or $1/f$) for low values of M against $\ln M$. According to eq 2.38 (or eq 2.33), the resulting plot should be linear and its slope and intercept (at $\ln M = 0$) allow M_L and d to be estimated. Next, with the parameter values so obtained, we look for a value of the remaining parameter q which leads to a best agreement between the $[\eta]$ (or f) data at high M and the corresponding theory of Yamakawa et al.

By applying this method to the triple–stranded helix of schizophyllan (an extracellular β–1, 3–glucan) in water, Yanaki et al. [36] obtained essentially the same values for the above three parameters ($M_L = 2150$ nm^{-1}, $d = 2.6$ nm, and $q = 200$ nm) from $[\eta]$ and f measured over a wide range of M. Figure 5–7 shows that their experimental results can be described with these parameter values. This M_L value yields 0.3 nm for h, which happens to equal the pitch per main chain glucose residue of the schizophyllan triple helix in the crystalline state [37].

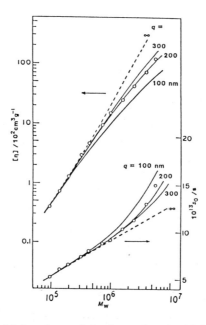

Fig. 5–7. Molecular weight dependence of $[\eta]$ and s_0 (sedimentation coefficient) for schizophyllan triple helix in water [36]. Curve calculated for an unperturbed wormlike cylinder with $q = 200$ nm, $M_L = 2150$ nm^{-1}, and $d = 2.6$ nm gives best fits to both $[\eta]$ and s_0 data.

Bushin et al. [38] and independently Bohdanecký [39] showed that the Yamakawa–Fujii–Yoshizaki theory for $[\eta]$ can be put in an approximate form as

$$(M^2/[\eta])^{1/3} = I + SM^{1/2} \tag{3.5}$$

with

$$I = 1.516 \times 10^{-8} A_0 M_L \quad (\mathrm{g}^{1/3}/\mathrm{cm}) \tag{3.6}$$

$$S = 1.516 \times 10^{-8} B_0 (M_L/2q)^{1/2} \tag{3.7}$$

Here, A_0 and B_0 are known functions of $d/2q$ and are tabulated in Bohdanecký's paper [39]. For typical stiff polymers for which $d/2q \sim 10^{-2}$, eq 3.5 is applicable over a wide range of $M/(2qM_L)$ from 0.4 to 300 and hence it should be quite useful for practical purposes; we note that $M/(2qM_L)$ is equal to the number of Kuhn's segments in the chain.

Bohdanecký found that published data on many semi-flexible polymers fit eq 3.5. Figure 5-8 shows this with the data [40] for poly (tere–phthalamide–p–benzohydrazide)(PPAH) in dimethyl sulfoxide (DMSO). The linear relation as observed here permits unequivocal estimation of I and S. If we have one more relation among q, M_L, and d, it becomes possible to determine these three unknowns. Molecular weight dependence data of $\langle S^2 \rangle$ or f may be utilized as such an additional relation. Bohdanecký [39] proposed the use of the relation

$$d = (4v M_L/\pi N_A)^{1/2} \tag{3.8}$$

where v is the partial specific volume of the polymer (at infinite dilution). This equation assumes that the hydrodynamic volume occupied by one gram of wormlike cylinder is equal to v. If it is applied to PPAH in DMSO (v = 0.676 cm^3/g), we obtain $q = 11.2$ nm, $M_L = 190$ nm^{-1}, and $d = 0.5$ nm. These values can be favorably compared with $q = 11.2$ nm, $M_L = 185$ nm^{-1}, and $d = 0.48$ nm which have been derived by Sakurai et al. [40] from molecular weight dependence data of $[\eta]$ and $\langle S^2 \rangle$. However, when the same method is applied to PHIC and triple–stranded schizophyllan, the resulting d values are much smaller than those either estimated from the combination of $[\eta]$ and $\langle S^2 \rangle$ data or expected from their helical structures.

3.2.2 Tests of the Yamakawa–Fujii–Yoshizaki Theory

Bohdanecký's finding that many stiff polymers obey eq 3.5 substantiates the predictability of the Yamakawa–Fujii–Yoshizaki theory for the M dependence of $[\eta]$ over a broad range of M. In Figure 5-9, this theory is compared with the data of Kuwata et al. [41] for PHIC in butyl chloride, which cover a very extended range of M_w. The solid line has been calculated for a hemisphere–capped wormlike cylinder whose q, M_L, and d are 35 nm, 760 nm^{-1}, and 1.5 nm, respectively. The dashed line represents behavior at the rod limit. Accurate fit of the former to the plotted points is worth noting. The M_L value of 760 nm^{-1} is close to 715 nm^{-1} in hexane (see the preceding section) and the d value of 1.5 nm is in the range 1.3–1.8 nm which can be estimated from crystallographic

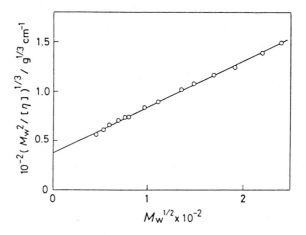

Fig. 5–8. Bushin–Bohdanecký plot for PPAH in DMSO [40].

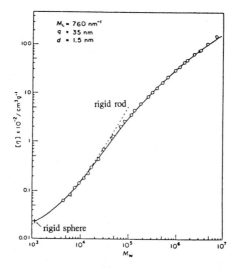

Fig. 5–9. Fit of the Yamakawa–Fujii–Yoshizaki theory to $[\eta]$ data [41] for PHIC in butyl chloride. Solid line, for M_L = 760 nm^{-1}, q = 35 nm, and d = 1.5 nm. Dashed line, for rigid cylinders with M_L = 760 nm^{-1} and d = 1.5 nm. Cross, for a rigid sphere with d = 1.5 nm.

data on poly(butyl isocyanate) and the chemical structure of PHIC. Finally, the q value in butyl chloride (35 nm) is significantly smaller than that in hexane (42 nm), suggesting that q should be considered a solvent–dependent quantity.

A more critical test of the hydrodynamic theories of Yamakawa et al. may be made on PHIC in hexane, since for this system reliable values of q and M_L are known separately (see Figures 5-3 and 5-4) so that only d is left as an adjustable parameter. With q and M_L chosen to be 42 (± 2) nm and 715 (± 15), respectively, Murakami et al. [27] examined (i) whether there exists a unique value of d which allows the theory to fit their M dependence data of $[\eta]$ and s_0 and (ii) whether if it exists, such a d falls in the range 1.3–1.8 nm expected from the chemical structure of PHIC.

The best value of d consistent with the $[\eta]$ data was found to be 1.6 (± 0.2) nm. The solid line A in Figure 5-10, calculated with $q = 42$ nm, $M_L = 715$ nm^{-1}, and $d = 1.6$ nm, is seen to fit the data points very accurately. Its close fit even for M_w above 3×10^6 appears to indicate that $[\eta]$ undergoes less excluded-volume effect than does $\langle S^2 \rangle$ (compare with Figure 5-3). However, this result is an accident due to the defect of the Yamakawa–Fujii–Yoshizaki theory which overestimates $[\eta]$ near the coil limit.

The solid line B in Figure 5-10 shows s_0 calculated with $q = 42$ nm, $M_L = 715$ nm^{-1}, and $d = 1.6$ nm. Interestingly, it deviates appreciably from the data points in the region of M_w below 7×10^5. The situation cannot be improved with minor adjustments of q, M_L, and d (being allowed to vary in the ranges ± 1 nm, ± 15 nm^{-1}, and ± 0.2 nm, respectively). The dashed line B' leading to a closer agreement is obtained only if d is increased to 2.5 nm. In contrast to the d value deduced from the $[\eta]$ data, this d is far off the expected range 1.3–1.8 nm. Furthermore, if $[\eta]$ is calculated with $q = 42$ nm, $M_L = 715$ nm^{-1}, and $d = 2.5$ nm, the dashed line A' is obtained, and it deviates from the data points more than experimental errors. Thus we find that there exists no unique d compatible with both $[\eta]$ and s_0 data and that the diameter d_s from s_0 is significantly larger than the diameter d_η from $[\eta]$, with the latter being in the expected range.

Murakami et al. were not the first to obtain such inconsistent d_s and d_η. Goldfrey and Eisenberg [31] had already found a similar disagreement from very precise hydrodynamic measurements on double–stranded DNA, another typical stiff chain. In fact, they obtained d_s (2.6 nm) about twice as large as d_η (1.2 nm), but, differing from the case of PHIC, d_s was in good accordance with the diameter of the established DNA double helix. For relatively short α–helical poly (γ–benzyl-L-glutamate) (PBLG) Itou et al. [42] found d_s = 2.2 nm and d_η = 1.7 nm, which were much larger than the diameter 1.4 nm of the PBLG α helix. Despite these findings, we have to observe the fact that there are many other polymers for which consistent d_s and d_η values have been obtained by application of the Yamakawa–Fujii–Yoshizaki theory (see Figure 5-7 as an example). Summarizing, we may tentatively conclude that the wormlike chain hydrodynamics due to Yamakawa et al. leads to $d_s \geq d_\eta$. According

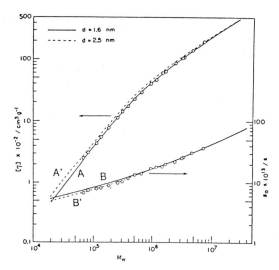

Fig. 5–10. Tests of the Yamakawa–Fujii–Yoshizaki theory with experimental data of Murakami et al. [27] on PHIC in hexane. Lines A and B, calculated with $q = 42$ nm, $M_L = 715$ nm^{-1}, and $d = 1.6$ nm. Lines A' and B', calculated with the same q and M_L as for A and B but $d = 2.5$ nm.

to Yamakawa [22], this inequality has something to do with the approximation which replaces actual polymer molecules with irregular surfaces by smooth uniform cylinders. However, one may be skeptical as to whether his idea will get a general acceptance. As Figure 5–10 illustrates, the hydrodynamic behavior of a polymer in dilute solution is quite insensitive to the chain thickness. This fact implies that only the combination of very accurate experiment and theory allows rigorous determination of the hydrodynamic diameter d of the wormlike cylinder to be made.

Unquestionably, Yamakawa and collaborators have made a substantial contribution to the understanding of transport behavior of semi–flexible polymers in dilute solution. However, their theories still leave something to be desired, as revealed by the recent careful experiments mentioned above. Their formulation is essentially the combination of the the Kirkwood–Riseman hydrodynamics and the statistics of wormlike chains. As mentioned in Chapter 2, this hydrodynamics fails to be good for flexible chains, but we have seen that it seems to work well for stiff chains. The reason is that the Kirkwood–Riseman formalism gives the exact solution in the limit of rigid rods.

For the sake of reference we summarize reported values of q and M_L for some typical semi–flexible polymers in the Appendix.

4. Excluded-volume Effects on Chain Dimensions

4.1 Experimental Information

Chain stiffness prevents the monomer units of a polymer molecule from having contact with one another. Thus, we may expect that there exists for a semi-flexible polymer a certain contour length L_c below which the excluded-volume effect on the chain dimensions disappears (in a statisitical sense). This prediction is borne out by the data of Figure 5-3, which show that $\langle S^2 \rangle/M$ follows the curve for an unperturbed wormlike chain until M_w reaches 3×10^6 ($L \sim 4 \times 10^3$). Obviously, L_c ought to be larger for a stiffer polymer, i.e., one with larger persistence length q.

Norisuye and Fujita [43] were the first to ask what correlation exists between L_c and q. Figure 5-11 summarizes log-log plots of $\langle S^2 \rangle/(2q)^2$ vs. n_K for several polymer + solvent systems which cover a range of q from 1 to 150. Here, n_K is equal to $M/(2qM_L)$, which gives the number of Kuhn's segments in a wormlike chain. We note that eq 2.4 can be rewritten

$$\langle S^2 \rangle/(2q)^2 = (n_K/6) - (1/4) + (1/4n_K)$$
$$- (1/8n_K^2)[1 - \exp(-2n_K)] \qquad (4.1)$$

The solid line in Figure 5-11 represents eq 4.1 and hereafter is referred to as the Benoit-Doty curve after the authors [50] who first derived eq 2.4.

It can be seen that all the data points for $n_K < 50$ fall on the Benoit-Doty curve. This finding implies that regardless of stiffness, chains with n_K smaller than 50 are essentially in the unperturbed state. Except for polystyrene in cyclohexane at θ, the data points start deviating upward from the solid line in a region of n_K near 50 and approximately follow the indicated straight lines at higher n_K. More precisely, n_K for the onset of the deviation depends on the system, varying from 30 to 80, as can be seen from Figure 5-12.

On the basis of these observations Norisuye and Fujita [43] concluded that the excluded-volume effect on chain dimensions becomes experimentally visible when n_K reaches about 50, roughly independent of q. Since the chain length L is related to n_K by $L = 2qn_K$, this conclusion means

$$L_c \sim 100q \qquad (4.2)$$

As expected, L_c is larger for a stiffer polymer.

4.2 Comparison with Theoretical Predictions

For a touched-bead chain whose backbone behaves like a wormlike chain Yamakawa and Stockmayer [51] calculated perturbatively the end distance expansion factor α_R and derived

$$\alpha_R^2 = 1 + K(n_K)z^* + \cdots \qquad (4.3)$$

159

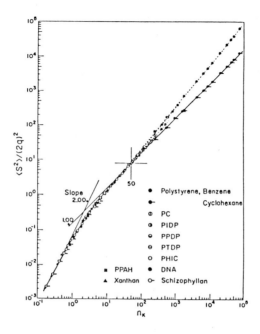

Fig. 5-11. Reduced mean–square radii of gyration $\langle S^2 \rangle / (2q)^2$ plotted against the number of Kuhn's segments n_K for various polymer + solvent systems [43]. Solid line, Benoit–Doty curve. Points, Experimental. PC, bisphenol A polycarbonate in tetrahydrofuran [44]; PIDP, poly(isophthaloyl–trans–2,5–dimethyl piperazine) in trifluoroethanol [45]; PPDP, poly (phthaloyl–trans–2,5–dimethyl piperazine) in chloroform and N–methyl–2–pyrrolidone [46]; PTDP, poly(terephthaloyl–trans–2,5–dimethyl piperazine) in trifluoroethanol [47]; PPAH, poly (terephthalamide–p–benzohydrazide) in DMSO [40]; DNA, in 0.2 M NaCl + 2 mM EDTA + 2 mM NaPO$_4$ [31]; Schizophyllan, in 0.01 N aqueous NaOH [48]; xanthan, in 0.1 M aqueous NaCl [49].

where

$$z^\star = (3/2\pi)^{3/2} B n_K^{1/2} \tag{4.4}$$

with

$$B = \beta/(2qa^2) \tag{4.5}$$

In these equations, β is the excluded–volume strength, a the bond length (equal to the diameter of one bead), and $z^\star$ an excluded–volume variable, which reduces to z defined previously when $2q$ is set equal to a. The coefficient $K(n_K)$ is essentially zero for $n_K < 1$ and increases to the familiar coil value 4/3 with

160

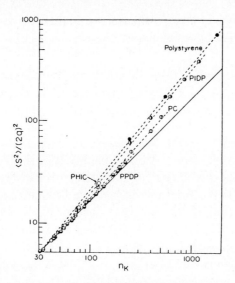

Fig. 5-12. Enlarged graph of Fig. 5-11 for n_K above 30.
The data for PS in CH are not shown.

increasing n_K. Assuming that the first-order coefficient in the squared radius expansion factor $\alpha_S{}^2$ has the same n_K dependence as $K(n_K)$ but approaches the coil value 134/105 as $n_K \to \infty$, we obtain

$$\alpha_S{}^2 = 1 + [67K(n_K)/70]z^\star + \ldots \tag{4.6}$$

Since $K(n_K)$ begins to depart from zero at n_K near unity, it follows from eq 4.6 that the theoretical critical contour length $L_c{}^\star$ for the onset of the excluded-volume effect on chain dimensions is approximately given by

$$L_c{}^\star \sim 2q \tag{4.7}$$

Comparing with eq 4.2, we see that $L_c{}^\star$ is at least one order of magnitude smaller than the experimentally estimated critical contour length L_c. This pronounced difference does not seem attributable to the fact that eq 4.6 is correct only to first order in $z^\star$.

To estimate $L_c{}^\star$ more accurately Yamakawa and Shimada [52] computed $\langle S^2 \rangle$ of polymethylene–like rotational isomeric state models with excluded volume by a Monte Carlo method. They considered that bead–bead interactions higher than ones leading to the "pentane" effect contribute to excluded–volume effects. They found that the simulated polymer in the unperturbed state can be modeled by a wormlike chain and that $\langle S^2 \rangle / (4q^2 n_K)$ of perturbed chains begins to deviate from those of unperturbed ones at n_K of 3–5. Thus, the $L_c{}^\star$ by Yamakawa and

161

Shimada is several times larger than that by Yamakawa and Stockmayer, but still much smaller than L_c estimated by Norisuye and Fujita from experimental data.

In Figure 5–13, the solid curve illustrates the Yamakawa–Shimada theoretical relation between $\log(6\langle S^2 \rangle/4q^2 n_K)$ and $\log n_K$ determined so as to fit the asymptotic behavior (experimentally determined) of polystyrene in benzene at large n_K. On the other hand, the dashed one indicates the Benoit–Doty curve for the unperturbed wormlike chain. When n_K reaches about 3 the former begins to separate upward from the latter. The $L_c{}^*$ predicted by the Yamakawa–Shimada theory corresponds to this critical n_K. However, it can be seen that the separation of the solid curve from the dashed one becomes appreciable after n_K exceeds about 50. Thus if we have $\langle S^2 \rangle$ data available only in the region of high M where they exhibit asymptotic behavior, we will extrapolate them to the solid curve and conclude from the point of intersection that chain dimensions begin undergoing volume exclusion at $n_K \sim 50$. It is the critical contour length so obtained that Norisuye and Fujita identified as L_c. Thus, we can understand why L_c and $L_c{}^*$ for polystyrene in benzene differ so markedly. To find the "true" critical length $L_c{}^*$ the measurement of $\langle S^2 \rangle$ has to be extended down to very short chains, and for this purpose it is advantageous to use small–angle X–ray or neutron scattering.

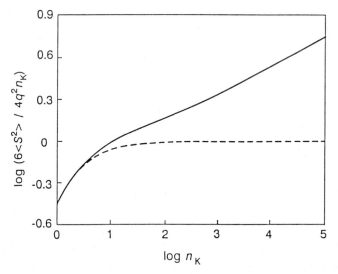

Fig. 5–13. Solid line, calculated relation between $\log(6\langle S^2 \rangle/4q^2 n_K)$ and $\log n_K$ for perturbed wormlike chains with $B = 0.23$ [52]. Dashed line, unperturbed wormlike chains.

The PHIC data of Figure 5–3 exhibit no discernible deviation from the Benoit–Doty curve up to $n_K = 30-50$, in contrast to the prediction from the Yamakawa–Shimada theory. Although whether this feature is characteristic or not of very stiff polymers such as PHIC remains to be seen, the following remark may be in order. The computation of Yamakawa and Shimada refers to a particular polymethylene–like chain in which bead–bead interactions higher than the pentane one are treated as responsible for excluded–volume effects. Though it sounds reasonable, this treatment seems to require further investigation.

Transport coefficients of actual polymers should also show the onset of the excluded–volume effect at certain critical contour lengths. It is interesting to study how such lengths can be estimated experimentally and what relations exist between them and L_c or $L_c{}^\star$. However, these problems remain almost unexplored.

5. Optical Anisotropy

5.1 Theory of Nagai

Stiff chains are usually optically anisotropic. Hence, the light scattering determination of their M and $\langle S^2 \rangle$ has to be made with this fact in mind. Nagai [53] was the first to present a theory of anisotropic light scattering from dilute solutions of a wormlike chain. He evaluated the scattering components $R_{Vv}, R_{Hv}(= R_{Vh})$, and R_{Hh} for this chain with cylindrically symmetric polarizabilities up to first order in k^2, using the point scatterer approximation. Here, k denotes the magnitude of the scattering vector, and, for example, R_{Hv} the reduced excess intensity of scattered light measured for vertically polarized incident light (signified by the subscript v) with an analyzer set in the horizontal direction (signified by the subscript H). Later, Yamakawa et al. [54] extended the calculation to higher orders in k^2 and also to helical wormlike chains. Here we outline Nagai's theory and its applications.

The reduced excess intensity $R_{\mu\nu}$ of scattered light ($\mu = V$ or H and $\nu = v$ or h) may be expressed by

$$R_{\mu\nu} = KcMI_{\mu\nu}/\Gamma^2 \tag{5.1}$$

provided that the polymer mass concentration c is so low that any intermolecular interactions can be neglected. In eq 5.1, K is the familiar optical constant in the theory of light scattering, $I_{\mu\nu}$ and Γ are the excess scattering intensity and the excess polarizability of the entire polymer chain, respectively, which are given by

$$I_{\mu\nu} = \int_0^L \int_0^L \langle (\mathbf{n}_\mu \mathbf{\Gamma}(s)\mathbf{n}_\nu{}^T)(\mathbf{n}_\mu \mathbf{\Gamma}(s')\mathbf{n}_\nu{}^T) \exp\left[i\mathbf{k} \cdot \mathbf{r}(s, s')\right] \rangle \, ds \, ds' \tag{5.2}$$

163

$$\Gamma = \int_0^L \langle \mathbf{n}_\mu \Gamma(s) \mathbf{n}_\nu{}^\mathrm{T} \rangle ds = \int_0^L (1/3)[\text{trace of } \Gamma(s)] \, ds \qquad (5.3)$$

In these, $\mathbf{n}_\mu$ and $\mathbf{n}_\nu$ are the unit column vectors in the polarization directions of scattered and incident light, respectively, $\mathbf{r}(s, s')$ the vector between contour points s and s', $\Gamma(s)$ the excess polarizability tensor per unit length at contour point s relative to a laboratory–fixed coordinate system, and the superscript T the transpose of the column vector.

For a wormlike chain having a polarizability α_1 along the chain contour and a polarizability α_2 normal to the chain, both per unit contour length, we may express $\Gamma(s)$ as follows:

$$\Gamma(s) = (\alpha_1 - \alpha_2)\mathbf{u}(s)^\mathrm{T}\mathbf{u}(s) + \alpha_2 \mathbf{E} \qquad (5.4)$$

with $\mathbf{u}(s)$ the unit tangent vector of the chain at contour point s and $\mathbf{E}$ a unit tensor. With eq 5.4 substituted into eq 5.3 we obtain

$$\Gamma = (1/3)(\alpha_1 + 2\alpha_2)L \qquad (5.5)$$

The average $\langle \cdots \rangle$ above is taken first over the orientation of the chain of fixed conformation relative to the laboratory–fixed coordinate system and then over all conformations of the chain. The resulting expressions for R_Vv and R_Hv correct to first order in k^2 and c are

$$R_\mathrm{Vv} = KcM\{1 + 4\delta - (k^2/3)[\langle S^2 \rangle - (4\delta_0 q^2/45)(g_1 - 4\delta_0 qg_2/63L)]\} \qquad (5.6)$$

$$R_\mathrm{Hv} = R_\mathrm{Vh} = KcM[3\delta - (\delta_0{}^2 q^3 k^2/630L)(11g_3/9 + g_4 \cos\psi)] \qquad (5.7)$$

The expression for R_Hh is lengthy and not shown here. In these, the $g_i(i = 1, 2, 3, 4)$ are known functions of L/q which tend to unity in the coil limit $(L/q = \infty)$, ψ the scattering angle, and δ and δ_0 the anisotropy parameters defined by

$$\delta = (2\delta_0{}^2 q/135L)[1 - (q/3L)(1 - e^{-3L/q})] \qquad (5.8)$$

$$\delta_0 = 3(\alpha_1 - \alpha_2)/(\alpha_1 + 2\alpha_2) \qquad (5.9)$$

In the rod limit $(L/q = 0)$, δ reduces to $\delta_0{}^2/45$, and eq 5.6 and 5.7 to Horn's results [55] for rigid rods with axially symmetric polarizabilities.

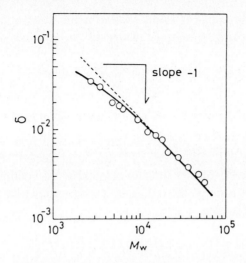

Fig. 5-14. Molecular weight dependence of δ for PPAH in DMSO [40]. Curve, calculated from eq 5.8 with $q = 11.2$ nm, $M_{\mathrm{L}} = 185$ nm^{-1}, and $\delta_0 = 2.1$.

5.2 Applications to Data Analysis

At the zero scattering angle ($k = 0$) eq 5.6 and 5.7 give

$$R_{\mathrm{Hv}}/R_{\mathrm{Vv}} = 3\delta/(1 + 4\delta) \quad (c \to 0) \tag{5.10}$$

$$R_{\mathrm{Hv}}/R_{\mathrm{Uv}} = 3\delta/(1 + 7\delta) \quad (c \to 0) \tag{5.11}$$

where

$$R_{\mathrm{Uv}} = R_{\mathrm{Hv}} + R_{\mathrm{Vv}} \tag{5.12}$$

Either eq 5.10 or 5.11 can be used to evaluate δ at infinite dilution. Once δ is known, the true molecular weight M may be determined from

$$R_{\mathrm{Vv}} = KcM(1+4\delta) \quad \text{or} \quad R_{\mathrm{Uv}} = KcM(1 + 7\delta)$$
$$(c \to 0, k \to 0) \tag{5.13}$$

Sakurai et al. [40] showed for poly(terephthalamide–p–benzohydrazide) (PPAH) in DMSO that the M values determined by eq 5.13 agreed with those evaluated by sedimentation equilibrium. Figure 5–14 depicts the M dependence of δ for this system. The curve shown appears to have a limiting slope of -1 at high M, as predicted by eq 5.8.

For an optically anisotropic polymer the conventional light scattering analysis, in which the initial slope of a Kc/R_{Vv} vs. k^2 plot at infinite dilution

165

determined, gives an apparent mean–square radius of gyration $\langle S^2 \rangle_{\mathrm{app}}$. With eq 5.6 we find that this quantity is related to the true mean–square radius of gyration $\langle S^2 \rangle$ by

$$\langle S^2 \rangle_{\mathrm{app}} = \langle S^2 \rangle_{\mathrm{Vv}}^{\star}/(1 + 4\delta) \tag{5.14}$$

where

$$\langle S^2 \rangle_{\mathrm{Vv}}^{\star} = \langle S^2 \rangle - (4\delta_0 q^2/45)(g_1 - 4\delta_0 q g_2/63L) \tag{5.15}$$

In the coil limit where $\langle S^2 \rangle = qL/3$ as can be derived from eq 2.5, the second term on the right–hand side of eq 5.15 can be neglected. Hence, it follows from eq 5.14 that $\langle S^2 \rangle$ in this limit is obtainable by multiplying $\langle S^2 \rangle_{\mathrm{app}}$ by $1 + 4\delta$. On the other hand, in the rod limit, it can be shown by use of the known functions of g_1 and g_2 that

$$\langle S^2 \rangle_{\mathrm{app}} = [1 - (4\delta_0/15) + (4\delta_0{}^2/63)]\,\langle S^2 \rangle/(1 + 4\delta) \tag{5.16}$$

Since, as noted above, $\delta = \delta_0{}^2/45$ in this limit, $\langle S^2 \rangle$ can be estimated from experimental $\langle S^2 \rangle_{\mathrm{app}}$ and δ if the sign of δ_0 is known by analyzing the data for R_{Hh} with the aid of Horn's method [55].

For wormlike chains at least two of the three parameters q, M_{L}, and δ_0 need be known to determine $\langle S^2 \rangle$ from $\langle S^2 \rangle_{\mathrm{app}}$ and δ. Sakurai et al. [40] proposed an iteration method which permits all the unknowns $\langle S^2 \rangle, q, M_{\mathrm{L}}$, and δ_0 to be estimated at one time if data for $\langle S^2 \rangle_{\mathrm{app}}$ and δ are available as functions of M and the sign of δ_0 is known in advance. A similar iteration method can be developed also for the evaluation of these unknowns from the measured M dependence of $\langle S^2 \rangle_{\mathrm{Uvapp}}$ and δ. The solid curve in Figure 5–14 is actually the final result obtained by its application; the convergent values of q, M_{L}, and δ_0 are shown in the caption. Sakurai et al. [40] found that when substituted into the theory of Yamakawa et al. for $[\eta]$ [3, 4], these q and M_{L} values along with a diameter value of 0.48 nm lead to a close fit to the data plotted in Figure 5–8. The M_{L} value of 185 nm^{-1} is favorably compared with 190 nm^{-1} from crystallographic data [56] and 180–190 nm^{-1} from SAXS measurements [57].

5.3 Remarks

According to Flory and coworkers [58, 59], experimentally measured R_{Hv} contains a contribution $R_{\mathrm{Hv}}(\mathrm{col})$ from transient, collision–induced depolarized scattering. At infinite dilution this contribution arises from the collision of individual polymer chains with solvent molecules. Hence it has nothing to do with the intrinsic optical anisotropy of the solute polymer.

The intensity of collision–induced scattering depends on the frequency of scattered light so that it could be detected as a spectral distribution as in quasi-elastic light scattering. However, elastic light scattering gives only an integrated intensity. Carlson and Flory [58] proposed the use of two interference filters with different band widths for eliminating $R_{Hv}(\text{col})$ from elastic scattering data for R_{Hv}. With this method applied to polystyrene and its oligomers in carbon tetrachloride Suter and Flory [59] found that $R_{Hv}(\text{col})$ amounted to 0.46–0.33 of R_{Hv}, depending on the degree of polymerization. This finding is striking, since it implies that the values of δ directly estimated from R_{Hv} are 30–100% larger than those from collision–corrected R_{Hv}.

No data for $R_{Hv}(\text{col})$ are yet available for stiff polymers. If the δ values for PPAH shown in Figure 5–14 contain appreciable collision contributions, the δ_0 value of 2.1 derived from them can no longer be related to the intrinsic optical anisotropy of PPAH. Fortunately, we may use uncorrected R_{Hv} data for the evaluation of true M and $\langle S^2 \rangle$.

6.   Helical Wormlike Chains

6.1   Introduction

The ideally flexible continuous chain generated from the spring–bead chain retains no microscopic feature of actual polymer molecules and hence it is the most abstract of polymer models. The wormlike chain, though a continuous chain, is more realistic since through the parameter q it allows for the stiffness possessed by actual molecules. Up to this point we have seen a number of examples which substantiate the usefulness of these chain models for the quantitative description of global behavior of polymers in dilute solution. But this never means that no other chain model need be considered.

The familiar rotational isomeric state (RIS) chain model takes account of the structural features of individual polymers in considerable detail, and thus is favored by many polymer physical chemists. This model should reveal polymer properties that are not predicted by the conventional more coarse–grained chain models. In fact, while the characteristic ratio $\langle R^2 \rangle / L$ of a wormlike chain always increases monotonically to the Gaussian coil limit with an increase in chain length L, it has been shown [60] that this ratio of some unperturbed RIS chains attains a maximum at an intermediate chain length. Furthermore, in other RIS chains, the characteristic ratio approaches the Gaussian coil limit much faster (i.e., at a shorter chain length) than does that of the wormlike chain [61]. Probably these theoretical findings motivated Yamakawa [5] to propose and investigate a new polymer model which he named the **helical wormlike chain** (HW chain). This is a continuous chain model more general than the wormlike chain in the sense that the chain can store torsional energy as well as bending energy.

In a series of recent papers Yamakawa and collaborators have formulated various static and dynamic properties of the HW chain in great detail and derived many interesting results (see Ref. [6] for an overview). Their contributions indeed mark a milestone in recent developments of polymer solution theory. However, since the HW chain needs for its characterization three more parameters than the wormlike chain it is not a simple matter to evaluate all the parameters from experimental results. Thus, much remains to be worked out to make this novel polymer model appealing to experimentalists. Here we confine ourselves to touching on its basic features.

6.2 Definitions

We fix a local Cartesian coordinate system (e_ξ, e_η, e_ζ) to every contour point of an elastic wire of finite thickness. In so doing, the ζ axis is taken in the direction of the unit tangent vector u of the chain centroid (or contour), and the ξ and η axes in the directions of the principal axes of inertia for the cross-section of the wire at the same point. When the wire is deformed (bent and twisted), the local coordinate system at contour point $s + ds$ is related to that at contour point s by an infinitesimal rotation $d\Omega$. The deformed state of the wire is then represented by a vector $\Theta(\Theta_\xi, \Theta_\eta, \Theta_\zeta)$ as a function of s, where Θ is defined by $d\Omega/ds$.

We assume that the cross-section of the wire remains circular when the wire is deformed and that its principal moments of inertia are equal. Then, if the centroid of the wire in the unstressed state is straight, the elastic energy $U(s)$ stored per unit contour length of the deformed wire at contour point s is expressed by [62]

$$U(s) = (\alpha_0/2)(\Theta_\xi{}^2 + \Theta_\eta{}^2) + (\beta_0/2)\Theta_\zeta{}^2 \tag{6.1}$$

provided that local deformations are sufficiently small. Here, α_0 and β_0 are the force constants for bending and twisting, respectively, and are assumed to be independent of s. If the centroid of the wire in the unstressed state takes the form of a regular helix, eq 6.1 has to be replaced by [62]

$$U(s) = (\alpha_0/2)[\Theta_\xi{}^2 + (\Theta_\eta - \kappa_0)^2] + (\beta_0/2)(\Theta_\zeta - \tau_0)^2 \tag{6.2}$$

where κ_0 and τ_0 are constants characterizing the centroid helix. In fact, the pitch and diameter of the helix are expressed by $2\pi|\tau_0|/(\kappa_0{}^2 + \tau_0{}^2)$ and $2\kappa_0/(\kappa_0{}^2 + \tau_0{}^2)$, respectively, with the helix sense being right-handed for $\tau_0 > 0$ and left-handed for $\tau_0 < 0$. What Yamakawa has called the HW chain is an elastic wire whose $U(s)$ is given by eq 6.2. Thus this chain stores both bending and torsional energies and when unstressed its centroid assumes a regular helix. In terms of Θ its unstressed state is expressed by $(0, \kappa_0, \tau_0)$. Its local stressed state is therefore represented by deviations of Θ_ξ, Θ_η, and Θ_ζ from 0, κ_0, and τ_0, respectively.

6.3 Basic Parameters

Statistical properties of an unperturbed HW chain in equilibrium are determined with five parameters: chain length L, $\alpha_0, \beta_0, \kappa_0$, and τ_0 (the latter four characterize $U(s)$ of the chain). This should be contrasted to the fact that only two parameters L and q are needed for the description of these properties of an unperturbed wormlike chain. In adapting the HW chain to actual polymers, the shift factor $M_{\rm L}(= M/L)$, instead of L, may be chosen as a parameter since M can be determined experimentally. Thanks to eq 2.10 the stiffness parameter $(2\lambda)^{-1}$ may be used for α_0. For HW chains we have no equation corresponding to eq 2.2. Hence $(2\lambda)^{-1}$ may not be equated to q according to eq 2.14. It should be noted that the persistence length q is the concept associated only with wormlike chains. The Poisson ratio σ_0 of the HW chain is expressed in terms of α_0 and β_0 as

$$\sigma_0 = (\alpha_0/\beta_0) - 1 \qquad (6.3)$$

Thus we see that $\lambda, \sigma_0, \kappa_0, \tau_0$, and $M_{\rm L}$ may be used as the basic parameters of the HW chain. These five parameters give the HW a greater flexibility than that obtained with the wormlike chain and should allow it to describe a wider range of physical properties of actual polymer molecules. However, in actuality, there would be considerable difficulty in evaluating so many parameters from comparisons between theory and experiment.

6.4 Special Cases

In the special case $\kappa_0 = 0$, the centroid helix of an unstressed HW chain reduces to a straight line and its pitch becomes infinitely long. Hence τ_0 also should vanish when $\kappa_0 = 0$. Then, eq 6.2 reduces to eq 6.1. However, the latter is not $U(s)$ for the wormlike chain because it contains the term associated with torsional energy ($U(s)$ consisting only of the first term of eq 6.1 defines the wormlike chain).

When $\alpha_0 = \beta_0 = 0$, eq 6.2 reduces to $U(s) = 0$, which corresponds to the case where the centroid of an unstressed HW chain behaves as a Gaussian chain.

The orientation Ω of the local coordinate system relative to a laboratory–fixed one may be expressed in terms of the Euler angles θ, ψ, and ϕ, with u represented by $(1, \theta, \psi)$. The distribution function $G(\mathbf{r}, \Omega|\Omega_0; s)$ for the centroid of an HW chain can then be defined in a way similar to $G(\mathbf{r}, \mathbf{u}|\mathbf{u}_0; s)$ for the wormlike chain. As may be anticipated from the definitions of Θ and $U(s)$, the trajectory of the HW chain centroid follows a Markoffian process. This means that $G(\mathbf{r}, \Omega|\Omega_0; s)$ obeys a Fokker–Planck equation.

Yamakawa and Fujii [63] showed that the Fokker–Planck equation for the HW chain reduces to eq 2.16 with $q = (2\lambda)^{-1} (= \alpha_0/k_{\rm B}T$, where $k_{\rm B}$ is the Boltzmann constant) when $G(\mathbf{r}, \Omega|\Omega_0; s)$ is integrated over ψ and then κ_0 is set equal to zero. As noted above, eq 6.2 reduces to eq 6.1 in the limit $\kappa_0 = 0$, but the latter is not $U(s)$ for the wormlike chain. The finding of Yamakawa and Fujii implies that despite this difference the HW chain with $\kappa_0 = 0$ behaves the same as the wormlike chain as far as the properties depending on contour length are concerned. Thus we may consider the wormlike chain to be a special case of the HW chain with zero κ_0 except when we deal with such properties as dipole moment and polarizability which depend on the local twisting of the chain.

According to Yamakawa and Shimada [64], the HW chain can be generated as a continuous limit of a certain Markoffian discrete chain. In this discrete chain, each bond is allowed to rotate about its own axis and such rotations of adjacent bonds are coupled. This coupling seems responsible for the feature that the HW chain can store torsional energy, but its physical picture is not clear to the author.

6.5 Chain Dimensions

Analytical expressions for $\langle R^2 \rangle$ and $\langle S^2 \rangle$ of unperturbed HW chains were obtained by Yamakawa and Fujii [63]. For example, $\langle R^2 \rangle$ in the case $\sigma_0 = 0$ may be written [65]

$$
\begin{aligned}
\lambda^2 \langle R^2 \rangle = {} & C_\infty(L\lambda) - (1/2)\hat{\tau}_0^2\nu^{-2} - 2\hat{\kappa}_0^2\nu^{-2}(4-\nu^2)(4+\nu^2)^{-2} \\
& + e^{-2L\lambda}\{(1/2)\hat{\tau}_0^2\nu^{-2} + 2\hat{\kappa}_02\nu^{-2}(4+\nu^2)^{-2}[(4-\nu^2) \\
& \times \cos(\nu L\lambda) - 4\nu \sin(\nu L\lambda)]\}
\end{aligned}
\tag{6.4}
$$

where $\hat{\kappa}_0 = \kappa_0/\lambda$, $\hat{\tau}_0 = \tau_0/\lambda$,

$$
\nu = (\hat{\kappa}_0{}^2 + \hat{\tau}_0{}^2)^{1/2}
\tag{6.5}
$$

and

$$
C_\infty = (4 + \hat{\tau}_0{}^2)/(4 + \hat{\kappa}_0{}^2 + \hat{\tau}_0{}^2)
\tag{6.6}
$$

It follows from eq 6.4 that

$$
\langle R^2 \rangle = C_\infty L/\lambda \quad (L\lambda \to \infty)
\tag{6.7}
$$
$$
\langle R^2 \rangle = L^2 \quad (L\lambda \to 0, \kappa_0/\lambda \ll 1)
\tag{6.8}
$$
$$
\langle R^2 \rangle = \tau_0{}^2 L^2/(\kappa_0{}^2 + \tau_0{}^2) \quad (L\lambda \to 0, L|\tau_0| \gg 1)
\tag{6.9}
$$

Equations 6.7, 6.8, and 6.9 give the well-known formulas for Gaussian coils, straight rods, and regular helices, respectively, indicating that the HW chain takes these three extreme conformations under the limiting conditions shown above in the parentheses. When κ_0 is zero, eq 6.4 reduces to eq 2.13 for the wormlike chain, as anticipated from what has been noted in the preceding section. Thus we can conclude that the HW chain takes a great variety of conformations consisting of coil, rod, and helix, depending on the values of its basic parameters. Figure 5-15 illustrates $\lambda \langle R^2 \rangle / (C_\infty L)$(a reduced characteristic ratio) as a function of λL (a reduced chain length) computed for HW chains with $\sigma = 0$ and different sets of κ_0 and τ_0, along with that for the wormlike chain (for which $(2\lambda)^{-1} = q$). The curve 1 markedly shifts to the left from the curve for the wormlike chain, while the curves 2 and 3 exhibit a maximum. Thus we see that the HW chain is capable of explaining what was mentioned in Section 6.1, i.e., the RIS model calculations which cannot be predicted by wormlike chains.

The appearance of a maximum in the chain length dependence of the characteristic ratio was first found by Yoon and Flory [60] in the RIS chain computations on atactic and syndiotactic poly(methyl methacrylate)(PMMA). Yamakawa and Fujii [63] showed that their formula for the HW chain with σ_0 can be fitted accurately to the computed results of Yoon and Flory over an almost entire range of chain length if the basic parameters of the HW chain are suitably chosen. A similar agreement in characteristic ratio between the HW ($\sigma_0 = 0$) and RIS chains was obtained for other flexible polymers. The choice of $\sigma_0 = 0$ seems odd, but, according to Yamakawa [6], $\langle R^2 \rangle$ of HW chains is relatively insensitive to σ_0 so that this choice does not matter. These results are striking, because the HW chain is Markoffian, whereas the RIS chain is not always so. Yamakawa and Shimada [64] tried to explore the relation between the RIS chain model and the discrete chain whose continuous limit is the HW chain.

The differences in the curves 1, 2, and 3 from the curve for the wormlike chain can be qualitatively explained by noting that the energy minimum ($U = 0$) is at the regular helix for the HW chain and at the straight rod for the wormlike chain. When L is sufficiently small, both chains are essentially rod–like; too short an HW chain cannot wind a helix. With increasing L, the HW chain tries to reduce its free energy by taking helical conformations because the helix is lower in energy than either coil or rod. On the other hand, the wormlike chain tends to coil itself because by so doing it gains more entropy and thus reduces the free energy even though its energy also increases owing to enhanced bending. Compared at the same chain length, the end–to–end distance is larger for a helix than for a coil unless the helix pitch is too small. Thus we can understand why the curves 1, 2, and 3 all appear above that for the wormlike chain. However, as L becomes larger, the HW chain finds it preferable to gain entropy by random coiling instead of reducing energy by further helix formation in order to decrease its free energy. The result is a decrease in chain dimensions. Thus, the characteristic ratio of the HW chain should asymptotically tend to that for a coil at sufficiently large L, as does that of the wormlike chain. Some

171

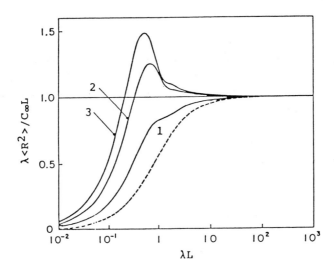

Fig. 5-15. Reduced characteristic ratio as a function of reduced chain length for HW chains with $\sigma_0 = 0$. Curve 1, $\kappa_0/\lambda = 3.0, \tau_0/\lambda = 2.0$; curve 2, $\kappa_0/\lambda = 4.0, \tau_0/\lambda = 1.0$; curve 3, $\kappa_0/\lambda = 5.0, \tau_0/\lambda = 1.0$. Dashed curve, wormlike chain ($\kappa_0 = 0$).

HW chains having a deep energy minimum may maintain dominantly helical conformations up to relatively large L. The characteristic ratios of such chains will rise above the limiting coil value in a certain range of L and exhibit a maximum, as seen in the curves 2 and 3.

6.6   Remarks

As exemplified by the curves of Figure 5-15, it is in the region of relatively small λL that the HW chain shows behavior distinctly different from that of the wormlike chain. This fact suggests that wormlike chains remain of use for the description of semi–flexible polymers down to low molecular weight unless the polymers studied have a special preference for assuming loose helical conformations. One of the distinct advantages of the HW chain over the wormlike one is that the former allows us to formulate the dynamics of polymers in which local dipole moments, polarizabilities, fluorescence probes, and so forth are distributed in an arbitrary fashion. In fact, Yamakawa et al. [73–77] developed detailed theories of dielectric and nuclear magnetic relaxation and fluorescence depolarization of the HW chain and compared the results with experiment. Their theories must contribute greatly to future studies of local motions of semi–flexible polymers.

172

Appendix

This appendix presents the reported values of persistence length q and shift factor M_L (molar mass per unit contour length) for some typical stiff polymers, along with the methods used for their estimation. These values are limited to ones determined from dilute solution data alone. Accumulated evidence shows that M_L can be evaluated approxinately using chemical structure or crystallographic data. As for the q values estimated on the basis of such an approximate M_L the reader is advised to see the review article of Tsvetkov and Andreeva [66].

Wormlike chain parameters for typical stiff polymers

Polymer	Solvent	q/nm	M_L/nm^{-1}	Method
schizophyllan	water	200	2150	$[\eta], s_0$ [36]
	0.01 N NaOH	150	2170	$\langle S^2 \rangle$ [48]
xanthan	0.1 M NaCl	120	1940	$\langle S^2 \rangle, [\eta], s_0$ [49]
DNA	0.2 M NaCl	58–68	1900–2000	$\langle S^2 \rangle, [\eta], s_0$ [3, 31, 67]
PBIC[a]	THF[g]	66	550	$\langle S^2 \rangle$ [68][j]
PHIC[b]	hexane	42	715	$\langle S^2 \rangle$ [27]
	butyl chloride	35	760	$[\eta]$ [41]
	toluene	37	740	$[\eta]$ [70]
CTN[c]	acetone	17	520	$[\eta], s_0$ [3][k]
pt-BD[d]	heptane	13	810	$[\eta], s_0$ [72]
PPAH[e]	DMSO[h]	11	185	$\langle S^2 \rangle, [\eta]$ [40]
PTDP[f]	TFE[i]	6.4	350	$\langle S^2 \rangle, [\eta]$ [47]

a: poly(butyl isocyanate).   b: poly(hexyl isocyanate).   c: cellulose trinitrate.
d: poly[trans–bis(tributylphosphine)platinum 1,4–butadiynediyl].
e: poly(terephthalamide–p–benzohydrazide).
f: poly(terephthaloyl–trans–2,5–dimethylpiperazine).
g: tetrahydrofuran.   h: dimethyl sulfoxide.   i: trifluoroethanol.   j: Data are due to Ambler et al. [69].   k: Data are due to Meyerhoff [71].

References

1. O. Kratky and G. Porod, Rec. Trav. Chim. Pay- Bas **68**, 1106 (1949).
2. H. Yamakawa and M. Fujii, Macromolecules **6**, 407 (1973).
3. H. Yamakawa and M. Fujii, Macromolecules **7**, 128 (1974).
4. H. Yamakawa and T. Yoshizaki, Macromolecules **13**, 633 (1980).
5. H. Yamakawa, Macromolecules **10**, 692 (1977).
6. H. Yamakawa, Ann. Rev. Phys. Chem. **35**, 23 (1984).
7. N. Saito, K. Takahashi, and Y. Yunoki, J. Phys. Soc. Japan **22**, 219 (1967).
8. L. D. Landau and E. M. Lifshitz, "Statistical Physics," Addison–Wesley, Reading, Mass., 1958.
9. W. Gobush, H. Yamakawa, W. H. Stockmayer, and W. S. Magee, J. Chem. Phys. **57**, 2839 (1972).
10. H. Yamakawa and M. Fujii, J. Chem. Phys. **59**, 6641 (1973).
11. J. J. Hermans and R. Ullman, Physica **18**, 951 (1952).
12. K. Nagai, Polym. J. **4**, 35 (1973).
13. H. Yamakawa, J. Chem. Phys. **59**, 3811 (1973).
14. J. Shimada and H. Yamakawa, J. Chem. Phys. **67**, 344 (1977).
15. H. Yamakawa, J. Shimada, and M. Fujii, J. Chem. Phys. **68**, 2140 (1978).
16. M. Fujii and H. Yamakawa, J. Chem. Phys. **72**, 6005 (1980).
17. K. Nagai, J. Chem. Phys. **38**, 924 (1963).
18. R. L. Jernigan and P. J. Flory, J. Chem. Phys. **50**, 4185 (1969).
19. P. Sharp and V. A. Bloomfield, Biopolymers **6**, 1201 (1968).
20. T. Norisuye, H. Murakami, and H. Fujita, Macromolecules **11**, 966 (1978).
21. T. Yoshizaki and H. Yamakawa, Macromolecules **13**, 1518 (1980).
22. H. Yamakawa, Macromolecules **16**, 1928 (1983).
23. T. Norisuye, M. Motowoka, and H. Fujita, Macromolecules **12**, 320 (1979).
24. J. E. Hearst and W. H. Stockmayer, J. Chem. Phys. **37**, 1425 (1962).
25. T. Yoshizaki and H. Yamakawa, J. Chem. Phys. **72**, 57 (1980).
26. H. Yamakawa, Macromolecules **8**, 339 (1975).
27. H. Murakami, T. Norisuye, and H. Fujita, Macromolecules **13**, 345 (1980).
28. L. Zhang, W. Liu, T. Norisuye, and H. Fujita, Biopolymers **26**, 333 (1987).
29. T. C. Troxell and H. A. Scheraga, Macromolecules **4**, 528 (1971).
30. U. Shmueli, W. Traub, and K. Rosenheck, J. Polym. Sci. Part A–2 **7**, 515 (1969).
31. J. E. Godfrey and H. Eisenberg, Biophy. Chem. **5**, 301 (1976).
32. R. G. Kirste and R. C. Oberthür, In "Small Angle X-Ray Scattering," Ed. O. Glatter and O. Kratky, Academic Press, London, 1982, Chap. 12, p. 387.
33. B. G. H. Ballard, M. G. Rayner, and J. Schelten, Polymer **17**, 349 (1976).
34. Y. Miyaki, Y. Einaga, and H. Fujita, Macromolecules **11**, 1180 (1979); Y. Miyaki, Ph. D. Thesis, Osaka University, 1981.

35. T. Oyama, K. Shiokawa, and K. Baba, Polym. J. **13**, 167 (1981).
36. T. Yanaki, T. Norisuye, and H. Fujita, Macromolecules **13**, 1462 (1980).
37. Y. Takahashi, T. Kobatake, and H. Suzuki, unpublished work; see T. Norisuye, Makromol. Chem. Suppl. **14**, 105 (1985).
38. S. V. Bushin, V. N. Tsvetkov, E. B. Lysenko, and V. N. Emel'yanov, Vysokomol. Soedin. A **23**, 2494 (1981).
39. M. Bohdanecký, Macromolecules **16**, 1483 (1983).
40. K. Sakurai, K. Ochi, T. Norisuye, and H. Fujita, Polym. J. **16**, 559 (1984).
41. M. Kuwata, H. Murakami, T. Norisuye, and H. Fujita, Macromolecules **17**, 2731 (1984).
42. S. Itou, N. Nishioka, T. Norisuye, and A. Teramoto, Macromolecules **14**, 904 (1981).
43. T. Norisuye and H. Fujita, Polym. J. **14**, 143 (1982).
44. T. Tsuji, T. Norisuye, and H. Fujita, Polym. J. **7**, 558 (1975).
45. J. Sadanobu, T. Norisuye, and H. Fujita, Polym. J. **13**, 75 (1981).
46. M. Motowoka, T. Norisuye, and H. Fujita, Polym. J. **9**, 613 (1977).
47. M. Motowoka, H. Fujita, and T. Norisuye, Polym. J. **10**, 331 (1978).
48. Y. Kashiwagi, T. Norisuye, and H. Fujita, Macromolecules **14**, 1220 (1981).
49. T. Sato, T. Norisuye, and H. Fujita, Macromolecules **17**, 2696 (1984).
50. H. Benoit and P. Doty, J. Phys. Chem. **57**, 958 (1953).
51. H. Yamakawa and W. H. Stockmayer, J. Chem. Phys. **57**, 2843 (1972).
52. H. Yamakawa and J. Shimada, J. Chem. Phys. **83**, 2607 (1985).
53. K. Nagai, Polym. J. **3**, 67 (1972).
54. H. Yamakawa, M. Fujii, and J. Shimada, J. Chem. Phys. **71**, 1611 (1979).
55. P. Horn, Ann. Phys. (Paris) **10**, 386 (1955).
56. V. F. Holland, J. Macromol. Sci.–Chem. A **7**, 173 (1973); R. L. Miller, J. Macromol. Sci.–Chem. A **7**, 183 (1973).
57. W. R. Krigbaum and S. Sasaki, J. Polym. Sci., Polym. Phys. Ed. **19**, 1339 (1981).
58. C. W. Carlson and P. J. Flory, J. Chem. Soc. Faraday II **73**, 1505 (1977).
59. U. W. Suter and P. J. Flory, J. Chem. Soc. Faraday II **73**, 1521 (1977).
60. D. Y. Yoon and P. J. Flory, Polymer **16**, 645 (1975).
61. P. J. Flory, "Statistical Mechanics of Chain Molecules," Interscience, New York, 1969.
62. A. E. H. Love, "A Treatise on the Mathematical Theory of Elasticity," Fourth Ed., Dover, New York, 1944, Chap. 18.
63. H. Yamakawa and M. Fujii, J. Chem. Phys. **64**, 5222 (1976).
64. H. Yamakawa and J. Shimada, J. Chem. Phys. **68**, 4722 (1978).
65. M. Fujii, K. Nagasaka, J. Shimada, and H. Yamakawa, Macromolecules **16**, 1613 (1983).
66. V. N. Tsvetkov and L. N. Andreeva, Adv. Polym. Sci. **39**, 95 (1981).

67. M. T. Record, Jr., C. P. Woodbury, and R. B. Inman, Biopolymers **14**, 393 (1975).
68. H. Yamakawa, J. Shimada, and K. Nagasaka, J. Chem. Phys. **71**, 3573 (1979).
69. M. R. Ambler, D. McIntyre, and L. J. Fetters, Macromolecules **11**, 300 (1978).
70. T. Ito, H. Chikiri, A. Teramoto, and S. M. Aharoni, Polym. J. **20**, 143 (1988).
71. G. Meyerhoff, J. Polym. Sci. **29**, 399 (1958).
72. M. Motowoka, T. Norisuye, A. Teramoto, and H. Fujita, Polym. J. **11**, 665 (1979).
73. T. Yoshizaki and H. Yamakawa, J. Chem. Phys. **81**, 982 (1984).
74. H. Yamakawa and M. Fujii, J. Chem. Phys. **81**, 997 (1984).
75. T. Yoshizaki, M. Fujii, and H. Yamakawa, J. Chem. Phys. **82**, 1003 (1985).
76. T. Yoshizaki and H. Yamakawa, J. Chem. Phys. **81**, 982 (1984).
77. H. Yamakawa, in "Molecular Conformation and Dynamics of Macromolecules in Condensed Systems," Ed. M. Nagasawa, Elsevier, Amsterdam, 1988, p.21.

PART II

CONCENTRATED SOLUTIONS

Chapter 6   Basic Concepts

1.   Concentration Regimes

1.1   Overlap Concentration

Sufficiently dilute polymer solutions may be viewed as systems in which "islands" of polymer coils scattered in the "sea" of a liquid solvent occasionally impinge and interpenetrate. Thus the spatial distribution of chain segments in them is quite heterogeneous and undergoes appreciable fluctuations from time to time. As the polymer concentration increases, the collision of the islands becomes more frequent and causes the chains to overlap and entangle in a complex fashion. As a result the segment distribution becomes less heterogeneous and its fluctuation is by and large suppressed. Increased chances of chain contact may lead to interchain association, and sometimes crystalline regions may be formed. However, in this book, we deal only with simple systems in which the formation of any higher-order structure does not take place.

One of the characteristic phenomena which occur when a polymer solution in a non-θ solvent is concentrated is the **screening effect** on chain dimensions. This is a change in $\langle S^2 \rangle(c)$ toward $\langle S^2 \rangle_\theta$ with the increase in polymer concentration. Here, $\langle S^2 \rangle(c)$ is the mean-square radius of gyration of a single polymer chain at a mass concentration c, and $\langle S^2 \rangle_\theta$ is its infinite-dilution value under the θ condition. Daoud et al. [1] were the first to observe the screening effect. Studying polystyrene in carbon disulfide by small-angle neutron scattering (SANS), they found $\langle S^2 \rangle(c)$ to decrease in proportion to $c^{-0.25}$. Figure 6-1 shows a more recent demonstration of the screening effect by King et al. [2] who investigated polystyrene in deuterated toluene by SANS; the abscissa n denotes the polymer concentration in terms of the mole fraction of styrene residues. The slope of the indicated line is -0.16, differing from -0.25 obtained by Daoud et al. The screening effect is a reflection of the decrease in intrachain segment-segment interactions caused by chain overlapping. Discussion on the mechanism is deferred to Section 3.

We look at a single chain in a polymer solution. The average number density of its segments, n_{coil}, i.e., their number contained in unit volume of the average space occupied by the chain, is given by

$$n_{\text{coil}} = \frac{3M/m}{4\pi[\langle S^2 \rangle(c)]^{3/2}} \tag{1.1}$$

where M is the molecular weight of the polymer and m the molar mass of one segment. On the other hand, if we denote the average number density of chain

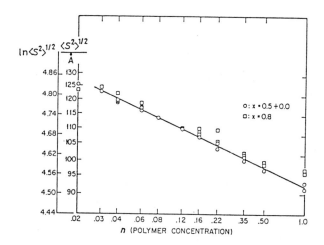

Fig. 6–1. Concentration dependence of coil radius for PS in d–toluene [2]. Line, least–squares fit to the sample pairs for $x = 0.50$ and 0.0. Here, n is the mole fraction of styrene residues, and x the mole fraction of d–styrene in a mixture of h– and d–styrenes.

segments in the entire solution by n_{soln}, we obtain

$$n_{\text{soln}} = (c/m)N_{\text{A}} \tag{1.2}$$

where N_{A} is the Avogadro constant. For dilute solutions in which the "islands" are separated by the solvent sea we have $n_{\text{soln}} < n_{\text{coil}}$ even though polymer coils occasionally overlap. With increasing c we reach the point at which n_{soln} catches up with n_{coil}. Denoting the c value for this solution by $c^{\star}$, we have the relation

$$c^{\star} = \frac{3M}{4\pi N_{\text{A}}[\langle S^2 \rangle(c^{\star})]^{3/2}} \tag{1.3}$$

Usually, $c^{\star}$ is called the **overlap concentration**. It corresponds to the situation where polymer coils would begin to touch one another throughout the solution if they behaved like mutually impenetrable spheres. However, having an open structure, actual polymer coils start interpenetrating as soon as the solution leaves the state of infinite dilution. Hence, it is not legitimate to consider $c^{\star}$ as if it marks the onset of coil overlapping. The term overlap concentration thus seems misleading. Nonetheless, it will be used in the ensuing presentation because it is almost settled in the current polymer literature.

Since $\langle S^2 \rangle(c)$ is not easy to measure, $c^{\star}$ is often approximated by $c^{\star}{}_0$ defined

180

as

$$c^{\star}_0 = \frac{3M}{4\pi N_A [\langle S^2 \rangle(0)]^{3/2}} \tag{1.4}$$

With $\langle S^2 \rangle(0) \sim M^{2\nu}$ valid unless M is low, this equation gives

$$c^{\star}_0 \sim M^{-(3\nu-1)} \tag{1.5}$$

which predicts

$$c^{\star}_0 \sim M^{-0.5} \quad (\theta \text{ solvents}) \tag{1.6}$$
$$c^{\star}_0 \sim M^{-0.8} \quad (\text{good solvents}) \tag{1.7}$$

Actually, there is no established consensus about the definition of the overlap concentration. For example, Hager and Berry [3] replace the factor $3/4\pi$ by $1/5$, and Graessley [4] by $1/8$, while Ying and Chu [5] drop this factor and use $\langle R^2 \rangle(0)/4$ for $\langle S^2 \rangle(0)$. Thus we have to be careful in comparing reported overlap concentrations.

1.2   Dilute and Concentrated Solutions

It is important to note that $c^{\star}$ is not a critical concentration. We should not expect that something special, such as sharp changes in the concentration dependence of some physical properties of the solution, takes place at this concentration. However, it seems certain that the macroscopic distribution of chain segments over the entire solution becomes essentially uniform when c passes through a relatively narrow region around $c^{\star}$. Thus we may take $c^{\star}$ as a measure of the crossover region where the island–sea heterogeneous structure of a polymer solution changes to the state of macroscopically uniform segment distribution. At concentrations higher than $c^{\star}$ the latter state remains unchanged, and the average segment density can be equated to c. However, when viewed microscopically, even the solutions above $c^{\star}$ are not uniform, the segment density in each volume element fluctuating from time to time about the mean value c and those in different volume elements at a given instant being different. Because of the chain connectivity of segments and their intra– and interchain interactions the density fluctuations at different places in the solution cannot take place independently. In other words, they are correlated. This correlation governs in various ways the physical properties of concentrated polymer solutions, and thus it is the key concept in the discussion of Part II.

In this book, for the convenience of description, we refer to a polymer solution as **dilute** or **concentrated** depending on whether c is below or above $c^{\star}$, though we are unable to draw a distinct boundary between these two regimes. The term

181

dilute used here should not be confused with that in Part I, which actually refers to "infinitely dilute." Some authors [6] call the region $c < c^\star$ the virial regime. Probably they consider that the virial expansion for osmotic pressure (eq 1–2.13) would diverge as c approaches $c^\star$. However, nothing is as yet known about the radius of convergence of this expansion.

1.3 Semi–dilute Regime

It follows from eq 1.4 that if M is sufficiently high, there is a rather wide range of concentration in which c is higher than $c^\star$ (to the approximation of regarding $c^\star$ as essentially equal to $c^\star_0$) but its absolute value is still low enough. Solutions of such concentrations are "dilute" in the sense that the volume fraction occupied by polymer coils is very small, but have to be regarded as "concentrated" according to the above definition of a concentrated solution. Thus they are referred to as **semi–dilute**. When the polymer molecular weight is infinitely large, even solutions very close to infinite dilution are "concentrated," and may be considered ideally semi–dilute.

In the semi–dilute regime, the segment density is so low that chain interpenetration or overlap is not yet extensive enough. Then, for the reason mentioned later in Section 3, the correlation of segment densities at different points is not sufficiently "screened" by chain overlapping, but remains significant until the distance between the points becomes comparable to the chain dimensions. Thus, the mean–field approximation ignoring density correlation may no longer be adequate for the formulation of semi–dilute solutions. This anticipation has led to many theoretical and experimental studies in recent years. Chapter 7 is devoted to a summary of typical findings from them.

The term "moderately" concentrated solution is sometimes used in the polymer literature. We may understand it as implying both dilute and semi–dilute solutions combined.

1.4 Highly Concentrated Regime

As the concentration increases, a polymer solution enters the regime subsequently called **highly concentrated**, in which chains overlap so extensively that the spatial correlation of density fluctuations is sufficiently screened and becomes effective only over distances comparable to the size of a few chain segments. The mean–field approximation would be valid in this regime. The crossover from semi–dilute to highly concentrated should occur over a range of concentration. Graessley [4] argued that an approximate location of this range may be estimated by a concentration $c^\dagger = 0.77/[\eta]^\dagger$, where $[\eta]^\dagger$ is the intrinsic viscosity for the molecular weight at which $[\eta]$ in the solvent considered begins to depart from $[\eta]_\theta$. Thus, Graessley's $c^\dagger$ is independent of M. Doi and Edwards [7] derived another expression for of $c^\dagger$, which also indicates that $c^\dagger$ is independent of M. However, as the data of Figure 6-1 show, the screening effect on chain dimensions persists up to the undiluted polymer. Hence, the crossover concentrations thus defined do not seem to have much physical significance.

In concentrated solutions and the melt of a flexible polymer, the chains entangle with each other in a very complex way, unless they are too short. Such an entangled system is often modeled by a pseudo–network with irregular meshes. As the concentration increases, the average size of the meshes diminishes monotonically, but density data suggest that it remains relatively large, of the order of a hundred monomer units, even in melts. The pseudo–network model is appealing to physical intuition, and has been widely used to formulate the viscoelastic behavior of polymer concentrates and melts [7]. However, for the formulation of thermodynamic behavior of concentrated solutions this model is not convenient, and non–network models are used. In the latter, polymer chains are distributed on a lattice fixed in space or uniformly in the system. Whatever model may be used, it is crucial to take the chain connectivity and interactions of polymer segments into statistical considerations.

1.5   Estimation of Overlap Concentration

Among several other methods proposed for the estimation of $c^\star$ the one using the concentration dependence of the sedimentation coefficient $s(c)$ appears to be most relevant. It has been thoroughly worked out by Roots and Nyström [8–10] and by Vidakovic et al. [11]

For a long time it has been well known that plots of $s(c)$ vs. c for different M merge to a single curve when c exceeds a certain value $\bar{c}$ characteristic of each M. Figure 6-2 illustrates this fact with the data of Vidakovic et al. [11] for polystyrene in cyclohexane at θ. The arrows point to $\bar{c}$. The usual assumption is that polymer chains in solutions of c above $\bar{c}$ sediment not individually but as an aggregate because of chain overlapping. Then it is reasonable to equate $\bar{c}$ to $c^\star$. With this idea Vidakovic et al. [11] derived from the data of Figure 6-2

$$c^\star = (6.75 \pm 0.09)M^{-0.5} \quad (\text{g/cm}^3) \tag{1.8}$$

which agrees in form with eq 1.6 (note that $c^\star$ agrees with $c^\star{}_0$ in a θ solvent). They also applied the same idea to their $s(c)$ data for polystyrene in benzene [11] and obtained

$$c^\star \sim M^{-0.66 \pm 0.03} \tag{1.9}$$

This relation differs considerably from eq 1.7, indicating that $\langle S^2 \rangle(c^\star)$ in eq 1.3 may not be approximated by $\langle S^2 \rangle(0)$, as should be expected. Thus, $c^\star{}_0$ is only

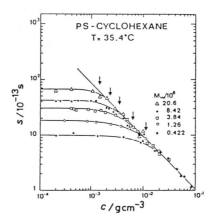

Fig. 6–2. Sedimentation coefficient s as a function of mass concentration c for PS in CH at θ [11]. Arrows indicate $c^\star$ at which the curves for different molecular weights converge to the indicated straight line. The slope of this line is -0.96.

an apparent overlap concentration, and for a rigorous discussion it may not be used as a substitute for $c^\star$.

Roots and Nyström [8, 9] derived for polystyrene

$$c^\star \sim M^{-0.58} \quad \text{(2–butanone at 25°C, marginal solvent)} \qquad (1.10)$$

$$c^\star \sim M^{-0.73} \quad \text{(toluene at 25°C, good solvent)} \qquad (1.11)$$

We note that the exponents in eq 1.9 and 1.11 differ by more than experimental uncertainty despite the fact that benzene and toluene are almost equally good solvents for polystyrene.

Nyström and Roots [10] investigated in detail the concentration dependence of s in the semi–dilute regime for various polymer + solvent systems. Some of the results are discussed in Chapter 7 along with the related theories.

Cornet [12] proposed equating $c^\star$ to c where the logarithm of zero shear viscosity plotted against c begins to follow a linear relation and obtained $c^\star = 0.614/[\eta]_\theta$ for polystyrene in cyclohexane. Since $[\eta]_\theta \sim M^{0.5}$ this relation agrees with eq 1.6. The author, however, does not know how Cornet's proposal can be justified.

Root and Nyström [13] tried to estimate $c^\star$ on the idea that fluorescent excimer formation would be enhanced when a polymer solution crosses over $c^\star$ because chances for interchain contact should increase by chain overlapping. Their fluorescence measurements gave data in support of the idea, but no definite determination of c^* was possible. There is still another idea proposed for the estimation of c^*. It utilizes the "cooperative" diffusion coefficient measured by quasi–elastic light scattering which becomes independent of M above a certain c [14] (see Section 2.1.4 of Chapter 7).

2.   Correlation Length

2.1   Distribution Functions

We consider an equilibrium homogeneous solution of volume V containing ω monodisperse polymer molecules, which are distinguished by a subscript σ. Each polymer consists of N ($\gg 1$) identical beads which are numbered with a subscript i. The position of bead i in polymer σ is thus expressed as $\mathbf{r}_{i(\sigma)}$. We introduce the symbols $[\sigma]$ and $d[\sigma]$ to represent the conformation of polymer σ and a volume element of its $3N$–fold conformation space, i.e.,

$$[\sigma] \equiv (\mathbf{r}_{1(\sigma)}, \mathbf{r}_{2(\sigma)}, \ldots, \mathbf{r}_{N(\sigma)}) \tag{2.1}$$

$$d[\sigma] \equiv d\mathbf{r}_{1(\sigma)}\, d\mathbf{r}_{2(\sigma)} \cdots d\mathbf{r}_{N(\sigma)} \tag{2.2}$$

The coordinates of one bead, for example, $\mathbf{r}_{1(\sigma)}$, can be used to determine the location of polymer σ in the solution, and the remaining ones $\mathbf{r}_{2(\sigma)}, \ldots$ to describe the conformation of the same polymer.

The probability of finding polymer σ in a state (position and shape) between $[\sigma]$ and $[\sigma] + d[\sigma]$ is written $V^{-1} F_1([\sigma]) d[\sigma]$, and that of finding polymer σ in a state between $[\sigma]$ and $[\sigma] + d[\sigma]$ and, at the same time, polymer τ in a state between $[\tau]$ and $[\tau] + d[\tau]$ is written $V^{-2} F_2([\sigma], [\tau])\, d[\sigma]\, d[\tau]$. We call F_1 and F_2 the one–body distribution function and the two–body distribution function, respectively. They satisfy the normalization conditions

$$V^{-1} \int_V F_1([\sigma])\, d[\sigma] = 1 \tag{2.3}$$

$$V^{-2} \int_V F_2([\sigma], [\tau])\, d[\sigma]\, d[\tau] = 1 \tag{2.4}$$

and F_1 can be expressed in terms of F_2 as follows:

$$F_1([\sigma]) = V^{-1} \int_V F_2([\sigma], [\tau])\, d[\tau] \tag{2.5}$$

We introduce $g_2([\sigma], [\tau])$ defined by

$$g_2([\sigma], [\tau]) = F_2([\sigma], [\tau]) - F_1([\sigma])\, F_1([\tau]) \tag{2.6}$$

185

This function represents how the states of polymers σ and τ at a given instant are correlated, and it is called the **polymer–pair correlation function.**

Next, if bead $i(\sigma)$ is fixed in a volume element $dr_{i(\sigma)}$ at the position $r_{i(\sigma)}$ and $V^{-1}F_1([\sigma])$ is integrated over all positions of the other beads of polymer σ, the result, written $V^{-1}f_1(r_{i(\sigma)})dr_{i(\sigma)}$, gives the probability that bead $i(\sigma)$ is found in the volume element $dr_{i(\sigma)}$. Since the solution is homogeneous, $V^{-1}f_1(r_{i(\sigma)})\,dr_{i(\sigma)}$ is simply equal to $V^{-1}\,dr_{i(\sigma)}$. From these facts we obtain

$$f_1(r_{i(\sigma)}) = \int_V F_1([\sigma])\,d[\sigma]/dr_{i(\sigma)} = 1 \tag{2.7}$$

If we integrate $F_1([\sigma])$ over all beads of polymer σ, excepting beads $i(\sigma)$ and $j(\sigma)$ fixed in space, and write the result as $f_2(r_{i(\sigma)}, r_{j(\sigma)})$, we have

$$f_2(r_{i(\sigma)}, r_{j(\sigma)}) = \int_V F_1([\sigma])d[\sigma]/dr_{i(\sigma)}\,dr_{j(\sigma)} \tag{2.8}$$

The function $f_2(r_{i(\sigma)}, r_{j(\sigma)})$ is the probability density for finding bead $i(\sigma)$ and bead $j(\sigma)$ simultaneously at $r_{i(\sigma)}$ and $r_{j(\sigma)}$, respectively. Hence it may be termed the intrachain two–body bead distribution function. This function may also be regarded as the product of the probability density for finding bead $i(\sigma)$ at $r_{i(\sigma)}$ and that for finding bead $j(\sigma)$ at a distance $r = r_{j(\sigma)} - r_{i(\sigma)}$ from $r_{i(\sigma)}$ when bead i is at $r_{i(\sigma)}$. If the latter is denoted by $P_{ij}(r)$, then we can write

$$f_2(r_{i(\sigma)}, r_{j(\sigma)}) = f_1(r_{i(\sigma)})P_{ij}(r) = P_{ij}(r) \tag{2.9}$$

because $f_1(r_{i(\sigma)}) = 1$. The quantity $P_{ij}(r)$ is called the **intrachain bead–pair radial distribution function.** Its form depends only on the bead numbers i and j, as indicated by the subscript ij. Since $f_1(r_{i(\sigma)}) = 1$, $f_2(r_{i(\sigma)}, r_{j(\sigma)})$ is sometimes confused with $P_{ij}(r)$, but the two functions have different physical meanings. It should be noted that $P_{ij}(r)$ is subject to the normalization condition

$$\int P_{ij}(r)\,dr = 1 \tag{2.10}$$

For beads belonging to different polymers σ and τ we may define the inter-chain two–body bead distribution function $f_2(r_{i(\sigma)}, r_{k(\tau)})$ and the **interchain bead–pair correlation function** $Q_{ik}(r_{i(\sigma)}, r_{k(\tau)})$ as follows:

$$f_2(r_{i(\sigma)}, r_{k(\tau)}) = \int_V F_2([\sigma], [\tau])\,d[\sigma]\,d[\tau]/d(r_{i(\sigma)})\,d(r_{k(\tau)}) \tag{2.11}$$

$$Q_{ik}(r_{i(\sigma)}, r_{k(\tau)}) = f_2(r_{i(\sigma)}, r_{k(\tau)}) - f_1(r_{i(\sigma)})\,f_1(r_{k(\tau)})$$
$$= f_2(r_{i(\sigma)}, r_{k(\tau)}) - 1 \tag{2.12}$$

It can be shown that the following relation holds

$$Q_{ik}(\mathbf{r}_{i(\sigma)}, \mathbf{r}_{k(\tau)}) = \int_V g_2([\sigma], [\tau]) \, d[\sigma] \, d[\tau] / d\mathbf{r}_{i(\sigma)} \, d\mathbf{r}_{k(\tau)} \qquad (2.13)$$

For a pair of chains sufficiently separated from each other, $g_2([\sigma], [\tau])$ vanishes, so that $Q_{ik}(\mathbf{r}_{i(\sigma)}, \mathbf{r}_{k(\tau)})$ vanishes for any i and k.

2.2 Correlation of Density Fluctuations

The number bead density at a position $\mathbf{r}^a$ is denoted by $n(\mathbf{r}^a)$. Then we can express it as

$$n(\mathbf{r}^a) = \sum_{\sigma=1}^{\omega} \sum_{i(\sigma)=1}^{N} \delta(\mathbf{r}_{i(\sigma)} - \mathbf{r}^a) \qquad (2.14)$$

where $\delta(\mathbf{r})$ is a three–dimensional delta function. If the ensemble average of $n(\mathbf{r}^a)$ is denoted by $\langle n(\mathbf{r}^a) \rangle$, it can be obtained as follows:

$$\langle n(\mathbf{r}^a) \rangle = V^{-1} \int_V n(\mathbf{r}^a) F_1([\sigma]) \, d[\sigma]$$

$$= V^{-1} \sum_{\sigma=1}^{\omega} \sum_{i(\sigma)=1}^{N} \int_V F_1([\sigma]) \, d[\sigma] / d\mathbf{r}_{i(\sigma)} \, |(\text{calc. at } \mathbf{r}_{i(\sigma)} = \mathbf{r}^a)$$

$$= \omega N / V \qquad (2.15)$$

As expected, $\langle n(\mathbf{r}^a) \rangle$ does not depend on the position $\mathbf{r}^a$. Hence it is designated by n in the ensuing discussion, i.e.,

$$n = \omega N / V \qquad (2.16)$$

The polymer mass concentration c is related to n by

$$c = mn / N_A \qquad (2.17)$$

where m is the molar mass of one bead.

Following Kurata [15], we can show that the ensemble average of $n(\mathbf{r}^a) n(\mathbf{r}^b)$ is given by

$$\langle n(\mathbf{r}^a) n(\mathbf{r}^b) \rangle = \frac{2n}{N} \sum_{i=1}^{N} \sum_{j>i}^{N} P_{ij}(\mathbf{r}^b - \mathbf{r}^a) + \frac{n^2}{N^2} \sum_{i=1}^{N} \sum_{k=1}^{N} [1 + Q_{ik}(\mathbf{r}^a, \mathbf{r}^b)] \quad (2.18)$$

The quantity $\langle n(\mathbf{r}^a)n(\mathbf{r}^b)\rangle$ represents the average correlation of bead densities at two points $\mathbf{r}^a$ and $\mathbf{r}^b$ and is expressed in terms of the intrachain bead–pair radial distribution function P_{ij} and the interchain bead–pair correlation function Q_{ik}.

The fluctuation of bead density at $\mathbf{r}^a$, $\Delta n(\mathbf{r}^a)$, is

$$\Delta n(\mathbf{r}^a) = n(\mathbf{r}^a) - n \tag{2.19}$$

It follows from eq 2.18 that $\langle \Delta n(\mathbf{r}^a)\Delta n(\mathbf{r}^b)\rangle$, the ensemble average of the correlation of density fluctuations at $\mathbf{r}^a$ and $\mathbf{r}^b$, can be expressed as

$$\langle \Delta n(\mathbf{r}^a)\Delta n(\mathbf{r}^b)\rangle = \frac{2n}{N}\sum_{i=1}^{N}\sum_{j>i}^{N} P_{ij}(\mathbf{r}^b - \mathbf{r}^a) + \frac{n^2}{N^2}\sum_{i=1}^{N}\sum_{k=1}^{N} Q_{ik}(\mathbf{r}^a, \mathbf{r}^b) \tag{2.20}$$

where the fact $\langle \Delta n(\mathbf{r}^a)\rangle = 0$ has been taken into account. For homogeneous solutions we may consider that $Q_{ik}(\mathbf{r}^a, \mathbf{r}^b)$ depends only on $\mathbf{r}^b - \mathbf{r}^a$ as does P_{ij} and that these functions of $\mathbf{r}^b - \mathbf{r}^a$ are actually spherically symmetric. Hence $\langle \Delta n(\mathbf{r}^a)\Delta n(\mathbf{r}^b)\rangle$ becomes a function of $r \equiv |\mathbf{r}^b - \mathbf{r}^a|$ only, and it follows from eq 2.20 that

$$h(r) = h_1(r) + (n/N)h_2(r) \tag{2.21}$$

where $h(r)$ is a reduced density fluctuation correlation function defined by

$$h(r) = \langle \Delta n(\mathbf{r}^a)\Delta n(\mathbf{r}^b)\rangle/(nN) \tag{2.22}$$

and $h_1(r)$ and $h_2(r)$ are given by

$$h_1(r) = \frac{2}{N^2}\sum_{i=1}^{N}\sum_{j>i}^{N} P_{ij}(r) \tag{2.23}$$

$$h_2(r) = \frac{1}{N^2}\sum_{i=1}^{N}\sum_{k=1}^{N} Q_{ik}(r) \tag{2.24}$$

We consider volume elements A and B separated by a distance r, and denote a test chain by L and any other coexisting chain by L'. The function $h(r)$ is proportional to the average density fluctuation produced in B when there is a bead of L in A. Equation 2.21 shows that it consists of the intrachain contribution $h_1(r)$ and the interchain contribution $(n/N)h_2(r)$. The function $h_1(r)$ depends on the interactions between paired beads of L in the presence of all L'. We note that these interactions include the constraint that the beads of L are chain-connected. In a θ solvent, $h_1(r)$ depends only on this constraint. On the other hand, the function $h_2(r)$ is concerned with how strongly the presence of a bead of L in A affects that of a bead of L' in B. This effect depends on the interaction transmitted from A to B by direct contact between L and L' and also by indirect contact between them via other intervening chains. In a θ solvent, no chain–chain interaction acts, so that $h_2(r)$ vanishes.

Summarizing, we see that $h(r)$ reflects the average interaction acting between a pair of beads separated by a distance r. Thus, it should have something to do with equilibrium behavior of polymer solutions. In fact, the osmotic pressure Π of a binary solution is related to $h(r)$ by (see the Appendix at the end of this chapter)

$$\frac{RT/M}{(\partial \Pi/\partial c)_{T,\mu_0}} = \int h(r)\, d\mathbf{r} \qquad (2.25)$$

where μ_0 is the chemical potential of the solvent and RT has the usual meaning. This relation is valid only for binary solutions.

It is obvious from their physical meaning that both $h_1(r)$ and $h_2(r)$ decay to zero and so does $h(r)$ with increasing r. A parameter called the correlation length is introduced in the next section to characterize the rate at which $h(r)$ tends to vanish.

2.3 Structure Factor

2.3.1 Intrachain and Interchain Interference Factors

We denote the excess structure factor per mass cocncentration of solution by $S(k)$ an simply call it the structure factor. Here by excess we mean the value for a solution minus that for its solvent. As before, k is the magnitude of the scattering vector. Textbooks of physical chemistry explain how $S(k)$ can be determined from measurements of the intensity of scattered light or X-ray or neutrons. For the polymer solution under consideration it can be shown that $S(k)$ is related to $h(r)$ by [15]

$$S(k) = KMH(k) \qquad (2.26)$$

where K is an experimental constant which depends on the kind of radiation used and $H(k)$ is the Fourier transform of $h(r)$, i.e.,

$$H(k) = \int h(r) \exp(i\mathbf{k} \cdot \mathbf{r})\, d\mathbf{r} \qquad (2.27)$$

With eq 2.21 substituted into eq 2.27, we obtain

189

$$H(k) = H_1(k) + (n/N) H_2(k) \tag{2.28}$$

where

$$H_1(k) = \int h_1(\mathbf{r}) \exp (i\mathbf{k} \cdot \mathbf{r}) \, dr \tag{2.29}$$

$$H_2(k) = \int h_2(\mathbf{r}) \exp (i\mathbf{k} \cdot \mathbf{r}) \, dr \tag{2.30}$$

Usually, $H_1(k)$ and $H_2(k)$ are called the intrachain interference factor and the interchain interference factor, respectively. At infinite dilution, the former reduces to the particle scattering factor $P(k)$ that appeared in Part I. Further we note the following:

(1) At the limit of zero scattering angle (i.e., $k = 0$) $H_1(k)$ converges to unity regardless of c, as can be verified with eq 2.10.

(2) For Gaussian chains Zimm [16] showed that

$$H_2(k) = [H_1(k)]^2 H_2(0) \tag{2.31}$$

(3) According to Zimm [16], we have

$$\lim_{c \to 0} H_2(0) = -(2M^2/N_A) A_2 \tag{2.32}$$

where A_2 is the second virial coefficient of the solution.

2.3.2 Definition of the Correlation Length

It follows from eq 2.27 that $H(k)$ (and hence $S(k)$) is an even function of k. Hence $S(k)$ for small k can be expressed in the form

$$S(k) = \frac{S(0)}{1 + \xi^2 k^2} \quad \text{(for small } k\text{)} \tag{2.33}$$

Thus, when experimental data for $S(k)$ are given, two parameters $S(0)$ and ξ can be determined from the intercept and initial slope of $1/S(k)$ plotted against k^2. If this determination of $S(0)$ is repeated at a series of c, the resulting data allow the concentration dependence of Π to be calculated from

$$S(0) = \frac{KRT}{(\partial \Pi/\partial c)_{T,\mu_0}} \tag{2.34}$$

which follows from eq 2.25 through eq 2.27. The quantity ξ determines how fast $S(k)$ decays with k in the region of small k.

If eq 2.33 happens to hold over the entire range of k, eq 2.26 and 2.27 give

$$1/(1 + \xi^2 k^2) = \int h(r) \exp(i\mathbf{k} \cdot \mathbf{r}) \, d\mathbf{r} / \int h(r) \, d\mathbf{r} \qquad (2.35)$$

which can be solved for $h(r)$ to give

$$h(r) = \frac{A}{r\xi^2} \exp(-r/\xi) \qquad (2.36)$$

where A is a constant. Actually, A can be determined from eq 2.25, giving

$$A = \frac{RT/M}{4\pi(\partial\Pi/\partial c)_{T,\mu_0}} \qquad (2.37)$$

Usually, $h(r)$ represented by eq 2.36 is called the Ornstein–Zernike form. It shows that $h(r)$ decays faster with r as ξ becomes smaller. Thus, ξ may be taken as a measure of the range of r in which density fluctuations are effectively correlated. It is called the **correlation length** for density fluctuation [17]. In general, eq 2.33 holds only for small k so that $h(r)$ for $r < \xi$ is no longer the Zernike–Ornstein form. This means that ξ characterizes the decay rate of $h(r)$ in the tail region ($r > \xi$). To calculate $h(r)$ valid over the entire range of r (hence $S(k)$ or $H(k)$ valid for all k) we have to face a very difficult problem. An approximate solution by Benoit and Benmouna is described in Section 2.4.

2.3.3   Correlation Length under the θ Condition

Under the θ condition the intrachain interference factor $H_1(k)$ is not affected by the presence of other chains, so that its value $H_{1\theta}(k)$ at any concentration is equal to that at $c = 0$. For a Gaussian spring–bead chain $P_{ij}(r)$ at this limit is given exactly by [15]

$$P_{ij}(r) = (\gamma_{ij}/\pi^{1/2})^3 \exp(-\gamma_{ij}^2 r^2) \qquad (2.38)$$

where

$$\gamma_{ij}^2 = 3/2|i - j| a^2 \qquad (2.39)$$

With eq 2.38 it can be shown that

$$H_{1\theta}(k) = \frac{2}{\sigma^2}(\sigma - 1 + e^{-\sigma}) \qquad (2.40)$$

where

$$\sigma = k^2 N a^2 / 6 = k^2 \langle S^2 \rangle_\theta \qquad (2.41)$$

The right–hand side of eq 2.40 is usually called the Debye function. Actually, it has already appeared in the preceding chapters. For small k, i.e., $\sigma \ll 1$, eq 2.40 yields

$$\frac{S(k)}{S(0)} = \frac{1}{1 + (\sigma/3)} \qquad (2.42)$$

Comparing with eq 2.33, we find

$$\xi = \langle S^2 \rangle_\theta^{1/2} / 3^{1/2} \qquad (2.43)$$

which indicates that density fluctuations in an isolated Gaussian chain have effective correlation up to distances comparable to the chain dimensions. This density correlation arises from the chain connectivity of beads, so that it is unique to polymer systems.

2.4 Theory of Benoit and Benmouna

In Figure 6–3a, i(1) and i'(1) are beads on a chain 1, while k(2) and k'(2) are beads on another chain 2. The definition of the interchain bead–pair correlation function allows $Q_{i(1)k(2)}(r)$ to be expressed as

$$Q_{i(1)k(2)}(r) = V^2 P_{i(1)i'(1)}(r_1) P_{k'(2)k(2)}(r_2) Q_{i'(1)k'(2)}(r_3) \qquad (2.44)$$

where r_1, r_2, and r_3 are the distances as shown in the figure, $P_{ii'}$ is the intrachain bead–pair radial distribution function, and V the volume of the solution. Fourier transforming eq 2.44 and summing the result over all beads of both chains, we obtain

$$H_2(k) = N^2 J(k) [H_1(k)]^2 \qquad (2.45)$$

where

$$J(k) = 4 \sum_{i(1)=1}^{N} \sum_{i'(1)>i(i)}^{N} \sum_{k(2)=1}^{N} \sum_{k'(2)>k(2)}^{N} \frac{\hat{P}_{i(1)i'(1)} \hat{P}_{k'(2)k(2)} \hat{Q}_{i'(1)k'(2)}}{[H_1(k)]^2} \qquad (2.46)$$

with the caret indicating the Fourier transform.

Benoit and Benmouna [18] applied the Ornstein–Zernike relation [19], first proposed for simple liquids, to polymer solutions. According to it, $Q_{i(1)k(2)}$ can be expressed as follows:

$$Q_{i(1)k(2)}(r) = D_{i(1)k(2)}(r) + (n/N) I_{i(1)k(2)}(r) \qquad (2.47)$$

192

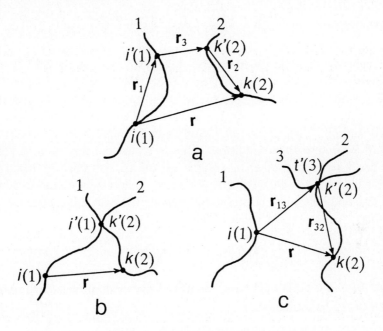

Fig. 6–3. a. Explanation of the vectors r_1, r_2, and r_3. b. Chain 1 is in direct contact with chain 2 at one point. c. Chain 3 is in contact with chain 2 at one point but not with chain 1.

with

$$I_{i(1)k(2)}(r) = \int Q_{i(1)t'(3)}(r_{13}) Q_{t'(3)k'(2)}(0) P_{k'(2)k(2)}(r_{32}) \, dr_{t'(3)} \qquad (2.48)$$

Here, $D_{i(1)k(2)}(r)$ is the contribution from the configuration of chains 1 and 2 shown in Figure 6–3b, in which beads i'(1) and k'(2) are in direct contact, and $I_{i(1)k(2)}(r)$ that from the configuration shown in Figure 6–3c, in which chain 3 is in contact with chain 2 directly at one point. Fourier transformation of eq 2.47 followed by summation over all beads of chains 1 and 2 yields

$$H_2(k) = \hat{D}(k) + (n/N)\hat{I}(k) \qquad (2.49)$$

where

$$\hat{D}(k) = \sum_{i(1)=1}^{N} \sum_{k(2)=1}^{N} \hat{D}_{i(1)k(2)}(k) \qquad (2.50)$$

193

$$\hat{I}(k) = \sum_{i(1)=1}^{N} \sum_{k(2)=1}^{N} \hat{I}_{i(1)k(2)}(k) \tag{2.51}$$

The quantity $\hat{D}(k)$ is given by the right–hand side of eq 2.45 if $J(k)$ is evaluated for the configuration of Figure 6–2b. We have $Q_{i'(1)k'(2)} = -\beta\delta(\mathbf{r})$ (where β is the excluded–volume strength) for beads i'(1) and k'(2) which are in direct contact. However, we have to introduce an approximation in order to proceed further. Thus, Benoit and Benmouna invoked what is usually called the **random phase approximation**. This is to assume that even in solutions where beads repel each other by volume exclusion, each polymer behaves as a Gaussian chain. Then the vectors $\mathbf{r}_1, \mathbf{r}_2, \mathbf{r}_3, \mathbf{r}_{13}$, and $\mathbf{r}_{32}$ in Figure 6–3 have no correlation and hence can be varied independently. With this maneuver we can show that $J(k) = -\beta$ and hence obtain

$$\hat{D}(k) = -\beta N^2 [H_1(k)]^2 \tag{2.52}$$

Though the proof is omitted here, the use of the random phase approximation allows $\hat{I}(k)$ to be evaluated as

$$\hat{I}(k) = -\beta N^4 J(k)[H_1(k)]^3 \tag{2.53}$$

Substituting eq 2.45, 2.52, and 2.53 into eq 2.49, we obtain

$$J(k) = -\beta[1 + (nN)J(k)H_1(k)]$$

which is solved for $J(k)$ to give

$$J(k) = -\beta/[1 + (nN)\beta H_1(k)] \tag{2.54}$$

When this is substituted back into eq 2.45 we obtain

$$H_2(k) = -\frac{\beta N^2 [H_1(k)]^2}{1 + (nN)\beta H_1(k)} \tag{2.55}$$

Thus, eq 2.28 yields the following formula of Benoit and Benmouna [18]:

$$H(k) = \frac{H_1(k)}{1 + (nN)\beta H_1(k)} \tag{2.56}$$

Interestingly, $H(k)$ is expressed in terms of $H_1(k)$ only.

Equation 2.56 gives no concrete information about $H(k)$ unless $H_1(k)$ can be calculated theoretically. Under the random phase approximation used to derive eq 2.56, $H_1(k)$ is actually equal to $H_{1\theta}(k)$. Hence, eq 2.56 should read

$$H(k) = \frac{H_{1\theta}(k)}{1 + (nN)\beta H_{1\theta}(k)} \qquad (2.57)$$

This formula was also derived by Jannink and de Gennes [20] and Daoud et al. [1] by using different methods but invoking the random phase approximation. To the author the theory of Benoit and Benmouna appears most transparent.

After eq 2.40 is substituted for $H_{1\theta}(k)$, eq 2.57 combined with eq 2.26 and 2.33 yields

$$\xi^2 = \frac{\langle S^2 \rangle_\theta}{3(1 + \beta c M N_A / m)} \qquad (2.58)$$

where m is the molar mass of one bead. This equation indicates that the correlation length ξ in a non-θ solvent ($\beta \neq 0$) starts at $\langle S^2 \rangle_\theta^{1/2}/(3)^{1/2}$ and decreases monotonically as the polymer concentration increases. Though the derivation is not given here, it can be shown that the correct correlation length at $c = 0$ is $\langle S^2 \rangle^{1/2}/(3)^{1/2}$, where $\langle S^2 \rangle$ is the mean–square radius of gyration (at infinite dilution) in a given non-θ solvent. Thus, eq 2.58 is incorrect. As shown in Chapter 7, this equation also fails to predict the experimentally measured concentration dependence of ξ. If not all, these shortcomings of eq 2.58 should originate from the use of the random phase approximation in deriving eq 2.57. Calculation of $H(k)$ with no recourse to this approximation still remains a challenging problem.

Figure 6–4 illustrates the ξ data of Kinugasa et al. [21] for semi–dilute solutions of polystyrene in cyclohexane at the θ temperature. Interestingly, these data do not support the prediction of eq 2.58, according to which ξ under the θ condition should be independent of c. The discrepancy suggests that the Benoit–Benmouna theory contains a shortcoming other than the use of the random phase approximation, which holds exactly at θ.

3. Effects of Concentration

3.1 Screening Effects

3.1.1 Excluded–volume Interaction

We consider what happens to a test chain in a polymer solution when chain overlap takes place. With increasing concentration, each bead of the test chain sees in its neighbor more beads of other chains, so that chances of intrachain

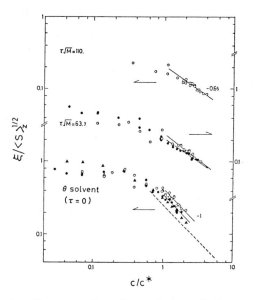

Fig. 6–4. Concentration dependence of the correlation length for PS in CH at three temperatures. Data for $\tau = 0$ are at 34.8°C (θ) [21]. Different marks are for different M. The dashed line is expected by the blob theory explained in Chapter 7.

bead–bead contact diminishes.[1] This phenomenon brings about the same effect on the test chain as weakening the strength of bead–bead repulsion, β (in the binary cluster approximation). Thus, as the concentration increases, it becomes possible for the test chain to take less swollen conformations, and its statistical radius shrinks. This explains why the screening effect on chain dimensions takes place.

To scrutinize the screening effect we look at a subchain (ij) between beads i and j of the test chain. As it becomes longer, the subchain is surrounded with more beads of other chains and hence undergoes a stronger screening effect. Therefore, if we denote the end–distance expansion factor for the subchain (ij) by $\alpha_R(ij)$, we can expect that $\alpha_R(ij)$ decreases monotonically to unity with increasing $|i - j|$. This decay behavior may be roughly described by a model in which subchains shorter than g_E are free from screening, while those longer than

---

[1] In solutions of a simple electrolyte, anions cluster around a cation owing to the attraction between them, and screen the repulsion between cations. In solutions of a neutral polymer, in which bead–bead interactions are repulsive, the clustering of beads of other chains around each bead of the test chain is caused by the thermodynamic force which tends to maximize the mixing of beads.

g_E are completely screened by other chains. Mathematically, $\alpha_R(ij) = \alpha_R(ij)^0$ for $|i - j| < g_E$ and $= 1$ for $|i - j| > g_E$, where the superscript 0 signifies the value at infinite dilution in a given solvent. Then we may speak of the strength of the screening effect on intrachain interactions in terms of g_E. In place of g_E we may use the root–mean–square radius of gyration, ξ_E, for the subchain characterized by g_E. In this book, we refer to ξ_E as the screening length for excluded–volume interaction. At infinite dilution, no screening occurs so that ξ_E can be equated to $[\langle S^2 \rangle(0)]^{1/2}$. As the concentration increases, ξ_E is supposed to diminish monotonically. Obviously, this behavior should depend on solvent quality.

3.1.2   Density Fluctuation Correlation

Since a shorter subchain is not very much screened by other chains, its $P_{ij}(r)$ is effectively equal to $P_{ij}{}^0(r)$. As the subchain gets longer, it is more strongly screened, so that $P_{ij}(r)$ of a longer subchain approaches $P_{ij}(r)_\theta$, which denotes the value of $P_{ij}(r)$ for the subchain in the unperturbed state. By definition $h_1(r)$ is the average of $P_{ij}(r)$ over all combinations of i and j. Hence, $h_1(r)$ should take a variety of intermediate shapes between $h_1(r)^0$ and $h_1(r)_\theta$, depending on the degree of chain overlap. Thus, it should change from the former to the latter with increasing concentration. The crossover behavior depends on solvent quality. In particular, under the θ condition, $h_1(r)$ remains unchanged over the entire range of concentration.

From what was explained in Section 2.2, we see that $h_2(r)$ is given by the sum of two terms, $h_{2d}(r)$ associated with direct contact between two chains and $h_{2i}(r)$ associated with indirect contact between them via other intervening chains. As the concentration increases, the chance of indirect contact is enhanced, so that $h_{2i}(r)$ begins to dominate the sum. The Benoit–Benmouna theory described in Section 2.4 gives an approximate picture of this change. It shows (see eq 2.55) that $h_2(r)$ tends to $-(N/n)h_1(r)$ at high concentration, regardless of solvent quality. Thus, with increasing c, the positive $h_1(r)$ is successively cancelled by the negative $(n/N)h_2(r)$, and $h(r)$ diminishes monotonically. In particular, at high enough concentrations, $h(r)$ may remain non–zero only for very small r, which may be comparable to the size of a few monomer units. This implies that, in the highly concentrated regime including melts, density fluctuations at different points become uncorrelated and the correlation length ξ diminishes to the order of a bond length.

As noted in Section 2.4, ξ at $c = 0$ is equal to $\langle S^2 \rangle^{1/2}/(3)^{1/2}$. Therefore, we see that the order of ξ varies from chain dimensions to a bond length as a polymer solution changes from the dilute regime to the highly concentrated one. This pronounced concentration dependence of ξ is a feature of polymer solutions. However, it would be prohibitively difficult task to predict it (more

197

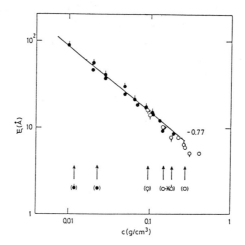

Fig. 6-5. SAXS correlation length data for PS in toluene at 25°C [22]. Different marks for different M. Arrows indicate $c^{\star}{}_0$ for respective M.

generally, the concentration dependence of $h(r)$) by theoretical calculation.

If the binary cluster approximation is valid, the transmission of interactions between different chains does not occur and hence $h_2(r)$ vanishes at any concentration in θ solvents. Then, $h(r)$ is equal to $[h_1(r)]_\theta$ and stays unchanged at any c. In other words, ξ undergoes no concentration effect under the θ condition. However, this consequence does not agree with the data of Figure 6-4, which indicate a definite decrease in ξ with c. The discrepancy again suggests that the binary cluster approximation fails in solutions near the θ condition.

Figure 6-5 shows the ξ data of Hamada et al. [22] for polystyrene in a good solvent toluene. It shows that, in the range of c studied, ξ is essentially independent of M^2 and that, as c increases, ξ diminishes from 10 nanometers to a few nanometers (the size of a few monomers) following a power $\xi \sim c^{-0.77}$. These features are the same as those found in the earlier SANS work of Daoud et al. [1] on polystyrene in carbon disulfide.

3.2 Chain Dimensions

In the early 1960s, Yamakawa [23], Eizner [24], and Grimley [25] showed by perturbation calculations valid to first order in c that the radius expansion factor α_S decreases with increasing c. Subsequently, Fixman and Peterson [26] predicted by an approximate theory that the shrinkage of chain dimensions occurs first rapidly and then gradually as the solution is concentrated. However,

---

[2] At low concentrations ξ depends on molecular weight because $\xi_0 \sim$ chain dimensions.

these highly mathematical theories do not give us a physical insight into the molecular mechanism responsible for the screening effect on chain dimensions.

According to eq 2–1.23 and the definition of $P_{ij}(r)$, $\langle S^2 \rangle$ of a single chain in a polymer solution is represented by

$$\langle S^2 \rangle = N^{-2} \sum_{i=1}^{N} \sum_{j>i}^{N} \int r^2 P_{ij}(r) d\mathbf{r} \tag{3.1}$$

With $\Delta_{ij}(r) \equiv P_{ij}{}^0(r) - P_{ij}(r)$, this equation may be rewritten

$$\langle S^2 \rangle = \langle S^2 \rangle(0) - \Delta \tag{3.2}$$

where

$$\Delta = N^{-2} \sum_{i=1}^{N} \sum_{j>i}^{N} \int r^2 \Delta_{ij}(r) d\mathbf{r} \tag{3.3}$$

For the reason mentioned in Section 3.1, $\Delta_{ij}(r)$ converges to $P_{ij}{}^0(r) - [P_{ij}{}^0(r)]_\theta$ at sufficiently high concentration. Hence, at this limit, eq 3.2 yields

$$\langle S^2 \rangle = \langle S^2 \rangle_\theta \tag{3.4}$$

which predicts that the chain dimensions in highly concentrated solutions become indistinguishable from those under the θ condition, regardless of solvent quality. Equation 3.4 is the consequence of the effect of intrachain excluded-volume interactions being just offset by that of interchain ones. The two effects cancel only partially at intermediate concentrations, and the degree of cancellation may be expressed in terms of how much the screening length ξ_E differs from zero. Since ξ_E decreases with increasing c, $\langle S^2 \rangle$ changes from $\langle S^2 \rangle(0)$ to $\langle S^2 \rangle_\theta$.

In his 1953 book, Flory [27] writes "It may be shown that when the polymer concentration is large, the perturbation (due to volume exclusion) tends to be less. In particular, in a bulk polymer containing no diluent α (= α_S) = 1 for the molecules of the polymer." His conjecture has long troubled many workers, but it can be easily justified by the consideration described above. Its first experimental verification was given by Cotton et al. [28] in the mid–1970s (see also Ref. [29, 30]). Their pioneering data are delineated in Figure 6–6. The success was due to the advent of SANS techniques which made it possible to measure $\langle S^2 \rangle$ of a single chain in solutions of finite dilutions.

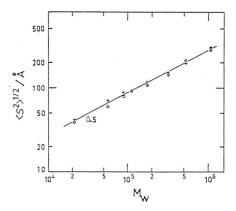

Fig. 6–6. SANS data of Cotton et al. [28] for radii of gyration of PS in CH at 36°C (θ). +, at infinite dilution; ⊙, in the bulk. The solid line has a slope of 0.5.

3.3 Hydrodynamic Screening

We look at various subchains of a test chain in a concentrated solution. Shorter ones are essentially free from the overlapping of other chains, so that they find themselves in an atmosphere similar to that at infinite dilution. Therefore, the motion of their segments is similar to that of the segments of an isolated chain, which is highly correlated owing not only to chain connectivity but also to intrachain hydrodynamic interactions. Longer subchains are more probably surrounded by other chains so that their motion undergoes the effects of a flow field generated by those other chains. Being more or less random, this field tends to make the subchain segments move in a less correlative fashion. In other words, chain overlap reduces the intrachain hydrodynamic interaction, just as it does the intrachain excluded–volume effect. This effect is called the hydrodynamic screening due to chain overlap. To express its strength we may introduce a parameter ξ_H analogous to ξ_E. For example, it may be defined in such a way that pairs of segments of a single chain separated by more than ξ_H essentially have no hydrodynamic interaction.

de Gennes [31] suggested equating ξ_H to ξ, but this maneuver does not seem legitimate since these characteristic lengths are associated with entirely different physical origins. Richter et al. [32] assumed that the hydrodynamic interaction between a pair of beads of a spring–bead test chain is represented by the Oseen tensor if the solvent viscosity η_0 in it is replaced by an r–dependent effective viscosity $\eta_{\text{eff}}(r)$ defined as

$$[\eta_{\text{eff}}(r)]^{-1} = (\eta_0^{-1} - \eta^{-1}) \exp\left(-r/\xi_H\right) - \eta^{-1} \tag{3.5}$$

200

where r is the distance between the two beads and η the viscosity of the solution. Equation 3.5, however, does not define ξ_H. In the theory of Richter et al. [32], ξ_H was treated as a concentration–dependent adjustable parameter. According to the approximate theory as described by Doi and Edwards [7] in their recent book, η_{eff} is given by eq 3.5 with no η^{-1} term. This theory predicts ξ_H at high concentration to be explicitly given by

$$\xi_H = 2/(\pi n a^2) \tag{3.6}$$

where n is the average number bead density in the solution and a the mean length of one spring. No idea has as yet been established to define ξ_H in terms of measurable quantities.

We have introduced three characteristic lengths ξ, ξ_E, and ξ_H to describe the effects of chain overlap on the density fluctuation correlation, the intrachain excluded–volume interaction, and the intrachain hydrodynamic interaction, respectively. In the following chapters, we will illustrate the important roles played by them in understanding the static and dynamic behavior of polymer solutions.

Appendix

Derivation of eq 2.25

In our notation, for a polymer + solvent binary solution in osmotic equilibrium with its solvent (i.e., a constant μ_0 system, where μ_0 is the chemical potential of the solvent component), eq 3.3.15 in Kurata's book [15] gives

$$S(0) = KRTM/c(\partial\mu_1/\partial c)_{T_0,\mu_0} \tag{6A-1}$$

Here, μ_1 is the chemical potential of the polymer component, and K is proportional to the square of the specific refractive increment of the solution. From eq 3.1.127 and 3.1.128 in the same book we get

$$(\partial\mu_1/\partial c)_{T,\mu_0} = (M/c)(\partial\Pi/\partial c)_{T,\mu_0} \tag{6A-2}$$

Hence, eq 6A-1 can be rewritten

$$S(0) = KRT/(\partial\Pi/\partial c)_{T,\mu_0} \tag{6A-3}$$

On the other hand, it follows from eq 2.26 and 2.27 of this chapter that

$$S(0) = KM \int h(r) \, d\mathbf{r} \tag{6A-4}$$

Equation 2.25 is obtained when eq 6A-4 is substituted into eq 6A-3. It should be noted that eq 2.25 does not hold for solutions containing more than one polymer component.

References

1. M. Daoud, J. P. Cotton, B. Farnoux, G. Jannink, G. Serma, H. Benoit, R. Duplessix, C. Picot, and P.-G. de Gennes, Macromolecules **8**, 804 (1975).
2. J. S. King, W. Boyer, G. D. Wignall, and R. Ullman, Macromolecules **18**, 709 (1985).
3. B. L. Hager and G. C. Berry, J. Polym. Sci., Polym. Phys. Ed. **20**, 911 (1982).
4. W. W. Graessley, Polymer **21**, 258 (1980).
5. Q. Ying and B. Chu, Macromolecules **20**, 362 (1987).
6. D. W. Schaefer and C. C. Han, in "Dynamic Light Scattering," Ed. R. Pecora, Plenum, New York, 1985, Chap. 5.
7. M. Doi and S. F. Edwards, "The Theory of Polymer Dynamics," Oxford Univ. Press, London, 1986, Chap. 5.
8. J. Roots and B. Nyström, J. Chem. Soc., Farad. Trans. I **77**, 947 (1981).
9. J. Roots and B. Nyström, J. Polym. Sci., Polym. Phys. Ed. **19**, 479 (1981).
10. B. Nyström and J. Roots, J. Macromol. Sci.–Rev. Macromol. Chem. **C19** (1), 35 (1980).
11. P. Vidakovic, C. Allain, and F. Rondelez, J. Phys. Lett. **42**, L-323 (1981); Macromolecules **15**, 1571 (1982).
12. C. F. Cornet, Polymer **6**, 373 (1965).
13. J. Roots and B. Nyström, Eur. Polym. J. **15**, 1127 (1979).
14. T. L. Yu, H. Reihanian, and A. M. Jamieson, J. Polym. Sci., Polym. Lett. Ed. **18**, 695 (1980).
15. M. Kurata, "Thermodynamics of Polymer Solutions," (translated by H. Fujita from the Japanese), Harwood Academic, Chur, 1982, Chap. 3.
16. B. H. Zimm, J. Chem. Phys. **16**, 1093 (1948).
17. S. F. Edwards, Proc. Phys. Soc. (London) **85**, 613 (1965). This paper appears to have been the first to introduce the correlation length.
18. H. Benoit and M. Benmouna, Polymer **25**, 1059 (1984).
19. L. S. Ornstein and F. Zernike, Proc. Acad. Sci., Amsterdam **17**, 793 (1914).
20. G. Jannink and P.-G. de Gennes, J. Chem. Phys. **48**, 2260 (1968).
21. S. Kinugasa, H. Hayashi, F. Hamada, and A. Nakajima, Macromolecules **19**, 2832 (1986).
22. F. Hamada, S. Kinugasa, H. Hayashi, and A. Nakajima, Macromolecules **18**, 2290 (1985).
23. H. Yamakawa, J. Chem. Phys. **34**, 1360 (1961).
24. Yu. E. Eizner, Vysokomol. Soedin. **3**, 748 (1961).
25. T. B. Grimley, Trans. Farad. Soc. **57**, 1974 (1961).
26. M. Fixman and J. M. Peterson, J. Am. Chem. Soc. **86**, 3524 (1964).

27.   P. J. Flory, "Principles of Polymer Chemistry," Cornell Univ. Press, Ithaca, 1953, p. 426.
28.   J. P. Cotton, D. Decker, H. Benoit, B. Farnoux, J. Higgins, G. Jannink, R. Ober, C. Picot, and J. des Cloizeaux, Macromolecules 7, 863 (1974).
29.   D. G. H. Ballard and M. G. Rayner, Polymer 17, 349 (1976).
30.   R. G. Kirste and B. R. Lebner, Makromol. Chem. 177, 1137 (1976).
31.   P.-G. de Gennes, Macromolecules 9, 587, 594 (1976).
32.   D. Richter, K. Binder, B. Ewen, and B. Stühn, J. Phys. Chem. 88, 6618 (1984).

Chapter 7   Semi–dilute Solutions

1.   Static Properties

1.1   Introduction

Solutions of a polymer at finite dilutions are many–body systems more complex than those of small molecules in that individual solute molecules have a high internal freedom because of their flexible chain structure. We therefore cannot hope to develop a theory for them which is realistic with respect to the structure and interaction of its repeat units. The best we can attempt is to study highly coarse–grained models of many–chain systems, hoping to find physical laws which may be obeyed in common by actual polymer solutions. The classic Flory–Huggins theory [1] based on the lattice model is the earliest example of such attempts, and the equation–of–state theories developed by Flory and also by Prigonine [2] in the 1960s are its important refinements. As is well known, these theories have contributed significantly to our understanding and interpretation of the thermodynamic behavior of polymer solutions. The point to note is that they all are based on the mean–field approximation, which neglects the spatial correlation of segment density fluctuations.

As has been mentioned in Chapter 6, the correlation length characterizing this correlation changes from the size of a single chain to that of a few monomer units as a polymer solution crosses over from the dilute to the highly concentrated regime via the semi–dilute regime. This fact casts doubt about the use of the mean–field approximation for semi–dilute solutions as well as for dilute ones. Thus, in recent years, there has arisen much interest in formulating concentrated polymer solutions, especially their semi–dilute regime, by non–mean–field methods. Typical results are the blob theory and the scaling theory due notably to de Gennes and his collaborators. These theories are based on bold assumptions, but allow us to predict in the form of power laws how the static and dynamic properties of semi–dilute solutions depend on molecular weight and concentration. In this chapter, we discuss the validity and limitations of the predicted relations on the basis of representative experimental data.

1.2   Osmotic Pressure

To begin with, we consider the osmotic pressure Π of a monodisperse polymer in a pure solvent. Usually it is assumed that $\Pi(c)$ at low polymer concentration c can be expanded in the virial series represented by eq 1–2.13. This series can be rewritten

$$\Pi_r(c) \equiv \Pi(c)/(RTc) = M^{-1} + A_2 c + A_3 c^2 + \ldots \tag{1.1}$$

The quantity $\Pi_r(c)$ is called the reduced osmotic pressure and has the dimensions of mol/g. Experimentally it is well known that $\Pi_r(c)$ increases rapidly with increasing c in θ solvents as well as in non-θ ones. This fact implies that knowledge of virial coefficients higher than the second and third ones becomes imperative for predicting $\Pi_r(c)$ at higher concentration. But either theoretical or experimental evaluation of such higher-order coefficients is not a simple matter. Probably, the series in eq 1.1 diverges as c approaches a certain value, which is believed to lie below the overlap concentration. Thus, various ideas have been proposed to fit $\Pi_r(c)$ data up to moderately high concentrations by an empirical closed form, but until recently, there was no guiding theory for this purpose.

Using a very sophisticated method based on field theory, des Cloizeaux [3] showed that $\Pi_r(c)$ in the semi-dilute regime can be expressed in the form

$$\Pi_r(c) = M^{-1}[1 + F(s)] \tag{1.2}$$

where s is a dimensionless variable defined as

$$s = \gamma c M^{3\nu - 1} \tag{1.3}$$

and F is a "universal" function. In eq 1.3, γ is a proportionality factor which depends on the combination of polymer and solvent, and ν the exponent to M in the asymptotic relation

$$\langle S^2 \rangle (0) \sim M^{2\nu} \quad (M \to \infty) \tag{1.4}$$

where (0) indicates the value at infinite dilution. As explained in Chapter 2, ν is 0.5 for θ solvents and may be taken to be 0.6 (Flory's value) for marginal as well as good solvents.

We have defined by eq 6-1.4 a concentration $c^\star_0$ analogous to the overalp concentration $c^\star$. This concentration is conveniently used in the analysis which follows. In passing, we note that, as explained in the Appendix to this chapter, $c^\star_0$ can be approximately estimated by

$$c^\star_0 = 1.46/[\eta] \tag{1.5}$$

With eq 6-1.4 and eq 1.4, eq 1.3 may be rewritten

$$s = \gamma(c/c^\star_0) \tag{1.6}$$

This γ is not the same as that in eq 1.3 (though both are proportional), but it is used below to save notation. Thus, we find that, according to des Cloizeaux's

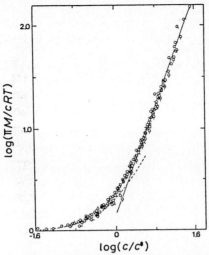

Fig. 7–1. Double–logarithmic plots of $M\Pi_r(c)$ vs. $c/c^{\star}{}_0$ for PαMS in toluene at 25°C. Different marks are for different M. The solid line has a slope of 1.33.

theory, when $M\Pi_r(c)$ data for different M are plotted against a reduced concentration $c/c^{\star}{}_0$, the data points in the semi–dilute regime should fall on a single curve. In modern jargon, $M\Pi_r(c)$ in this region is scaled by $c/c^{\star}{}_0$. It should be noted that, strictly speaking, eq 1.2 is valid asymptotically for very long chains. For such a chain, $c^{\star}{}_0$ is so small that there is a region of c in which the absolute value of c is sufficiently small but $c/c^{\star}{}_0$ can be much larger than unity.

Figure 7–1 illustrates, with the data of Noda et al. [4] on poly (α–methyl-styrene), that $M\Pi_r(c)$ is indeed scaled by $c/c^{\star}{}_0$. Schäfer [5] showed that earlier data on other polymers [6, 7] obeyed the same scaling law.

1.3 Expressions for $F(s)$

It would be very difficult to derive the exact form of $F(s)$. des Cloizeaux [3] showed that $F(s)$ for large s (which does not always mean $c/c^{\star}{}_0 \gg 1$, i.e., being deep in the semi–dilute regime) is asymptotically represented by

$$F(s) \sim s^{1/(3\nu-1)} \tag{1.7}$$

If Flory's value of 0.6 is used for ν, substitution of eq 1.7 into eq 1.2 gives for large s

$$\Pi_r(c) \sim M^{-1}(c^{\star}{}_0)^{-5/4}c^{5/4} \tag{1.8}$$

Since it follows from eq 6–1.4 and 1.3 that $c^{\star}{}_0 \sim M^{1-3\nu} = M^{-4/5}$, eq 1.8 can

207

be written

$$\Pi_r(c) \sim c^{5/4} \tag{1.9}$$

which predicts that $\Pi_r(c)$ in a good solvent is independent of M and increases in proportion to $c^{1.25}$ in the semi–dilute regime. If a more exact value of 0.588 is used for ν (see Section 1.4 of Chapter 2) we obtain 1.31 instead of 1.25 for the exponent to c in eq 1.9. This value is in excellent accordance with the data shown in Figure 7–1.

Using a renormalization group theory, Knoll et al. [8] derived a complex expression for $F(s)$, and Schäfer [5] showed that the numerical values it gives can be represented accurately by an empirical equation which reads

$$F(s) = \frac{s}{4} \left(\frac{40 + 27s + s^2}{22 + s} \right)^{0.3} [1 + 0.06(0.68)^s] \tag{1.10}$$

In fact, as he illustrated, reported osmotic pressure data on many polymer + solvent systems can be fitted precisely with eq 1.10 over a range including both dilute and semi–dilute solutions. Thus, Schäfer's equation should be quite useful for estimating Π of polymer solutions over a relatively wide concentration range.

Recently, Schulz and Stockmayer [9] have remarked that the empirical osmotic pressure equation proposed by one of them over 50 years ago, which reads

$$M\Pi_r(c) = 1/(1 - \mu c) \tag{1.11}$$

with

$$\mu = k_1 \Pi^{-q} \tag{1.12}$$

can be fitted to the data of Figure 7–1 if eq 1.12 is properly amended. Thus, they proposed in place of eq 1.11

$$1 - \frac{1}{M\Pi_r(c)} = \frac{k's}{(1 + k''sM\Pi_r(c))^q} \tag{1.13}$$

If the adjustable parameters k', k'', and q are chosen accordingly, eq 1.13 yields $F(s)$, the values of which agree closely with those given by eq 1.10 over a range of c including dilute and semi–dilute solutions. This exemplifies the wisdom of earlier scientists.

1.4 Blob Concept

de Gennes [10] considered that in solutions where chains overlap, not individual polymer chains but their subchains whose end distance is of the order of the correlation length ξ may behave independently of one another, and named the domain occupied by each of them the **blob**. His idea tacitly assumes that there is no thermodynamic interaction bewtween chain segments separated by more than ξ. It allows us to make some interesting predictions.

1.4.1 Correlation Length

According to the de Gennes idea, we may view a polymer solution as consisting of closely packed blobs which have no mutual interaction. Since the volume of each blob is proportional to ξ^3, the solution contains as many blobs as are proportional to ξ^{-3} per unit volume. The osmotic pressure of a solution is in direct proportion to the number density of thermodynamically independent particles. Therefore, we obtain the relation

$$\Pi \sim \xi^{-3} \tag{1.14}$$

Comparing this with $\Pi \sim c^{3\nu/(3\nu-1)}$, which can be derived from eq 1.2, 1.3, and 1.7 for $s \gg 1$, we find

$$\xi \sim c^{-\nu/(3\nu-1)} \tag{1.15}$$

where the proportionality factor is independent of M. For Flory's value of ν this gives

$$\xi \sim c^{-3/4} \tag{1.16}$$

which predicts that the correlation length in the semi–dilute regime does not depend on M and decreases with c in proportion to $c^{-0.75}$. This consequence was first confirmed by the pioneering SANS measurement of Daoud et al. [11] on polystyrene in carbon disulfide, who actually obtained $-(0.72 \pm 0.06)$ for the exponent to c. Another confirmation can be seen from the SAXS data of Hamada et al. [12] presented in Figure 6–5, where the exponent to c is -0.77. Recent SANS experiments of King et al. [13] on polystyrene in toluene led to $-(0.70 \pm 0.02)$ for the exponent. They took note of the earlier studies of Cotton et al. [14] and Benoit and Picot [15], which yielded exponents smaller than 0.75. Summarizing, we may say that, as far as ξ in the semi–dilute regime is concerned, the blob theory of de Gennes can predict experimental behavior to a good approximation.

1.4.2 Chain Dimensions

The Benoit–Benmouna theory described in Section 2.4 of Chapter 6 gives eq 6–2.58 for ξ, but, as noted in the same section, this equation does not give the correct ξ at infinite dilution. A simple ad hoc modification is to replace the term $\langle S^2 \rangle_\theta$ in it by $\langle S^2 \rangle(c)$ (the mean–square radius of gyration at concentration c). Then, with eq 1.16, the modified equation predicts the following concentration dependence of $\langle S^2 \rangle(c)$ in the semi–dilute regime:

$$\langle S^2 \rangle(c) \sim c^{-1/2} \tag{1.17}$$

209

This prediction does not agree with the SANS data of Daoud et al. [11], who found the exponent to c to be -0.25. More recent SANS data of King et al. [3] (Figure 6–1) yield a much smaller exponent of -0.16. These findings suggest that even after the above modification, eq 6–2.58 is still inaccurate. To make the modified equation consistent with their experiment, Daoud et al. [11] allowed the excluded–volume strength β to vary with c as $\beta \sim c^{1/4}$. But this ad hoc maneuver conflicts with the fact that as c increases, there occurs a screening of the excluded–volume effect, which is equivalent to a decrease in β with increasing c.

In place of defining a blob in terms of the correlation length, we may redefine it as the region occupied by a subchain consisting of g_E chain segments (see Section 3.1 of Chapter 6 for the definition of g_E). Each chain can then be regarded as a linear sequence of (N/g_E) blobs, where N is the number of segments in the chain. According to the definition of g_E, these blobs exert no excluded–volume interaction on one another, so that the chain made up of them behaves as an unperturbed Gaussian chain. Hence, we have

$$\langle S^2 \rangle \sim (N/g_E)\xi_E^2 \tag{1.18}$$

because the mean end distance of each blob is proportional (though not exactly) to the screening length ξ_E for the excluded–volume effect.

Again, using the definition of g_E, we assume that each blob is fully swollen by volume exclusion. Then, we have

$$\xi_E \sim g_E^{\nu} \tag{1.19}$$

To go further we boldly assume that the blobs may be treated as being thermo-dynamically independent. This allows us to derive the same equation as eq 1.15 for ξ_E, and substitution of eq 1.19 into it yields

$$g_E \sim c^{-1/(3\nu-1)} \tag{1.20}$$

With these formulas it follows from eq 1.18 that

$$\langle S^2 \rangle \sim c^{-(2\nu-1)/(3\nu-1)} \tag{1.21}$$

If Flory's value of 0.6 is used for ν, this relation predicts $\langle S^2 \rangle$ varying in proportion to $c^{-0.25}$, in agreement with the data of Daoud et al.

Since the blob theory based on ξ_E is not concerned with the correlation of density fluctuations, it gives no information about ξ. In their derivation of eq

210

1.21, Daoud et al. [11] implicitly equated ξ to ξ_E, but this is an ad hoc maneuver. We have as yet no clear–cut answer to the important question of what relation exists between the correlation length ξ and the screening length ξ_E.

The blob is a physical concept for defining an average working range of molecular interactions in polymer solutions. Hence, it is not always legitimate to visualize as if it were the entity consisting of a subchain, as has been done in the above discussion. We have to note that, though the same term is used, the blobs in the above theories for semi-dilute solutions are different from those in the theory of Weill and des Cloizeaux for dilute solutions (Section 1 of Chapter 4). The former do not interact, while the latter repel each other by volume exclusion. To distinguish between them, the former are sometimes referred to as concentration blobs and the latter as thermal blobs.

1.5   Remarks

1.   For θ solvents eq 1.15 gives

$$\xi \sim c^{-1} \tag{1.22}$$

The SAXS data of Kinugasa et al. illustrated in Figure 6–4 appear to substantiate this theoretical prediction. As has been noted in the preceding chapter, in the binary cluster approximation, ξ in a θ solvent should undergo no concentration effect.

2.   Strictly speaking, eq 1.9 is valid for semi–dilute solutions in which $s = \gamma(c/c^\star_0) \gg 1$. des Cloizeaux did not give the general expression for $F(s)$ in eq 1.2. As noted above, Knoll et al. [8] obtained it by renormalization group calculations, and their results were approximated accurately by eq 1.10 [5]. A renormalization group theory of Ohta and Oono [16] also derived an analytical expression for $F(s)$ in the good solvent limit, which yielded the following for $\partial\Pi/\partial c$:

$$(M/RT)(\partial\Pi/\partial c) = 1 + (8X)^{-1}\{9X^2 - 2X + 2\ln(1 + X)\}$$
$$\times \exp\{(2X)^{-2}[X + (X^2 - 1)\ln(1 + X)]\} \tag{1.23}$$

where $X = \alpha(c/c^\star_0)$ with α being constant for a given polymer + solvent pair. Wiltzius et al. [17] evaluated the quantity on the left–side of eq 1.23 by use of eq 6–2.34 from SANS measurements of $S(0)$ and found that the resulting values are scaled by X and agree fairly well with eq 1.23 as illustrated in Figure 7–2. In the limit of large X eq 1.23 reduces to

$$(M/RT)(\partial\Pi/\partial c) = (9/8)(\alpha c/c^\star_0)^{5/4} \tag{1.24}$$

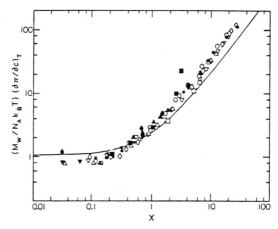

Fig. 7-2. Solid line, calculated from eq 1.23 with $X =$ $(16/9)A_2c$ (A_2 is the second virial coefficient). Points, derived from SANS data for PS in MEK and toluene [17] (different symbols are for different M).

Interestingly, this equation is consistent with eq 1.9. These results are rather surprising, because the theory of Ohta and Oono is correct only to first order in $\epsilon = 4 - d$, where d is the dimensionality of space. We note that the calculation of Knoll et al. was performed up to second order in ϵ.

3.   A concentrated polymer solution may be viewed as a transient network of entangled chains. In this visualization, the average end distance (size) of a network mesh is sometimes identified to the correlation length ξ, according to de Gennes [10]. If this is accepted, we have to interpret typical data of ξ (as presented in Ref. [11–13]) as implying that the mesh size diminishes to a few monomer units in highly concentrated solutions. However, as "plateau modulus" data from viscoelastic measurements indicate [18], even in the melt, the mesh size is at least one order of magnitude larger than these. Thus, it is misleading to envisage ξ as a network mesh size.

2.   Transport Coefficients

2.1   Friction Coefficients

2.1.1   Mutual Diffusion Coefficient

Transport of a substance caused by a thermodynamic force which arises from the gradient of its chemical potential is termed diffusion. It can be formulated phenomenologically by application of the thermodynamics of irreversible processes [19]. In what follows we confine ourselves to binary systems consisting of a monodisperse polymer and a pure solvent. The diffusion process in such a system can be described in terms of a single parameter D_m called the **mutual diffusion coefficient**. This parameter can be evaluated with the gradient method, in which a sharp boundary is initially formed between a test solution and its solvent and the subsequent change in the concentration (or concentration gradient) distribution is measured by an adequate optical means. It is crucial to recognize that D_m is a parameter of the system and hence not assignable to either polymer or solvent component. In general, D_m varies with polymer concentration c, temperature T, and pressure p, though the dependence on the last parameter is usually very weak. When we are concerned with a series of homologous polymers, we have to include the molecular weight M as an additional parameter.

The quantity directly associated with the translation of the polymer component relative to the solvent is not D_m, which is the volume–fixed diffusion coefficient, but the solvent–fixed diffusion coefficient D, which is defined by [19]

$$D = D_m/(1 - v_p c) \tag{2.1}$$

where v_p is the partial specific volume of the polymer component. The difference between D and D_m becomes important as the concentration increases. Sometimes it is convenient to discuss diffusion using a friction coefficient f of the polymer component defined by [20]

$$D = \left(\frac{M}{N_A f} \right) \left(\frac{\partial \Pi}{\partial c} \right)_{T,p} \tag{2.2}$$

This "phenomenological" f should not be confused with the "molecular" friction coefficient which is defined as a force needed to have a molecule translate steadily at unit velocity. The latter appears below when the self–diffusion coefficient is introduced.

Both v_p and f depend on c, T, p, and M, but the M dependence of the former may be neglected unless M is too small ($< 10^4$). Furthermore, as far as liquid solutions are concerned, the pressure dependence of f as well as v_p may be treated as negligible. The ensuing discussion tacitly takes these facts into account.

As seen from eq 2.2, the concentration dependence of D is an interplay of the hydrodynamic factor f and the thermodynamic factor $\partial \Pi/\partial c$ (the subscripts T and p attached to this derivative are omitted below). In estimating f from diffusion measurements, we have to correct D for the thermodynamic factor $\partial \Pi/\partial c$ according to eq 2.2.

2.1.2 Self-diffusion Coefficient

Thermal wriggling of the segments of polymer molecules gives rise to zig-zag migration of the centers of mass of the molecules (Figure 7–3), called self-diffusion. Until recently, the rate of this migration in concentrated solutions or melts was considered to be too slow to be measured experimentally, and little interest evolved in formulating it by molecular considerations. Motivated by de Gennes' novel idea explained later, however, the situation changed dramatically in the mid–1970s, and many papers have since been published on theoretical and experimental aspects of polymer self-diffusion. Since we have a lot to say about this subject, we defer its full discussion to the next chapter.

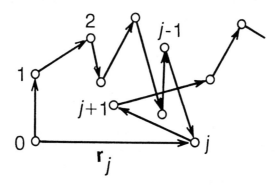

Figure 7–3. Trajectory of polymer self-diffusion. Each circle marks the center of mass of a migrating polymer chain at a certain instant.

In Figure 7–3, points 0, 1, ..., j, ... mark the positions of the center of mass of a self-diffusing polymer chain at times 0, $\Delta t, \ldots, j\Delta t, \ldots$, respectively, where Δt is a very short time interval. The vector from point 0 to point j is denoted by r_j, and the average of $r_j{}^2$ ($r_j \equiv |r_j|$) over all chains in the solution by $\langle r_j{}^2 \rangle$. According to the theory of Brownian motion, $\langle r_j{}^2 \rangle / j\Delta t$ converges to a constant as j increases, i.e., as time goes on. We designate this limiting value by $6D_s$, i.e.,

$$D_s = (1/6) \lim_{j \to \infty} \langle r_j{}^2 \rangle / j\Delta t \tag{2.3}$$

and call D_s the **self-diffusion coefficient** of the polymer. In recent years, many elaborate techniques have been developed for the determination of this diffusion coefficient with admirable ideas and efforts of experimentalists. We will touch upon them in Chapter 8.

The "molecular" friction coefficient mentioned above is f_s associated with D_s by the Einstein relation

$$D_s = k_B T / f_s \tag{2.4}$$

It can be shown [20] that f_s and f converge to the same value at infinite dilution, but, in general, they are not equal at finite dilutions.

2.1.3 Sedimentation Coefficient

Sedimentation velocity measurements on polymer + solvent binary solutions allow the volume–fixed sedimentation coefficient s_v of the polymer component to be evaluated. The solvent–fixed sedimentation coefficient s of the polymer component can then be calculated from s_v by use of the relation [21]

$$s = s_v(1 - v_p c) \tag{2.5}$$

According to the thermodynamics of irreversible processes, s is expressed in terms of f as

$$s = M(1 - v_p \rho)/(N_A f) \tag{2.6}$$

where ρ is the density of the solution. In many reports, the term $v_p c$ in eq 2.5 is ignored, but this approximation is good except in highly concentrated solutions with which sedimentation experiments are seldom concerned. Equation 2.6 may be taken as the definition of f.

The determination of f by gradient diffusion and sedimentation velocity is firmly established on the thermodynamics of irreversible processes. However, it needs time-consuming experiments, and has rapidly lost popularity in recent years. At present, few polymer laboratories in the world have the routine use of a gradient diffusion apparatus or an analytical ultracentrifuge.

2.1.4 Cooperative Diffusion Coefficient

QELS (quasi–elastic light scattering) experiments on concentrated solutions often reveal two dynamic structure factors (see Section 3 of Chapter 4 for the definition) which decay at distinctly different rates when the sampling time for current autocorrelation measurements is varied by orders of magnitude (beautiful evidence can be seen in a paper by Aims and Han [22]). The decays with fast and slow rates are referred to as the fast and slow relaxation modes, respectively. We denote the diffusion coefficient associated with the fast mode by $D(\text{fast})$ and that with the slow mode by $D(\text{slow})$. At low concentrations and small scattering angles, only the slow mode is observed. As c tends to zero, $D(\text{slow})$ converges to $D(\infty)$ defined in Section 3.5 of Chapter 4. Therefore, if a friction coefficient $f(\text{slow})$ is defined by $D(\text{slow}) = k_B T / f(\text{slow})$, it agrees with f_s and f at $c = 0$. In early days, it was concluded that $f(\text{slow})$ may be equated to f_s even at finite

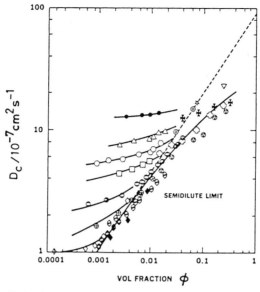

Fig. 7–4. Concentration dependence of the cooperative diffusion coefficient D_c for PS in THF at 25°C. Different marks are for different M and different data sources [25]. The dashed line is drawn with slope 0.65.

dilutions, but later experiments have shown that this conclusion is invalid, i.e., the slow mode has nothing to do with self-diffusion. It appears that the origin of the slow mode in concentrated polymer solutions still remains unknown [23].

It has been found that D(fast) increases with c and becomes independent of M when the solution enters the semi–dilute regime.[1] The data in Figure 7–4, collected by Schaefer and Han [25] from several sources, elegantly illustrate these characteristic properties of D(fast). Usually, D(fast) is termed the **cooperative diffusion coefficient** and denoted by D_c or D_{coop}. The M independence of D_c suggests that the fast mode is associated with the relaxation of local density fluctuations.

Another remarkable fact known experimentally is that D tends to merge with D_c as the solution approaches and enters the semi–dilute regime. This fact explains why QELS has recently replaced the time–consuming gradient method for routine D measurements on polymer solutions, concentrated as well as dilute. However, as far as the author is aware, no theoretical answer is given as yet to the important question of why D becomes indistinguishable from D_c as the solution crosses over the overlap concentration. This frustrating situation exemplifies our

---

[1] The latter behavior has been used to estimate the overlap concentration [24].

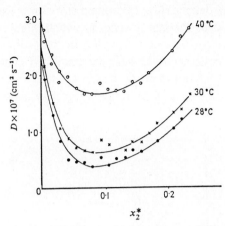

Fig. 7–5. Concentration and temperature dependence of D for PS in CH [26]. x_2^* denotes the mole fraction of styrene residues.

incomplete knowledge about the physics of polymer concentrates.

2.2 Experimental Information

2.2.1 Concentration Dependence of Diffusion Coefficient D

Equation 2.2 shows that D is determined by the interplay of the thermodynamic factor $\partial\Pi/\partial c$ and the friction factor f. In general, the friction factor is expected to increase monotonically with increasing c and decrease with rising temperature. On the other hand, as can be deduced from the known information about osmotic pressure, the thermodynamic factor as a function of c varies in complex ways with solvent quality and temperature. Thus, the concentration dependence of D for a given polymer should exhibit a variety of features depending on solvent conditions.

This prediction is well illustrated with the classic data of Rahage and collaborators [26]. Figure 7–5 shows those for polystyrene in cyclohexane above and below the θ temperature. At the indicated temperatures the system is close to the θ state (some authors call it the pseudo–ideal state), so that $\partial\Pi/\partial c$ stays nearly constant in the region of low concentration. Hence, the initial decline in D as seen in the figure can be attributed to an increase in f with c. The rise in D after a minimum indicates the suppression of a continuous increase in f by a sharp increase in $\partial\Pi/\partial c$.

Figure 7–6 shows the data for polystyrene in a good solvent ethylbenzene (EB). Here, the concentration dependence of D is opposite to that seen in Figure 7–5. The initial rise in D is due to an increase in $\partial\Pi/\partial c$ suppressing an

217

increase in f, while the decline in D after the maximum is ascribable to the fact that the system approaches the glassy state, in which large-scale motion of the chains is so suppressed owing to the decrease in free volume that f increases appreciably. At high concentrations near the undiluted state, D is essentially controlled by f, and the latter depends on the average free volume present in the system [27]. Available data on D in the highly concentrated regime are still scant and fragmentary, however.

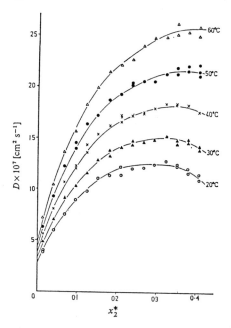

Fig. 7–6. Concentration and temperature dependence of D for PS in EB, a good solvent for PS [26].

In good solvents, $\partial\Pi/\partial c$ is virtually independent of T. Hence, the sharper initial increase in D with c at higher temperature as demonstrated in Figure 7–6 can be ascribed to a significant decrease in f with rising temperature.

For the reason mentioned above, the D_c data of Figure 7–4 show the behavior of D near and in the semi-dilute regime. The dashed line in the figure has been drawn to fit the data points for $0.002 < \phi < 0.03$. Its slope 0.65 is compared with the blob theory in Section 2.3.

2.2.2 Concentration Dependence of Sedimentation Coefficient s

Already in Figure 6–2 we have shown the data of Vidakovic et al. [28] for $s(c)$ of polystyrene in cyclohexane at θ. Interestingly, they do not agree with the data of Roots and Nyström [29] for polystyrene in another θ solvent cyclopentane

218

(20°C) depicted in Figure 7-7. Differing from the finding of Vidakovic et al., the latter give a system of $\log s$ vs. $\log c$ curves which asymptotically merge not to a straight line but a curve bent downward, whose slope changes continuously from −1.0 to −2.0 as c increases. It should be noted that Roots and Nyström measured s over a wider concentration range than did Vidakovic et al.

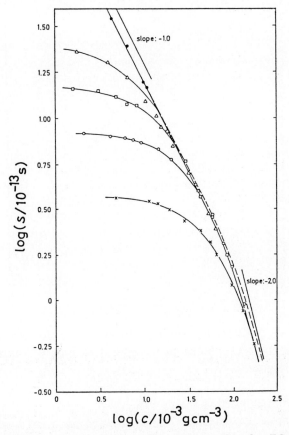

Fig. 7-7. Concentration dependence of s for PS in cyclopentane at $\theta(20°\text{C})$. Solid curves from top to bottom are for increasing molecular weights. Actually, these represent not s but s_v, and the dashed line indicates the former.

Figure 7-8 shows the concentration dependence of s for polystyrene in a good solvent, toluene at 25°C, with the data compiled by Roots and Nyström [29]. The behavior of the data points is similar to that in Figure 7-7, except that the slope of the asymptotic curve varies from −0.6 to −1.0 with increasing c. Roots and Nyström [30] also obtained s data similar to Figure 7-7 for polystyrene in

219

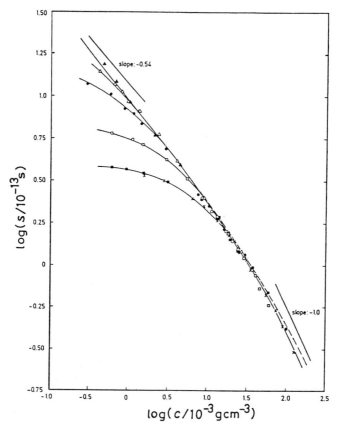

Fig. 7–8. Concentration dependence of s for PS in toluene at 25°C. The same remarks as in Fig. 7–7 apply to this figure.

a marginal solvent methyl ethyl ketone (MEK).

Summarizing these experimental studies, we see that s as a function of c becomes independent of M regardless of solvent quality when the solution enters the semi–dilute regime. This feature suggests that the unit of sedimentation in semi–dilute polymer solutions is not an individual polymer chain but its subchain.

2.2.3 Comparison of Friction Coefficients

It is the fundamental requirement of the thermodynamics of irreversible processes that for a two-component system f evaluated from D and Π agrees with that from s within experimental error. An excellent check of this requirement, described by Nyström and Roots [31], is reproduced in Figure 7–9. However,

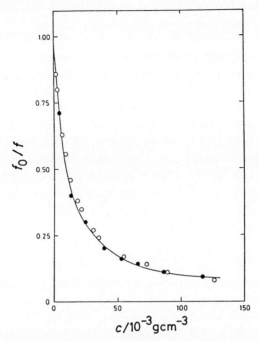

Fig. 7-9. Agreement of f from D and Π (•) and from s (○) for PS ($M = 1.1 \times 10^5$) in toluene. f_0 is the value of f extrapolated to infinite dilution.

Nyström and Roots [31] found for poly(ethylene oxide) in water that $1/f$ from D and Π decreased less rapidly with c than does $1/f$ from s. They attributed the discrepancy to a polydispersity of the polymer sample used.

2.2.4 Remarks

1. We now have ample evidence that s becomes independent of M when the solution enters the semi–dilute regime. Hence, with eq 2.6, we can conclude that $f \sim M$ in this regime. As mentioned in Section 1, Π in the same regime does not depend on M. Thus we can conclude with eq 2.2 that D too should become independent of M in semi–dilute solutions. This conclusion agrees with what we see in Figure 7–4.

2. The predicted proportionality between f and M in the semi–dilute regime differs from the finding from recent self–diffusion measurements (see Chapter 8) that the M dependence of f_s in the same regime is stronger than linear. This difference indicates that f in semi–dilute solutions is not associated with the Brownian motion of a single chain as a whole.

3. Unless a relation between f and f_s is established, gradient diffusion on a polymer + solvent binary solution is of no use for measurement of f_s. However, if applied to a labeled polymer dispersed very dilutely in an unlabeled polymer + solvent solution, it allows us to determine f_s. The experiment may be done as follows. We prepare a pair of solutions containing an unlabeled polymer at a desired concentration, disperse in one of them a labeled species very dilutely, allow the two solutions to contact at a sharp boundary in a diffusion cell, and then measure the resulting change in the concentration distribution of the labeled species as a function of time by an optical method. The data obtained gives D of the labeled species. The value of f calculated from this D (in this case, the correction for the thermodynamic factor is not needed because the concentration of the labeled species is very low) may be equated to f_s of the labeled species, because f and f_s agree at infinite dilution. This idea is the basis of the tracer diffusion method for self–diffusion.

2.3 Blob Theory

As illustrated above, s vs. c curves for homologous polymers with different M merge to a single line above the overlap concentrations characteristic of respective M. This behavior is not a recent finding but has long been noticed by ultracentrifuge workers [32]. They considered that, in concentrated solutions, polymer chains entangle with one another and hence sediment not individually but as a porous plug (or a sponge) when undergoing a centrifugal force. If this is the case, the sedimentation coefficient s should depend on the rate at which solvent molecules flow back through the porous plug by a pressure gradient Δp set up in the ultracentrifuge cell. Applying the thermodynamics of irreversible processes to this flow, Mijnlieff and Jaspers [33] derived

$$s = [1 - (v_{\mathrm{p}}/v_0)]c(1 - v_{\mathrm{p}}c)P/\eta_0 \qquad (2.7)$$

where v_0 is the partial specific volume of the solvent, η_0 the viscosity of the solvent, and P the permeability defined by Darcy's law:

$$\Delta p = -(\eta_0/P)V \qquad (2.8)$$

with V being the velocity of the solvent. Equation 2.7 simply converts the consideration of $s(c)$ to that of P as a function of c.

In Section 1, we have treated de Gennes' blob as a thermodynamically independent unit to formulate Π of semi–dilute solutions. Brochard and de Gennes [34] (see also Pouyet and Dayantis [35]) assumed that the blob also acts as a hydrodynamically independent unit and applied Stokes' law of resistance to it. Perhaps their idea may be more adequate when applied to the blob defined in

222

terms of the hydrodynamic screening length ξ_H (see Section 3.3 of Chapter 6). Another assumption in the Brochard–de Gennes theory is to model the porous plug by an assembly of blobs of radius ξ. It allows us to put the resistance experienced by the back flow of solvent from the porous plug proportional to $\eta_0 \xi V \times \xi^{-3}$ per unit volume of solution. Here, ξ^{-3} is proportional to the number density of blobs. In the steady state, this resistance balances Δp, leading to

$$\eta_0 \xi^{-2} = C \Delta p \qquad (2.9)$$

where C is a constant. Comparing with eq 2.8, we find

$$P \sim \xi^2 \qquad (2.10)$$

Hence eq 2.7 gives

$$s \sim c \xi^2 \qquad (2.11)$$

where $v_p c$ is neglected in comparison with unity.

Substituting eq 1.15 for ξ, we obtain from eq 2.11

$$s \sim c^{-(1-\nu)/(3\nu-1)} \qquad (2.12)$$

which predicts

$$s \sim c^{-0.50} \quad \text{(good solvents)} \qquad (2.13)$$
$$s \sim c^{-1.0} \quad (\theta \text{ solvents}) \qquad (2.14)$$

We note that the prefactors in these relations are independent of M and that the exponent to c in eq 2.13 is -0.54 if the "best" value of 0.588 is used for ν in place of Flory's 0.60.

Eliminating f from eq 2.2 and 2.6 and substituting eq 2.12 and $\Pi \sim c^{3\nu/(3\nu-1)}$ (obtained from eq 1.14 and 1.15), we obtain the following prediction for D in the semi–dilute regime:

$$D \sim c^{\nu/(3\nu-1)} \qquad (2.15)$$

where the prefactor is independent of M. This relation leads to

$$D \sim c^{0.75} \quad \text{(good solvents)} \qquad (2.16)$$
$$D \sim c^{1.0} \quad (\theta \text{ solvents}) \qquad (2.17)$$

223

2.4 Comparison with Experiment

2.4.1 Sedimentation Coefficient

Equation 2.14 may be compared favorably with the data of Vidakovic et al. already quoted in Figure 6-2. These authors gave -0.96 for the slope of the asymptote to $\log s$ vs. $\log c$ curves for different M. However, the data of Figure 7-7 do not conform to eq 2.14, giving a curved asymptote whose slope changes from -1.0 to -2.0 with increasing c. Equation 2.13 also fails to explain the data of Figure 7-8, which give a curved asymptote whose slope changes from -0.60 to -1.0.

2.4.2 Diffusion Coefficient

The solid curve fitting semi-dilute solution data in Figure 7-4 represents the asymptote to $D(c)$ for polystyrene + tetrahydrofuran, and is adequate for comparison with eq 2.16, since THF is a good solvent for polystyrene. First of all, it is not a straight line, thus inconsistent with the theoretical prediction. Its slope decreases from 0.65 to about 0.40 as the volume fraction ϕ changes from 0.001 to 0.20. Even the maximum slope 0.65 is significantly lower than 0.75 predicted by eq 2.16. The asymptote to D for polystyrene in a marginal solvent ethyl acetate [25] displays behavior similar to that for the same polymer in tetrahydrofuran. Its slope changes from 0.65 to about 0.35 with increasing ϕ. Nyström and Roots [31] showed that D data for polystyrene in toluene from different sources are consistent with one another on a log–log graph and can be fitted by straight lines over the semi–dilute range of c. However, the slopes of the lines were all close to 0.50, thus differing more markedly from the predicted 0.75, though Nyström and Roots [36, 31] themselves found a slope of 0.70 for the same system. Munch et al. [37] reported a slope of 0.68 for polystyrene in benzene, which closely agrees with 0.67 obtained independently by Adam and Delsanti [38]. However, according to the review of Schaefer and Han [25], the data of Munch et al. [37] for poly(dimethylsiloxane) in toluene are consistent with eq 2.16. The situation is indeed quite confusing.

Since D in the semi–dilute regime does not depend on M, its concentration dependence for monodisperse polymers and that for their blends should become coincident in this regime. Recent work of Nemoto et al. [39] on polystyrene in benzene confirmed the prediction. In fact, their D_c data for monodisperse and binary blended samples fell on a single line with slope 0.67 on a log–log graph. This slope value agrees with that obtained by Adam and Delsanti [38] for monodisperse samples.

According to Schaefer and Han [25], D_c data by QELS under the θ condition are still scanty. Roots and Nyström [40] found that their gradient diffusion data on polystyrene in cyclohexane at θ agree with eq 2.17 for $M = 8.6 \times 10^5$ but not for $M = 3.9 \times 10^5$.

Summarizing, we may conclude that reported sedimentation and diffusion data on semi–dilute solutions do not always lend support to the predictions of the Brochard–de Gennes blob theory. Certainly, this theory oversimplifies the complex hydrodynamic interactions involved in many–chain systems. Nonetheless we should not underestimate its merit that has sparked off many sedimentation and diffusion measurements on concentrated polymer solutions in recent years.

2.5   Viscosity

The (steady–state) viscosity η is one of the most practically important transport properties of polymer solutions. From the theoretical point of view it is one of the toughest objects in current polymer research. In an excellent book published in 1982, Bohdanecký and Kovář summarized [40] and discussed a number of contributions made by the end of the 1970s to η of both dilute and concentrated polymer solutions. This book gives a good description of the molecular mechanisms envisaged at that time to explain the molecular weight and concentration effects on η of polymer concentrates. Though many reports on η have since appeared, it does not seem that our understanding of these effects in terms of molecular concepts has made substantial progress for the past decade. In this connection, it would be interesting to ask why the blob concept has been of little use for the viscosity problem. In Chapter 8, we touch upon some recent ideas for approaching the molecular weight dependence of η.

3.   Other Theories

3.1   Scaling Theory

In Section 1.4, we explained that the blob concept enables us to derive expressions for the concentration dependence of static properties of concentrated polymer solutions. In the following, we show that the same results can be derived by a simple argument called the scaling theory. First, we consider $\langle S^2 \rangle(c)$, which, as before, denotes the mean–square radius of gyration of a test polymer (modeled by a spring–bead chain) in a solution of concentration c. Basic assumptions of the scaling theory are that a dimensionless factor H_S defined by

$$\langle S^2 \rangle(c) = \langle S^2 \rangle(0) H_S \tag{3.1}$$

depends on the number of beads per chain, N, and the "effective" volume occupied by the beads in unit volume of solution, ϕ, and that ϕ is proportional to na^3. Here, n is the number density of beads and a the mean spring length. Thus, the scaling theory assumes $H_S = H_S(N, na^3)$.

We group μ beads into one. This operation changes N, n, and a to N/μ, n/μ, and $ag(\mu)$, respectively. If N is large enough, we may write $\langle S^2 \rangle(0) \sim a^2 N^{2\nu}$, with ν defined by eq 1.4. Hence, $\langle S^2 \rangle(0)$ changes to $\sim a^2[g(\mu)]^2(N/\mu)^{2\nu}$ after the grouping. Being a global property, $\langle S^2 \rangle(c)$ should be independent of how many beads are grouped into one, i.e., the extent to which the chain is coarse–grained. This means that the following relation must hold regardless of μ:

$$a^2 N^{2\nu} H_S(N, na^3) = a^2[g(\mu)]^2 (N/\mu)^{2\nu} H_S(N/\mu, (n/\mu)a^3[g(\mu)]^3) \quad (3.2)$$

This relation is satisfied if

$$g(\mu) = \mu^\nu \quad (3.3)$$

and

$$H_S(x, y) = H_S(yx^{3\nu-1}) \quad (3.4)$$

Thus, we arrive at

$$\langle S^2 \rangle(c) = \langle S^2 \rangle(0) H_S(na^3 N^{3\nu-1}) \quad (3.5)$$

If eq 6–1.5 is used, this may be rewritten

$$\langle S^2 \rangle(c) = \langle S^2 \rangle(0) H_S(c/c^\star{}_0) \quad (3.6)$$

which shows that the statistical radius of a single chain as a function of c is scaled by a reduced concentration $c/c^\star{}_0$.

For concentrations at which $\langle S^2 \rangle(c)$ comes close to $\langle S^2 \rangle_\theta \sim N$ owing to the screening effect, eq 3.6 approaches

$$N^{-(2\nu-1)} \sim H_S(c/c^\star{}_0) \quad (3.7)$$

which, when eq 6–1.5 is substituted, allows us to find

$$H_S(x) \sim x^{-(2\nu-1)/(3\nu-1)} \quad (3.8)$$

Hence, we obtain an asymptotic relation:

$$\langle S^2 \rangle(c) \sim \langle S^2 \rangle(0)(c/c^\star{}_0)^{-(2\nu-1)/(3\nu-1)} \quad (3.9)$$

which is the same as eq 1.21 derived from the blob theory.

Using a similar argument and imposing the requirement that the osmotic pressure and the correlation length be independent of chain coarse–graining, we can derive the following scaling laws:

$$\Pi_r = M^{-1} H_\pi(c/c^\star{}_0) \to c^{1/(3\nu-1)} \quad (3.10)$$

$$\xi^2 = \langle S^2 \rangle(0) H_\xi(c/c^\star{}_0) \rightarrow c^{-2\nu/(3\nu-1)} \tag{3.11}$$

where $\rightarrow$ indicates the approach to asymptotic behavior valid for solutions in which Π_r and ξ are independent of N. These asymptotic formulas agree with those obtained from eq 1.14 and 1.15.

The SAXS data of Hamada et al. shown in Figure 6–5 clearly illustrate that, as eq 3.11 predicts, $\xi/[\langle S^2 \rangle(0)]^{1/2}$ is scaled by the reduced concentration $c/c^\star{}_0$. Similar scaling behavior can be seen in the SANS data of Wiltzius et al. [17].

Kosmas and Freed [41] presented another approach to scaling laws. Differing from the theory described above, it starts from the partition function for a solution of continuous chains which interact subject to the binary cluster approximation. For example, their theory derives for osmotic pressure (in three dimensions) a general scaling law which, in our notation, may be written

$$M\Pi_r = H_\pi(nb^3 N^{1/2}, \beta_c b^{-3} N^{1/2}) \tag{3.12}$$

where β_c is the excluded–volume strength defined by eq 1–3.19 and b the Kuhn segment length defined by eq 1–3.14. In agreement with the theory of des Cloizeaux [3], Kosmas and Freed introduced the hypothesis that H_π depends only on $c/c^\star{}_0$. Unless this hypothesis can be deduced from the Hamiltonian that they assumed for the solution, their theory is not self–sufficient. Anyway, the hypothesis allows us to deduce that H_π depends on a single variable $nb^3 N^{1/2}(\beta_c b^{-3} N^{1/2})^{3(2\nu-1)}$ formed by combining its two variables $nb^3 N^{1/2}$ and $\beta_c b^{-3} N^{1/2}$. Thus we have

$$M\Pi_r = H_\pi(nN^{3\nu-1}\beta_c{}^{3(2\nu-1)}b^{6(2-3\nu)}) \tag{3.13}$$

If, as before, we require that Π_r becomes independent of N at high concentrations we can derive from eq 3.13

$$\Pi_r \sim c^{1/(3\nu-1)}\beta_c{}^{3(2\nu-1)/(3\nu-1)}b^{6(2-3\nu)/(3\nu-1)} \tag{3.14}$$

as an asymptotic relation valid for $c/c^\star{}_0 \gg 1$. For Flory's ν value of 0.6 this gives

$$\Pi_r \sim c^{5/4}\beta_c{}^{3/4}b^{3/2} \tag{3.15}$$

which predicts the same concentration dependence as eq 3.10. Using the same hypothesis as mentioned above, Kosmas and Freed derived the following asymptotic relations:

$$\langle S^2 \rangle(c) \sim c^{-1/4}\beta_c{}^{1/4}b^{1/2} \tag{3.16}$$

$$\xi \sim c^{-3/4}\beta_c{}^{-1/4}b^{-1/2} \tag{3.17}$$

These agree with eq 3.9 and 3.11 for $\nu = 0.6$, respectively, as far as the concentration dependence is concerned. What is new in the Kosmas–Freed theory is that it predicts the dependence on b and β_c as well as on c.

3.2 Theory of Muthukumar and Edwards

Either the blob theory or the scaling theory predicts only in proportionality form the dependence of various static properties of a polymer solution on concentration and molecular weight. Naturally it is more desirable to have a theory which is capable of predicting the prefactors in the proportionality relations. Efforts toward such a theory have been made notably by Edwards and collaborators, starting from Edwards' paper [42] in 1966, but the theories reported so far leave much to be desired.

Here, omitting the mathematical details, we sketch a relatively new theory due to Muthukumar and Edwards [43]. The M–E theory considers a uniform solution consisting of a pure solvent and the Edwards continuous chains with a Kuhn segment length b. It is formulated on the assumption that the Hamiltonian H of the solution is given by

$$
\frac{H}{k_B T} = \frac{3}{2b} \sum_{i=1}^{\mathcal{N}} \int_0^L \left[\frac{\partial \mathbf{r}(s_i)}{\partial s_i} \right]^2 ds_i
$$

$$
+ \frac{\beta_c}{2b^2} \sum_{i=1}^{\mathcal{N}} \sum_{j=1}^{\mathcal{N}} \int_0^L \int_0^L \delta \left[\mathbf{r}(s_i) - \mathbf{r}(s_j) \right] ds_i \, ds_j \qquad (3.18)
$$

where $\mathcal{N}$ is the number of chains present in the solution, s_i the contour length along chain i, and all other symbols the same as those defined in Section 3 of Chapter 1. In fact, eq 3.18 is a generalization of eq 1–3.23 to a many–chain system, though the cut–off is not introduced here.[2] Muthukumar and Edwards introduced an effective (or "renormalized") Kuhn segment length b_1 defined by

$$
\langle R^2 \rangle (c) = L b_1 \qquad (3.19)
$$

where the left–hand side represents the mean–square end distance of a test chain at concentration c. The parameter b_1 absorbs all effects on $\langle R^2 \rangle (c)$ from inter– and intrachain interactions. The intrachain interaction is affected by the presence of other chains (the screening effect). According to Muthukumar and Edwards, the modified intrachain interaction is described by a "renormalized" excluded–volume strength $\Delta(\mathbf{r})$, which is represented by

$$
\Delta(\mathbf{r}) = \beta_c [\delta(\mathbf{r}) - (4\pi \xi^2 r)^{-1} \exp(-r/\xi)] \quad (r = |\mathbf{r}|) \qquad (3.20)
$$

---

[2] The Kosmas–Freed theory mentioned above also starts with eq 3.18.

where ξ is the correlation length. They showed that b_1 and ξ are coupled by integral equations. In general, these quantities are functions of L, c, β_c, b, and k, where k is the Fourier conjugate variable to the position variable which labels the chain mode. In the limit $L \to \infty$, if b_1 and ξ are independent of k, the coupled integral equations are replaced by coupled algebraic equations as

$$\frac{1}{\xi^2} = \frac{6wn}{(\frac{b_1}{b})^2 + \frac{27w\xi}{8\pi b^2}} \left(\frac{b_1}{b}\right) \tag{3.21}$$

$$\left(\frac{b_1}{b}\right)^3 - \left(\frac{b_1}{b}\right)^2 = \alpha \left(\frac{w\xi}{b^2}\right) \tag{3.22}$$

where α is a numerical factor of the order unity that is left as an adjustable parameter in the M–E theory, n the number density of monomer units, and $w = \beta_c/b^2$.

Since $c^* \to 0$ for $L \to \infty$, the system for which eq 3.21 and 3.22 hold lacks the dilute regime. If $n \sim 0$, it corresponds to ideally semi–dilute solutions, and these equations lead to

$$b_1 = \left(\frac{9\alpha}{16\pi}\right)^{1/4} \beta_c^{1/4} n^{-1/4} b^{-1/2} \tag{3.23}$$

$$\xi = \left(\frac{9}{16\pi\alpha^{1/3}}\right)^{3/4} \beta_c^{-1/4} n^{-3/4} b^{-1/2} \tag{3.24}$$

Interestingly, eq 3.24 agrees with eq 3.17 as far as the concentration dependence is concerned.

If $n \gg 1$, eq 3.21 and 3.22 give

$$b_1 = b + \frac{\alpha\beta_c^{1/2}}{6^{1/2}b^2 n^{1/2}} + \cdots \tag{3.25}$$

$$\xi = \frac{b}{(6\beta_c n)^{1/2}} \tag{3.26}$$

Hence, as the concentration increases, b_1 approaches b (which means that the chain dimensions tend to the values for a Gaussian chain) and ξ diminishes to zero in proportion to $c^{-1/2}$ (which was obtained earlier by Edwards [42]).

Though limited to sufficiently long chains, eq 3.21 and 3.22 can be used to evaluate β_c from experimental determinations of ξ and b_1/b, provided that α is known. To a good approximation we may equate b_1/b to $\langle S^2 \rangle(c)/\langle S^2 \rangle_\theta$. Hence both ξ and b_1/b as functions of c can be evaluated by SANS. Muthukumar and

Edwards [43] deduced that α can be related to $(b_1/b)_0$, the value of b_1/b at infinite dilution, by

$$\alpha = \frac{2(6\pi)^{1/2}b^{7/2}}{\beta_c L^{1/2}}\left[\left(\frac{b_1}{b}\right)_0^{5/2} - \left(\frac{b_1}{b}\right)_0^{3/2}\right] \tag{3.27}$$

With the approximation $(b_1/b)_0 = \langle S^2\rangle(0)/\langle S^2\rangle_\theta = \alpha_S^2$ and the aid of eq 2–1.12, this equation may be rewritten

$$\alpha = (9/\pi z)(\alpha_S^5 - \alpha_S^3) \tag{3.28}$$

where z is the excluded–volume variable introduced in Chapter 2. If Flory's original equation, eq 2–1.15, is used (with α_S is approximated by α_R), eq 3.28 gives $\alpha = 7.7$.

King et al. [13] analyzed their SANS data for polystyrene in toluene by the above method and found the calculated β_c at different concentrations to stay approximately constant.

Appendix

Notes on $c^\star{}_0$

Usually, $[\eta]$ is expressed as

$$[\eta] = 6^{3/2}\Phi[\langle S^2\rangle(0)]^{3/2}/M \qquad (7A\text{-}1)$$

and Φ is called the Flory viscosity factor. Under the θ condition, Φ agrees with Φ_θ defined by eq 2-3.19 and does not depend on M. In good solvents, this factor decreases slowly with increasing M, but, to a first approximation, it may be regarded as a constant whose value is about 2.5×10^{23} (mol^{-1}) according to well-documented experiments. Then, with eq 7A-1, eq 6-1.4 yields

$$c^\star{}_0 = \frac{3 \times 6^{3/2} \times 2.5 \times 10^{23}}{4\pi N_A [\eta]} = \frac{1.46}{[\eta]}$$

which gives eq 1.5. The numerical factor 1.46 should not be taken seriously, because it varies with the definition of $c^\star{}_0$.

If we combine eq 6-1.4 with eq 2-2.5, we obtain

$$c^\star{}_0 = \frac{3\pi^{1/2}\Psi}{A_2 M} \qquad (7A\text{-}2)$$

In good solvents, the penetration function Ψ is nearly independent of M unless the chain is too short, and roughly equals 0.22. Then eq 7A-2 gives

$$c^\star{}_0 = 1.17/(A_2 M) \qquad (7A\text{-}3)$$

This relation is also sometimes used for calculating the reduced concentration $c/c^\star{}_0$. Again, the factor 1.17 should not be taken seriously.

References

1. P. J. Flory, "Principles of Polymer Chemistry," Cornell Univ. Press, Ithaca, 1953.
2. I. Prigogine, "The Molecular Theory of Solutions," North Holland, Amsterdam, 1957; P. J. Flory, Disc. Farad. Soc. **49**, 7 (1970).
3. J. des Cloizeaux, J. Phys. (Paris) **36**, 281 (1975).
4. I. Noda, N. Kato, T. Kitano, and M. Nagasawa, Macromolecules **14**, 668 (1981).
5. L. Schäfer, Macromolecules **15**, 652 (1982).
6. T. G Fox, J. B. Kinsinger, H. F. Mason, and E. M. Schuele, Polymer **3**, 71 (1962).
7. N. Kuwahara, T. Okazawa, and M. Kaneko, J. Polym. Sci., Part C **23**, 543 (1968); J. Chem. Phys. **47**, 3357 (1967).
8. A. Knoll, L. Schäfer, and T. A. Witten, J. Phys. (Paris) **42**, 767 (1981).
9. G. V. Schulz and W. H. Stockmayer, Makromol. Chem. **116**, 25 (1986).
10. P.-G. de Gennes, "Scaling Concepts in Polymer Physics," Cornell Univ. Press, Ithaca, 1979.
11. M. Daoud, J. P. Cotton, B. Farnoux, G. Jannink, G. Serma, H. Benoit, R. Duplessix, C. Picot, and P.-G. de Gennes, Macromolecules **8**, 804 (1975).
12. F. Hamada, S. Kinugasa, H. Hayashi, and A. Nakajima, Macromolecules **18**, 2290 (1985).
13. J. S. King, W. Boyer, G. D. Wignall, and R. Ullman, Macromolecules **18**, 709 (1985).
14. J. P. Cotton, B. Farnoux, and G. Jannink, J. Chem. Phys. **57**, 290 (1972).
15. H. Benoit and C. Picot, Pure & Appl. Chem. **12**, 545 (1966).
16. T. Ohta and Y. Oono, Phys. Lett. A **89**, 460 (1982).
17. P. Wiltzius, H. R. Haller, D. S. Cannell, and D. W. Schaefer, Phys. Rev. Lett. **51**, 1183 (1983).
18. J. D. Ferry, "Viscoelastic Properties of Polymers," Wiley, New York, 1980.
19. D. D. Fitts, "Thermodynamics of Irreversible Processes," McGraw–Hill, New York, 1962.
20. A. R. Altenberger and M. Tirrell, J. Polym. Sci., Polym. Phys. Ed. **22**, 909 (1984).
21. H. Fujita, "Foundations of Ultracentrifugal Analysis," Wiley, New York, 1975.
22. E. J. Amis and C. C. Han, Polym. Commun. **21**, 1403 (1982).
23. S. Balloge and M. Tirrell, Macromolecules **18**, 817 (1985).
24. T. L. Yu, H. Reihanian, and A. M. Jamieson, J. Polym. Sci., Polym. Lett. Ed. **18**, 695 (1980).

25. D. W. Schaefer and C. C. Han, in "Dynamic Light Scattering," Ed. R. Pecora, Plenum, New York, 1985, Chap. 5.
26. G. Rehage and O. Ernst, Koll.-Z. Z. Polym. 197, 64 (1964); G. Rehage, O. Ernst, and J. Fuhrmann, Disc. Farad. Soc. 49, 208 (1970).
27. H. Fujita, Adv. Polym. Sci. 3, 1 (1961).
28. P. V. Vidakovic, C. Allain, and F. Rondelez, J. Phys. Lett. 42, L-323 (1981).
29. J. Roots and B. Nyström, J. Poly. Sci., Polym. Phys. Ed. 19, 479 (1981).
30. J. Roots and B. Nyström, J. Chem. Soc. Farad. Trans. I 77, 947 (1981).
31. B. Nyström and J. Roots, J. Macromol. Sci.–Rev. Macromol. Chem. C19(1), 35 (1980).
32. T. Svedberg and K. O. Pedersen, "The Ultracentrifuge," Clarendon Press, Oxford, 1940, Part IV, Chap. B.
33. P. F. Mijnlieff and W. J. M. Jaspers, Trans. Farad. Soc. 67, 1839 (1971).
34. F. Brochard and P.-G. de Gennes, Macromolecules 10, 1157 (1977).
35. G. Pouyet and J. Dayantis, Macromolecules 12, 293 (1979).
36. J. Roots and B. Nyström, Macromolecules 13, 1595 (1980).
37. J. P. Munch, P. L. Lemarichal, and S. Candau, J. Phys. (Paris) 41, 519 (1980).
38. M. Adam and M. Delsanti, Macromolecules 10, 1229 (1977).
39. N. Nemoto, Y. Makita, Y. Tsunashima, and M. Kurata, Macromolecules 17, 2619 (1984).
40. M. Bohdanecky and J. Kovář, "Viscosity of Polymer Solutions," Elsevier Science, Amsterdam, 1982.
41. M. K. Kosmas and K. F. Freed, J. Chem. Phys. 69, 3647 (1978).
42. S. F. Edwards, Proc. Phys. Soc. (London) 88, 265 (1966).
43. M. Muthukumar and S. F. Edwards, J. Chem. Phys. 76, 2720(1982); Polymer 23, 345 (1982).

Chapter 8   Polymer Self–Diffusion

1.   Definitions and Basic Relations

1.1   Introduction

Bulk polymer properties such as viscosity and elasticity are concerned with averaged responses of an assembly of polymer chains to external stimuli. On the other hand, the self–diffusion coefficient D_s has something to do with the average speed of translation of the centers–of–mass of individual chains. Thus its study should give us a clue to the clarification of the modes of Brownian motion of a single chain on long timescales. This expectation must have been in the mind of polymer workers for many years, but, except in dilute solutions, few measurements of D_s were undertaken until recently, probably on the one hand because of experimental difficulties and on the other because of the lack of an adequate guiding theory.

The situation has changed dramatically since de Gennes [1] proposed a novel idea in 1971. He advocated that when the concentration of a polymer solution is high enough to produce dense chain entanglements, each chain is forced to wriggle, on a long timescale, in an anisotropic (curvilinear) mode that he named "reptation." For reasons mentioned later the idea of reptation soon fascinated workers who were concerned with the dynamic behavior of polymer concentrates, and high research activity on polymer self–diffusion has evolved. Worthy of special mention is the phenomenal progress that has been made in measuring techniques for D_s in order to test various theoretical predictions.

In an excellent review article, Tirrell [2] summarized and discussed most theoretical and experimental contributions made up to 1984 to polymer self–diffusion in concentrated solutions and melts. Although his conclusion seemed to lean toward the reptation theory, the data then available were apparently not sufficient to support it with sheer certainty. Over the past few years further data on self–diffusion and tracer diffusion coefficients (see Section 1.3 for the latter) have become available and various ideas for interpreting them have been set out. Nonetheless, there is yet no established agreement as to the long timescale Brownian motion of polymer chains in concentrated systems. Some prefer reptation and others advocate essentially isotropic motion. Unfortunately, we are unable to see the chain motion directly. In what follows, we review current challenges to this controversial problem by referring to the experimental data which the author believes are of basic importance.

1.2   Bead Friction Coefficient and Topological Factor

The friction coefficient f_s introduced in eq 7–2.4 is the force required to make a polymer molecule translate as a whole at unit velocity relative to its immediate environment which consists of other chains and coexisting solvent molecules. We model the probe polymer by a spring–bead chain made of N identical beads and define the bead friction coefficient ζ as the average force needed to pull one isolated bead at unit velocity in the solution considered. Then the factor F_s defined by

$$f_s = N\zeta F_s \tag{1.1}$$

combines all effects due to the chain connectivity of beads and to dynamic interactions of the probe chain with the surrounding chains. In what follows, we call it the topological factor [2]. In terms of F_s, eq 7–2.4 can be written

$$D_s = \frac{k_B T}{N\zeta F_s} \tag{1.2}$$

Usually, ζ is assumed to depend on polymer mass concentration c and temperature T, unless the chain is too short. It can be adequately formulated by free volume theory [3]. On the other hand, F_s is assumed to be a function of c and N (or molecular weight M). The central theme in polymer self–diffusion studies is to evaluate the latter by theory or experiment for a variety of chain architectures and solvent conditions.

1.3 Tracer Diffusion

When we referred above to self–diffusion, we tacitly meant the random motion of the center of mass of a probe polymer in a strictly binary solution made up of a monodisperse polymer and a pure solvent. However, it has been recognized as useful to examine polymer self–diffusion in more complex systems as well as in binary solutions. One example is quasi–binary solutions which consist of a solvent and two homologous monodisperse polymers differing in chain length. The other is ternary solutions consisting of a solvent and two chemically different monodisperse polymers. Usually, one of the two polymers is chosen as the probe and dispersed at a very low concentration in solutions of the other polymer. Thus it is called the tracer component. The other polymer is usually called the matrix component, even when its concentration is not high enough to form a pseudo–network. Though somewhat confusing, the self–diffusion coefficient of the tracer component is named the **tracer diffusion cocfficient** and is designated by D_{tr}, while the term self–diffusion coefficient and the notation D_s are reserved for the diffusion of the probe polymer in strictly binary solutions. In general, D_{tr} is treated as a function of c_p, T, N, and P, since the probe concentration is chosen to be very low. Here, c_p and P denote the concentration and chain length of the matrix component, respectively. For special quasi–binary solutions in which $P = N$, D_{tr} reduces to D_s.

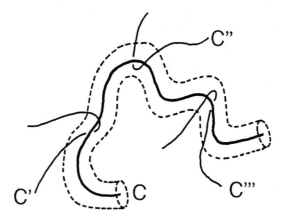

Fig. 8-1. Probe chain C is forced to slide along the centroid of a tube (shown by a dashed line) which visualizes dynamic effects of entangled chains C', C", ...

2. Theoretical Aspects

2.1 Reptation

Since it emerged, the idea of reptation motion has rapidly gained great popularity in the polymer community. In fact, it makes testable predictions about a wide range of dynamic properties of polymer concentrates, and also the reptation motion is appealing to physical intuition. Details of the reptation theory and its applications can be found in the book of Doi and Edwards [4].

Polymer chains are uncrossable with one another. Hence, in concentrated solutions where polymer chains are entangled as sketched in Figure 8-1, the lateral displacement of each chain would be largely impeded by the neighboring chains. This effect due to chain uncrossability is called the **topological constraint** of entangled chains. If it is strong enough, the thermal motion of the chain would be dominantly restricted to wriggling in the direction of the chain contour. Thus, there emerged an idea [1] which replaces the chain by a wire trapped in a curvilinear tube fixed in space and only allowed to slide back and forth along the centroid of the tube. Edwards [5] called this wire the primitive chain or path, and de Gennes [1] gave its slithering motion the name **reptation**. In what follows, we use the terms reptation model and tube model synonymously to describe a chain forced to move reptatively.

2.2 Predictions of the Doi–Edwards Theory

2.2.1 *Displacement of the Center of Mass*

The tube mentioned above is a conceptual thing which represents the topological contraints exerted on a probe chain by its surroundings. We have to consider that it is not extended beyond the ends of the primitive chain (see Figure 8–1). Thus, when escaping out of the tube, the end of the primitive chain receives no topological constraint so that it can take any direction with equal probability. Thanks to this mechanism, the primitive chain can change the conformation at its ends, and hence its conformation successively departs from the initial one as the chain repeats reptation. The process is illustrated in Figure 8–2. We may view it as the disengagement of the chain from its initial tube.

Pure reptation is possible only under the very strong topological constraint that puts the instantaneous orientations of all segments of a primitive chain, except ones at its ends, in a complete correlation. In actual entangled systems, since the constraint may not be that strong, the chains are likely to wriggle in modes other than reptation. We have no a priori reason to deny the possibility of such non–reptative chain motions.

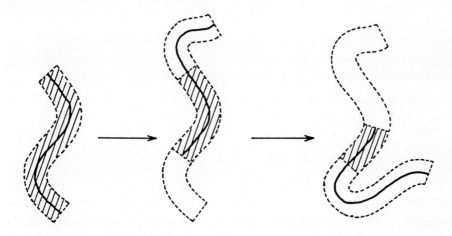

Fig. 8–2. Conformation changes of a primitive chain due to repeated reptation. The chain retains its initial conformation in shaded parts.

For a spring–bead primitive chain consisting of $N(\gg 1)$ springs of length a, we consider the motion of its center–of–mass $\mathbf{r}_G(t)$. Doi and Edwards [6] showed that

$$\langle[\mathbf{r}_G(t+\Delta t)-\mathbf{r}_G(t)]^2\rangle = \langle s(\Delta t)^2\rangle/N \qquad (2.1)$$

237

where $s(\Delta t)$ is the distance slithered by the chain in a time interval Δt and the notation $\langle \cdots \rangle$ signifies the average taken over all chains present in the solution under consideration. According to eq 2.1, while the primitive chain slithers a distance $\xi(\Delta t) \equiv [\langle s(\Delta t)^2 \rangle]^{1/2}$ along its contour, its center-of-mass displaces a distance $\xi(\Delta t)/N^{1/2}$. With the help of this equation it is possible to derive

$$\langle [\mathbf{r}_G(t) - \mathbf{r}_G(0)]^2 \rangle = \frac{[\xi(\Delta t)]^2 t}{N \Delta t} \tag{2.2}$$

Equation 7-2.3 defining the self-diffusion coefficient D_s may be written in terms of $\mathbf{r}_G(r)$ as

$$D_s = (1/6)\langle [\mathbf{r}_G(t) - \mathbf{r}_G(0)]^2 \rangle / t \quad (t \to \infty) \tag{2.3}$$

Substitution of eq 2.2 gives

$$D_s = \frac{[\xi(\Delta t)]^2}{6N \Delta t} \tag{2.4}$$

It can be shown that

$$\langle \mathbf{r}_G(t) - \mathbf{r}_G(0) \rangle = 0 \tag{2.5}$$

which means that the center-of-mass of the system does not move while those of individual chains migrate randomly with time. This is an expected consequence, since we are considering the solution which undergoes no external force.

2.2.2   Molecular Weight Dependence of D_s

Reptation of the primitive chain along the centroid of a curved tube is a curvilinear diffusion process. It can be shown [6] that the diffusion coefficient D_p for this process is defined by

$$D_p = (1/2)[\xi(\Delta t)]^2/\Delta t \tag{2.6}$$

Hence, eq 2.4 may be written

$$D_s = D_p/(3N) \tag{2.7}$$

A new friction coefficient f_p may be introduced by

$$D_p = k_B T/f_p \tag{2.8}$$

Then f_p represents the force required to pull the primitive chain through the tube at unit velocity. Doi and Edwards [6] set it equal to $N\zeta$. This maneuver tacitly assumes that no other chains are dragged when the chain in question is pulled. In other words, the dynamic coupling of entangled chains is neglected. This assumption for f_p and the existence of the tube are the key ingredients of the Doi–Edwards theory. With $f_p = N\zeta$, it follows from eq 2.7 and 2.8 that

$$D_s = k_B T/(3\zeta N^2) \tag{2.9}$$

which predicts that D_s is inversely proportional to N^2. When this equation is compared with eq 1.2, it follows that the topological factor F_s is equal to $3N$. These simple predictions have sparked off many recent experiments on polymer self-diffusion in concentrated solutions and melts.

2.2.3 Changes in Primitive Chain Conformation

A primitive chain successively changes its conformation as it repeats disengagement from the previous tubes. To formulate this process it is convenient to use a continuous chain model. We define $g(s, s'; t)$ by

$$g(s, s'; t) = \langle [\mathbf{r}(s, t) - \mathbf{r}(s', 0)]^2 \rangle \tag{2.10}$$

where s is the contour length from one end of the chain and $\mathbf{r}(s, t)$ the position of a contour point s at time t. Then, $g(s, s; t)$ represents the mean square of the distance traveled by the contour point s in a time interval t. In the continuous chain limit, we have to make D_p infinitely large in such a way that a new diffusion coefficient D_p' defined by

$$D_p' = \lim_{\Delta s \to 0} D_p \Delta s / b \tag{2.11}$$

remains finite, where Δs denotes the contour length of one spring in the spring–bead chain and b the Kuhn segment length (defined by eq 1–3.14).

Following the theory of Doi and Edwards [6], it can be shown that

$$g(s, s'; t) = b|s - s'| + (2bD_p'/L)t$$
$$+ \sum_{k=1}^{\infty} \frac{4bL}{\pi^2 k^2} \cos\left(\frac{\pi k s}{L}\right) \cos\left(\frac{\pi k s'}{L}\right) [1 - \exp(-\lambda_k t)] \tag{2.12}$$

which gives

$$g(s, s; t) = (2bD_p'/L)t$$
$$+ \sum_{k=1}^{\infty} \frac{4bL}{\pi^2 k^2} \cos^2\left(\frac{\pi k s}{L}\right) [1 - \exp(-\lambda_k t)] \tag{2.13}$$

where λ_k is given by

$$\lambda_k = k^2/\tau_d \tag{2.14}$$

with a time constant τ_d defined by

$$\tau_d = \frac{L^2}{\pi^2 D_p'} \tag{2.15}$$

If we define $g(t)$ by

$$g(t) = \frac{1}{L} \int_0^L g(s, s; t)\, ds \tag{2.16}$$

it represents the average mean–square displacement of a point on the primitive chain for a time interval t. Substitution from eq 2.13 gives

$$g(t) = \left(\frac{2bD_p'}{L}\right) t + \frac{Lb}{3} - \sum_{k=1}^{\infty} \frac{2Lb}{\pi^2 k^2} \exp\left(-\lambda_k t\right) \tag{2.17}$$

For t much larger than the time constant τ_d, this equation reduces to

$$g(t) = (2bD_p'/L)t + (Lb/3) \tag{2.18}$$

It can be shown that in the continuous chain limit eq 2.7 becomes

$$D_s = \frac{bD_p'}{3L} \tag{2.19}$$

Hence, for $t \gg \tau_d$, we obtain

$$g(t) = 6D_s t \tag{2.20}$$

Comparison of this with eq 2.3 gives the following important conclusion: In the region of time much larger than τ_d, the mean–square displacement of any point on the primitive chain changes with time in exactly the same way as does the mean–square displacement of the chain's center–of–mass.

For t much smaller than τ_d it follows from eq 2.17 that

$$g(t) = \frac{2b(D_p')^{1/2}}{\pi^{1/2}} t^{1/2} + \frac{\pi^2 b(D_p')^2}{3L^3} t^2 + \dots \tag{2.21}$$

Thus, $g(t)$ increases in proportion to $t^{1/2}$ for $t \ll \tau_d$.

Summarizing, we see that the time dependence of $g(t)$ for a reptating primitive chain changes from the $(t)^{1/2}$ type to the t type roughly at the characteristic

240

time τ_d as t increases. The assumption $D_p \sim N^{-1}$ by Doi and Edwards leads to $D_p' \sim L^{-1}$, which is introduced into eq 2.15 to give

$$\tau_d \sim L^3 \tag{2.22}$$

Thus, if the chain is very long, τ_d becomes so large that only the $t^{1/2}$ type of $g(t)$ appears in the timescale accessible to diffusion experiments, and we are unable to measure D_s experimentally.

2.2.4 Disengagement Time

For sufficiently small t the primitive chain would stay in the initial tube except for small parts near its ends. In this state the following relation should hold for any s not close to 0 and L:

$$\langle [\mathbf{r}(s,t) - \mathbf{r}(s,0)]^2 \rangle = b\langle |\Delta(t)| \rangle \tag{2.23}$$

where $\Delta(t)$ is the contour length slithered by any contour point in the time interval t, being positive for forward and negative for backward slithering.[1] Since $\Delta(t)$ is a random variable, its probability density $P(\Delta(t))$ may be assumed to be one–dimensional Gaussian. Then we find

$$\langle |\Delta(t)| \rangle = \{2\langle \Delta(t)^2 \rangle / \pi\}^{1/2} \tag{2.24}$$

We may equate $\langle \Delta(t)^2 \rangle$ and t to $[\xi(\Delta t)]^2$ and Δt in eq 2.6, respectively, if D_p is replaced by D_p'. Thus

$$\langle \Delta(t)^2 \rangle = 2D_p't \tag{2.25}$$

With eq 2.24 and 2.25, eq 2.23 can be written

$$g(s,s;t) = \frac{2b(D_p')^{1/2}}{\pi^{1/2}} t^{1/2} \tag{2.26}$$

which is introduced into eq 2.16 to give

$$g(t) = \frac{2b(D_p')^{1/2}}{\pi^{1/2}} t^{1/2} \tag{2.27}$$

This agrees with the leading term in eq 2.21. Thus, we may consider that while $g(t)$ follows the $t^{1/2}$ type, the primitive chain essentially stays in the initial tube. In other words, for $t > \tau_d$ the primitive chain is surrounded by tubes which are substantially different from the initial one. For this reason τ_d is called the disengagement time. It is also the time at which the chain begins to lose the memory of its initial tube.

---

[1] This relation can be derived from the assumption that the centroid of the tube obeys Gaussian statistics. Though not mentioned explicitly in the above discussion, the Doi–Edwards theory invokes this assumption.

2.3 Tube Renewal

The crucial assumption of the Doi–Edwards theory is that the primitive chain reptates in a tube fixed in space. However, in monodisperse solutions, all chains are wriggling simultaneously, so that the tube around each chain is never fixed but successively renewed by different chains. Hence, the Doi–Edwards theory is not self–consistent. This fact has given rise to recent measurements of the tracer diffusion coefficient as a function of the molecular weight and concentration of the matrix component.

When tube renewal takes place, the topological constraint is weakened, and it should become possible for the chain to move sideways over a distance. This may be pictured as the leakage of the chain out of its confining tube. Lateral chain displacements can occur in a variety of modes other than reptation. However, sticking to the reptation model, Klein [7], Daoud and de Gennes [8] and Graessley [9] formulated the tube renewal and derived expressions which approximately correct the Doi–Edwards theory for this process. For example, Graessley derived the following expression for D_s:

$$D_s = D_s{}^0 \left[1 + \frac{48}{25} z \left(\frac{12}{\pi^2} \right)^{z-1} \frac{n_e{}^2}{N^2} \right] \tag{2.28}$$

where $D_s{}^0$ denotes D_s for the fixed tube (eq 2.9), n_e the average number of beads of a subchain between entanglements, and z a factor whose largest physically allowable value is 5.[2] The second term in the brackets represents the tube renewal effect. It becomes negligible when the primitive chain is entangled with about five other chains.

2.4 Concentration Dependence

We assume (see Ref. [2]) that F_s in the dilute regime is independent of c and that in the semi–dilute regime follows a power law as

$$F_s \sim h(N) c^x \tag{2.29}$$

where $h(N)$ is an unknown function of N and x an unknown numerical parameter. From this assumption it follows that

$$F_{s0} \sim h(N)(c^\star{}_0)^x \tag{2.30}$$

in the approximation of replacing $c^\star$ by $c^\star{}_0$. Here, F_{s0} indicates the value of F_s at infinite dilution. Referring back to eq 2-3.3 for D_0 ($\sim (F_{s0}N)^{-1}$), we may express F_{s0} as

$$F_{s0} \sim N^{\mu-1} \tag{2.31}$$

---

[2] Graessley's theory [9] predicts that eq 2.28 holds for tracer diffusion coefficients from experiments with quasi–binary solutions if the term $1/N^2$ is replaced by N/P^3.

where μ is an empirical parameter; μ is 0.5 in θ solvents and increases as the solvent improves. Substituting eq 2.31 and eq 6-1.5 into eq 2.30 and assuming that $h(N)$ is expressed by a power law as

$$h(N) \sim N^\gamma \tag{2.32}$$

we find that x must equal $(1 + \gamma - \mu)/(3\nu - 1)$. Thus we obtain the scaling law:

$$F_s \sim N^\gamma c^{(1+\gamma-\mu)/(3\nu-1)} \tag{2.33}$$

If the reptation model is valid in the semi–dilute regime, γ can be taken to be unity (see Section 2.2.2), and eq 2.33 yields

$$F_s \sim N c^{(2-\mu)/(3\nu-1)} \tag{2.34}$$

In θ solvents, $\mu = \nu = 0.5$. Hence

$$F_s \sim c^3 \quad (\theta \text{ solvents}) \tag{2.35}$$

On the other hand, in good solvents where $\nu = 0.6$, we obtain

$$F_s \sim c^{1.25(2-\mu)} \quad (\text{good solvents}) \tag{2.36}$$

Usually, μ in good solvents is taken to be 0.6. Then eq 2.36 predicts

$$F_s \sim c^{1.75} \tag{2.37}$$

We note, however, that this choice of μ is established neither theoretically nor experimentally. For example, a well–documented QELS study of Adam and Delsanti [10] on polystyrene in benzene yielded $\mu = 0.55 \pm 0.02$. Equation 2.34 with μ equated to ν was derived by Brochard and de Gennes [11] using the blob concept.

Equation 2.34 assumes the validity of reptation in the semi–dilute regime, but it is unlikely that such an assumption holds in the region near c^*, where chain entanglements are supposed to be sparse and weak.

2.5 Possibility of Rouse–like Motion

One of the features of a free flexible polymer is that any pair of its segments virtually have no orientation correlation unless they are situated very close on the chain. Its segments thus tend to wriggle almost isotropically. In order to have a flexible chain reptate, this tendency of the segments has to be suppressed so that all the segments displace along a curve at the same speed. A naive question is whether the constraint due to chain uncrossability is strong enough to induce such highly correlated motion.

Subchains between neighboring entanglements are as long as the order of a hundred repeating units even in undiluted polymers. Thus the pseudo–networks formed in concentrated systems are relatively coarse. Their knots are not fixed but fluctuate owing to tube renewal. If these facts are considered, it is reasonable to suspect that chain segments in entangled systems, except a few near entanglements, may feel little topological constraint from the tube wall and hence each of them may wriggle almost isotropically. Hydrodynamic interactions are largely screened out at high concentrations. Thus, the motion of an individual chain in concentrated solutions may well be compared to that of a free–draining flexible polymer chain, i.e., the so–called Rouse chain, in dilute solutions. However, since the two motions could not be exactly the same, we call the former Rouse–like in the subsequent discussion. The idea of the Rouse–like chain motion in entangled systems stands at the extreme opposite to the reptation model. Although, at present, there is a dominant preference for reptation, some recent data quoted below demand a serious reconsideration of non–reptative motion.

As the concentration is lowered, chains entangle more coarsely with one another, and below a certain concentration, no chain entanglement is formed. In the latter region, while the concentration is still so high that the hydrodynamic screenig of segments is effective, chains should behave Rouse–like so that D_s varies linearly with N^{-1} as expected by the theory of an isolated free–draining chain (the Rouse model)[12]. Upon further dilution, the hydrodynamic interaction becomes less strongly screened, and the N dependence of D gradually approaches $D \sim N^{-\mu}$ $(0.5 < \mu < 0.6)$ valid for an isolated non–draining chain (the Zimm model). Thus, in the non–entangled regime, the molecular weight dependence of D_s is expected to change from the Rouse–model type to the Zimm–model one, as the concentration is decreased.

2.6 Theories of Hess and Skolnick et al.

The reptation idea premises that the topological constraint exists and is sufficiently strong. In principle, we should be able to determine whether it is right or not by solving the equations of motion governing the long-timescale dynamics of many–chain systems. Recently, some attempts to this very difficult problem have appeared. Here we mention two of them.

Hess [13] neglected the hydrodynamic interactions among chain beads and treated the global motions of different chains as uncorrelated (this is to assume a small number of chain–chain contacts and thus to focus on the semi–dilute regime). He deduced that polymer self–diffusion consists of both lateral and longitudinal modes of chain motion until the "entanglement parameter" $\psi(c, N)$ reaches unity, but it is dominated by the latter (i.e., chains move reptatively) for $\psi \gg 1$. Here $\psi(c, N)$ is a quantity which may be equated to the number of chain entanglements per chain, i.e.,

$$\psi(c, N) = N/n_e \qquad (2.38)$$

Hess showed that F_s is expressed by

$$F_s = 1/[1 - (2/3)(N/n_e)] \quad (\psi \leq 1)$$
$$F_s = 1 + 2(N/n_e) \quad (\psi \geq 1) \qquad (2.39)$$

Thus we see that the N dependence of D_s changes from the Rouse type $D_s \sim 1/N$ to the reptation type $D_s \sim 1/N^2$ as the chain length increases and crosses over n_e. However, it is too early to conclude with this result that the premise of the reptation idea has been justified theoretically, because Skolnick et al. [14] have shown that an equation similar to eq 2.39 can be obtained from a different dynamic model, as explained below.

The basic idea of Skolnick is that for an entanglement to have an important effect on self–diffusion of a chain, two chains have to remain in contact for a sufficiently long time (which they call the terminal relaxation time t_r). Thus they distinguish between static and dynamic entanglements. The former are contacts between the chains randomly diffusing apart in timescales shorter than t_r. Most of the chain entanglements are supposed to be static ones. Occasionally, two chains are moving in the same direction, keeping their contacts for timescales of the order of t_r. Skolnick et al. call such a long–lived chain contact the dynamic entanglement, and consider that the rate of self–diffusion of an individual chain is primarily controlled by how strongly one chains drags another through dynamic entanglements. Although dynamic entanglements are not permanent and sooner or later undone, in steady self–diffusion, an equal number of dynamic entanglements will be formed somewhere on the same chain. Thus, what happens to a chain self–diffusing steadily in an entangled system may be pictured as a continuous series of forming dynamic entanglements, dragging another chain for a while, and then breaking up the entanglements.

The above consideration suggests that dynamic entanglements per chain are few even in concentrated solutions. Hence, in timescales comparable to t_r, the motion of individual chains in such a system may be hardly affected by topological constraint and essentially Rouse-like. Even so, it should differ from the motion of a Rouse chain in dilute solution, because each chain drags another through dynamic entanglements. Thus, D_s for polymer concentrates may be expressed by the Rouse diffusion coefficient $D_s = k_B T/\zeta N$ if ζ is corrected for the dragging effect. Skolnick et al. assumed ζ to be proportional

245

to N/n_{ed}, where n_{ed} is the average number of beads between neighboring dynamic entanglements. Then, this assumption gives

$$F_s = 1 + \gamma(N/n_{ed}) \tag{2.40}$$

where γ is a factor standing for the strength of interaction at a dynamic entanglement. This equation predicts the N dependence of D_s essentially similar to that given by the theory of Hess.

The key difference between the above two theories is as follows. The theory of Skolnick et al. distinguishes between static and dynamic entanglements according to timescales and considers that the motion of individual chains is controlled by mutual dragging through dynamic entanglements. On the other hand, the theory of Hess treats all entanglements equally and regards them as obstacles to chain wriggling. Thus, in effect, the latter is akin to the original reptation idea of de Gennes. It should be noted that any mechanism leading to $F_s \sim N$ gives the reptation prediction $D_s \sim N^{-2}$. Thus, this N dependence of D_s is not unique to the reptation model.

3. Typical Experimental Results

3.1 Measuring Techniques

Measurements of D_s in entangled polymer systems were initiated by Bueche et al. [15] as early as 1952, who studied self–diffusion in plasticized polystyrene and poly (n–butyl acrylate) by the radioactive tracer method and obtained D_s values of the order of $10^{-11} - 10^{-12}$ cm^2 s^{-1}. Values of this quantity for polymers in dilute solutions being of the order of $10^{-6} - 10^{-8}$ cm^2 s^{-1}, we see how slow polymer self–diffusion processes in concentrated systems are. Probably owing to the foreseeable difficulty in tracing such very slow processes and also to the lack of an incentive from the theoretical side, only fragmentary measurements of D_s on polymer concentrates were reported occasionally until about a decade ago. Then, triggered by de Gennes' reptation theory, on the one hand, and aided by the advent of various high technologies, on the other hand, there has occurred remarkable progress in measuring techniques for self–diffusion in concentrated solutions and melts. In an excellent review, Tirrell [2] in 1984 outlined the underlying principles as well as the scope and limitations of the new methods, along with a critical evaluation of the experimental data then available. Here we add more recent data and discuss how much our comprehension of polymer self–diffusion in concentrated systems has advanced. It will be seen that despite the great efforts of many workers, the mechanism controlling this dynamic process still remains controversial.

The average time t taken by a polymer chain to self–diffuse a distance x is given approximately by (see eq 7–2.3)

$$t = x^2/6D_s \qquad (3.1)$$

The following table shows values of t for x = 0.01 cm and D_s which covers the range measurable by current techniques.

D_s (in cm^2 s^{-1})	10^{-7}	10^{-9}	10^{-11}	10^{-13}
t (in hours)	0.05	5	5×10^2	5×10^4

We see that, for example, it takes as long as 20 days for a polymer with D_s of 10^{-11} cm^2 s^{-1} to travel 0.01 cm. Thus, recent efforts of experimentalists have focused on developing methods which allow D_s to be estimated from chain migration over distances of the micron level.

Many such methods are similar in their principles to the tracer diffusion method. They are concerned with measuring the relaxation of an appropriate physical property that occurs when labeled chains diffuse through a bulk of unlabeled (identical or different) chains. To evaluate D_s correctly it is necessary that the concentration of the labeled species is sufficiently low, since otherwise we would deal with diffusion not of the center–of–mass of a single chain but of an assembly of interacting chains. Labeling is done by deuteration, radio–labeling, dye–labeling, and so on. Of course, the labeling should not significantly alter the structure and properties (e.g., solubility) of the host polymer.

Table 8–1 lists the techniques that have been developed to date to measure D_s or D_{tr} in polymer concentrates. The reader is advised to see Tirrell's concise account [2] for the principles, experimental problems, and related references.

Table 8–1. Techniques for measurement of D_s

Techniques	For melts	For solutions	Lower limit of D_s in cm^2 s^{-1}
Radioactive tracer	x		
Infrared microdensitometry	x		10^{-10}
Nuclear magnetic resonance (NMR)[a]	x	x	10^{-10}
Quasi–elastic light scattering (QELS)		x	10^{-12}
Neutron scattering[b]	x		

Forced Rayleigh scattering (FRS)	x	x	10^{-14}
Fluorescence recovery after pattern photobleaching (FRAPP)	x	x	10^{-10}
Rutherford backscattering from interfacial marker	x		10^{-15}
Forward recoil spectroscopy	x		10^{-15}

a: More correctly, this includes the steady gradient spin echo method and the pulse gradient spin echo method. Both require no labeling.

b: This method is more relevant to the study of chain motions faster than the center–of–mass movement, since neutrons have much shorter wavelengths than light.

As the glass transition temperature T_g is approached, the friction coefficient ζ sharply increases [3] and hence D_s becomes too small to be measured. Thus, self–diffusion measurements on undiluted polymers are usually made at temperatures far above T_g. For example, all the reported data on polystyrene melts ($T_g \sim 100°C$) were taken in the range 150 – 250°C. Working at such high temperatures, however, is not simple for various technical reasons including polymer degradation. It is therefore advantageous to study polymers with T_g far below room temperature if nearly monodisperse samples are available (use of polydisperse samples should be avoided for basic research). Examples of such polymers are poly(isoprene) and polybutadiene.

Addition of a solvent (or diluent) often lowers T_g. Hence even for a polymer whose T_g is high, its self–diffusion in solutions can be studied at room temperature, though this is limited to concentrations c at which T_g of the solution is still far below the measuring temperature. The upper limit of c could be made higher by experimenting at a higher temperature, but this would make it difficult to handle the solution if the solvent is volatile. Furthermore, preparing homogeneous polymer solutions becomes increasingly difficult as c increases, especially when the solvent is thermodynamically poor. Thus it is only feasible for a polymer with high T_g (such as polystyrene) to measure D_s in solution over a range of relatively low concentration.

3.2   Melts

3.2.1   Monodisperse Melts

To date, the most extensively studied polymer, both in solutions and in the melt, is polystyrene. Kumagai et al. [16] were the first to measure D_s for the melt of this polymer as a function of molecular weight M. Using a radioactive

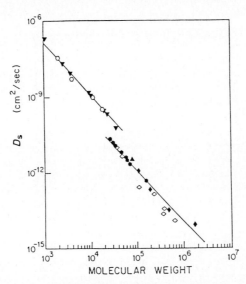

Fig. 8-3. Self-diffusion coefficient data for PS in the melt at nearly the same temperature (175°C. ▼, Bachus and Kimmich [17] (NMR); ○, Fleisher [18] (NMR); ●, Antonietti et al. [19] (FRS); ◇, Kramer [20] (Rutherford backscattering); ◆, Kumagai et al. [16] (radioactive tracer); ▲, Bueche [21] (radioactive tracer). The lines are drawn with slope −2.

tracer method, they found that $D_s \sim M^{-2.7}$, which does not conform to the reptation prediction. Figure 8-3 was prepared by Tirrell [2], who collected polystyrene melt D_s data available to 1984. The data points below $M \sim 10^5$ closely fit the indicated straight lines of slope −2, thus favoring the reptation theory. However, excepting one point for the highest M studied by Kumagai et al., those above $M \sim 2 \times 10^5$ appear to follow a somewhat steeper line. Regardless of whether real or not, this deviation from the reptation exponent suggested that more experiments should be done for low D_s at high M. Another point of note is that the upper line fitting the NMR data is significantly above the lower one fitting the data by other methods.

Self-diffusion in the melt of polyethylene (PE) (or hydrogenated polybutadiene) has also been investigated rather extensively. Figure 8-4, taken from Tirrell's review [2], shows typical experimental data. Each set of data points is fitted closely by a straight line of slope −2, consistent with the reptation theory. However, we see again that the NMR data [17, 18, 22, 23] appear above those from tracer diffusion [24, 25] and neutron scattering [26]. The reason for the discrepancy was left unexplained in Tirrell's paper.

249

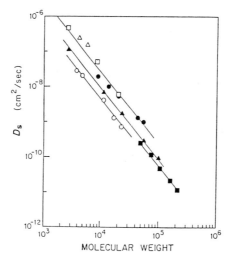

Fig. 8–4. Self–diffusion coefficient data for PE in the melt at 176 °C. $\triangle$, McCall et al. [22] (NMR); $\square$, Bachus and Kimmich [17] (NMR); $\bullet$, Fleisher [18, 23] (NMR); $\circ$, Klein and Briscoe [24] (IR microdensitometry); $\blacktriangle$, Klein et al. [25] (IR microdensitometry); $\blacksquare$, Bartels et al. [26] (neutron scattering). The lines are drawn with slope –2.

3.2.2   Matrix Chain Effects

The data shown above were taken on monodisperse systems (pure melts), in which all chains were thermally migrating and had to be treated on an equal footing. Hence, they are not adequate for a meaningful test of de Gennes' original reptation theory or the Doi–Edwards theory, which assumes that the tube trapping a moving chain is fixed in space. To obtain D_{s} good for such a test we may go as follows. We measure the tracer diffusion coefficient D_{tr} of a labeled polymer (see Section 1.3 for the definition) in the matrix of the host polymer as a function of the latter's chain length P and estimate D_{tr} for $P = \infty$ by extrapolation. Since at this limit the matrix polymer cannot move as a whole, the extrapolated D_{tr} (denoted $D_{\mathrm{tr}}{}^{\infty}$) is considered to meet the condition of fixed tube and thus may be compared meaningfully with the reptation theory.

As early as 1968 when this theory did not exist, Bueche [21] found the same value of D_{tr} for 8×10^4 molecular weight polystyrene in three matrices of molecular weights $18.6 \times 10^4, 41.0 \times 10^4$, and 106×10^4. Thus, Bueche's D_{tr} was actually equal to $D_{\mathrm{tr}}{}^{\infty}$. In fact, it is included in Figure 8–3.

Probably, Tanner [27] was the first to determine D_{tr} as a function of P. His

250

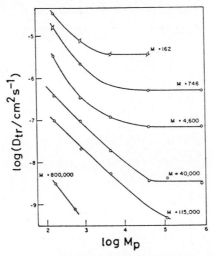

Fig. 8–5. Tracer diffusion coefficient data for blended PDMS melts consisting of 10 wt.% N chains and 90 wt.% P chains [27]. The two species are here distinguished in terms of molecular weights M and M_p instead of N and P, respectively.

NMR data on poly(dimethylsiloxane) are shown in Figure 8–5. Typical features are that, as P increases, D_{tr} for a fixed N decreases sharply and tends to be insensitive to P for P somewhat larger than N, unless N is in the region of oligomers. Those who are in favor of the reptation theory would explain the observed N dependence of D_{tr} in terms of the concept of tube renewal, while Tanner gave it a different interpretation. Klein [28] confirmed the insensitivity of D_{tr} to P above N with blended polyethylene melts.

Recently, more extensive work on the matrix chain effect on D_{tr} was reported by Antonietti et al. [19] (by FRS) and by Green and Kramer [29] (by Rutherford backscattering), who both studied polystyrene melts. Figure 8–6 shows the data of Green and Kramer. The values of D_s and D_{tr}^{∞} at 212°C reported by Antonietti et al. are reproduced in Figure 8–7. Here, the abscissa M/m is the polymerization degree of the tracer component (hence, proportional to N). The line fitting the D_{tr}^{∞} has a slope of -2, and this fact appears to to substantiate the reptation idea, since no tube renewal is expected to occur in the limit $P = \infty$. Th data for D_s also follow a straight line, but the slope is $-(2.45 \pm 0.05)$ according to Antonientti et al. The data of Green and Kramer give similar dependences of D_{tr}^{∞} and D_s on M, with $-(2.3 \pm 0.1)$ for the slope of the D_s line.

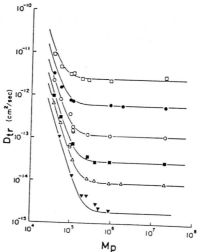

Fig. 8-6. Tracer diffusion coefficient data for blended PS melts at 174°C [29]. $\square$, $M = 5.5 \times 10^3$; $\bullet$, $M = 11 \times 10^4$; $\circ$, $M = 25.5 \times 10^4$; $\blacksquare$, $M = 52 \times 10^4$; $\triangle$, $M = 91.5 \times 10^4$; $\blacktriangledown$, $M = 200 \times 10^4$. Lines, calculated by Graessley's theory [9].

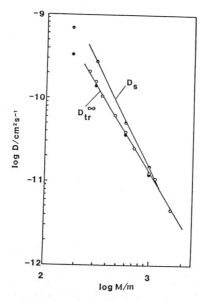

Fig. 8-7. M dependence of D_{tr}^∞ and D_s for PS in the melt at 212°C [19]. The data points for D_{tr}^∞ are actually for $P = 2150$

What is puzzling in Figure 8-7 is that the indicated two lines intersect at $M/m(\sim N) \sim 10^3$. This phenomenon also occurs in the data of Green and Kramer, though the intersection of the lines appears at a much higher value of N (this difference is also puzzling, because the two groups studied the same polymer species). It follows that the values of D_s and D_{tr}^{∞} for a given M should coincide for all M above M' at the intersection, since otherwise we are led to a physically unreasonable conclusion that D_s for $M > M'$ falls below D_{tr}^{∞}, i.e., the tracer chain can diffuse more easily when trapped in a matrix of longer chains. We may consider two modes for the coincidence of the two lines. In one, the line for D_s breaks toward the line for D_{tr}^{∞} at $M = M'$, so that D_s varies as M^{-2} for $M > M'$. In the other, the line for D_{tr}^{∞} bends toward the line for D_s at $M = M'$, so that D_{tr}^{∞} varies as $M^{-\alpha}$, with $\alpha = 2.3 - 2.5$, for $M > M'$. If the former occurs, it suggests that when its length exceeds a certain value, each chain in a monodisperse melt self-diffuses in the pure reptation mode as if the surrounding chains are frozen, i.e., tube renewal has little to do with the motion of individual chains. On the other hand, if the latter is the case, it implies that even the matrix of infinitely long chains cannot have a topological constraint strong enough to force a tracer chain to reptate. At present, the former idea seems to be more preferred by many authors, but it has to be confirmed by further experiments. For this purpose, it is desirable to determine $D_{tr}(M, M_p)$ accureately up to as high M as possible. Green and Kramer extended the experiment up to $M = 200 \times 10^4$, which is the highest reported so far in the self-diffusion measurements on polymer melts. As can be seen from Figure 8-6, D_{tr}^{∞} for such a high M is of the order of 10^{-15} cm^2 s^{-1}. However, it appears that none of the currently available techniques for self-diffusion can proceed with precision below this limit. Thus, the interest of workers has been directed to solutions of adequate concentrations, in which polymer segments become more mobile than in the melt and thus it should be possible to determine the M dependence of D_{tr} up to higher molecular weights.

3.3 Concentrated Solutions

3.3.1 Binary Solutions

A dozen measurements of D_s on concentrated polymer solutions have been carried out with well-characterized polymers in solvents of different quality, with the aim of checking the concentration dependence predicted by scaling arguments (eq 2.35 for θ solvents and eq 2.37 for good solvents) as well as the molecular weight dependence. As can be seen from Tirrell's review [2], it was evident already in 1984 that the topological factor F_s as a function of c does not follow a simple power law, while its M dependence appears to be consistent with the reptation theory over a wide range of c. However, it was

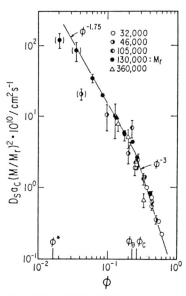

Fig. 8–8. D_s for PS in THF [30], corrected for the concentration dependence of ζ (the factor a_c) and for the molecular weight dependence predicted by the reptation theory. ϕ is the PS volume fraction and $\phi^\star = v_p c^\star_0$, where v_p is the PS specific volume.

puzzling that the reptation model was likely to hold down to the vicinity of overlap concentration $c^\star$ where polymer chains are not sufficiently entangled. In order to extract the concentration effect on F_s from measured D_s values we have to correct the latter for the concentration dependence of the bead friction coefficient ζ. However, this correction was not made to the data summarized by Tirrell [2].

In an FRS study on polystyrene in tetrahydrofuran, Nemoto et al. [30] estimated $\zeta(c)$ from the tracer diffusion coefficient of methyl red, an azo dye. Figure 8–8 shows their D_s data corrected for ζ and molecular weight dependence according to the reptation theory. The data points fall on a single curve, thus substantiating the reptation exponent –2 over the entire concentration range studied. The curve is reasonably approximated by broken straight lines having slopes –1.75 and –3. The former is consistent with eq 2.37 for good solvents and the latter with eq 2.35 for θ ones. This change in the slope is consistent with the prediction that, as the concentration increases, the effective solvent quality changes from good to poor owing to the increased screening of volume exclusion.

For semi–dilute solutions of polystyrene in carbon tetrachloride and deuter-

254

obenzene Callaghan and Pinder [31] measured D_s as a function of c and M by NMR and obtained data similar to those of Nemoto et al., though they applied no correction for the concentration dependence of ζ. In both solvents, the reptation exponent was found in the region of high concentration where the slope of $\log D_s$ vs. $\log c$ was about -3, i.e., the solvent displayed θ–like behavior.

3.3.2 Quasi–binary Solutions

Although the data quoted above appeared to establish the behavior of D_s in the semi–dilute regime and to verify the validity of the reptation idea, Yu and coworkers recognized that further experimental work sufficient in precision and range would be needed. Thus, Kim et al. [32] carried out extensive FRS measurements of the tracer diffusion coefficient D_{tr} in quasi–binary solutions consisting of labeled tracer polystyrene, unlabeled polystyrene, and toluene. The molecular weights of the tracer and matrix polymers were varied over wide ranges, and the concentration c_p of the matrix solution was changed from 0 to 15 wt.%, with the tracer concentration kept as low as about 0.05 wt.%.

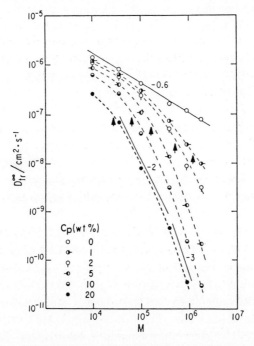

Fig. 8–9. Molecular weight dependence of $D_{tr}{}^\infty$ for PS in PS + toluene solutions of different c_p at 20°C [32]. See the text for the meaning of arrows.

Firstly, it was found that the dependence of D_{tr} on the matrix polymer molecular weight M_p persisted to M_p which was a factor of 3–5 higher than tracer molecular weight M. Secondly, D_{tr}^∞ at various c_p varied with M as displayed in Figure 8-9. In this graph, each of the arrows points to a molecular weight M^* at which the average segment density in the coils of tracer chains becomes equal to the overall segment concentration of a given matrix solution, c_p. Making the same consideration as in Section 1.1 of Chapter 6, we can show that the following relation holds approximately:

$$c_p \sim (M^*)^{1-3\nu} \tag{3.1}$$

Kim et al. estimated M^* by use of an empirical relation [33]

$$c_p(\text{wt.\%}) = 620(M^*)^{-0.785} \tag{3.2}$$

At a given c_p, as M exceeds M^*, tracer chains overlap the matrix chains and hence find themselves in an environment similar to a semi–dilute solution. Thus, except for the solid line for $c_p = 0$, all other curves contain at least one data point in the semi–dilute regime. It can be seen that, at $c_p = 5$, 10, and 20 wt.%, the exponent α in $D_{tr}^\infty \sim M^{-\alpha}$ changes smoothly from about 2 to 3 as the solution enters and goes deeper into the semi–dilute regime. Since D_{tr}^∞ concerns the limit of infinitely large M_p, it should correspond to the absence of tube renewal. Hence, if the reptation model is valid in the semi–dilute regime, the exponent α should remain equal to 2 as M is increased. The curves in Figure 8-9 do not follow this prediction and suggest that even the matrix of infinitely long chains is incapable of constraining long tracer chains to pure reptation.

Recently, Nemoto et al. [34] conducted FRS measurements of D_{tr} in quasi-binary solutions containing about 1 wt.% labeled polystyrene, 40 – 41 wt.% unlabeled polystyrene, and dibutyl phthalate (DBP), with M and M_p varied over broad ranges. Figure 8-10 illustrates D_{tr} for different M plotted against M_p, with all data reduced to 60°C. Consistent with the observation of Kim et al., the M_p dependence of D_{tr} does not vanish until M_p reaches more than five times M. In Figure 8-11, D_s (D_{tr} for $M_p = M$) and D_{tr}^∞ are plotted as functions of M. Differing from the melt data of Antonietti et al. and Green and Kramer discussed in Section 3.2.2, these functions can be fitted by lines which do not intersect at least in the range of measurement. Remarkable is the finding that the linear parts of the two lines have slopes clearly steeper than the reptation exponent –2; the slope for D_{tr}^∞ is about –2.5, and that for D_s is slightly larger. Though nothing more definite can be claimed than these, the two lines are likely to merge asymptotically to a straight line of slope ~ -2.5 as M becomes higher.

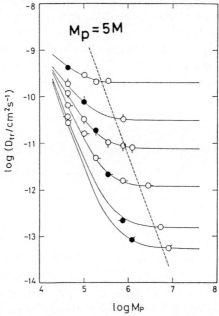

Fig. 8–10. M_p dependence of D_{tr} for PS in PS + DBP solutions of c_p (wt.%) $\sim$ 40% at 60°C [34]. M from top to bottom: $4.4 \times 10^4, 10.2 \times 10^4, 18.6 \times 10^4, 35.5 \times 10^4, 77.5 \times 10^4, 126 \times 10^4$.

The findings of Kim et al. and Nemoto et al. on the M dependence of D_s cannot be explained not only by de Gennes' original reptation theory but also by its current modifications taking the tube renewal into account. Although it is still too early to conclude that they are definite evidence for the anti–reptation opinion, the following remark may be of some interest.

In a recent report Yu [35] showed the D_s data of Landry [36] for poly(iso-prene) in bulk at 25 and 45°C, obtained by FRS and NMR. This polymer is advantageous for self–diffusion study in that its T_g is far below room temperature and its M_e (mean molecular weight between neighboring entanglements) is only one tenth of that in bulk polystyrene [3]. The latter feature allows us to reach well–entangled states with samples of relatively low M. At either temperature, Landry's data show that D_s tends to vary in proportion to the -2.8 power of M as M exceeds about 5×10^4. It seems fair not to ignore this significant deviation from the reptation exponent -2, even though its adequate explanation is not feasible at present.

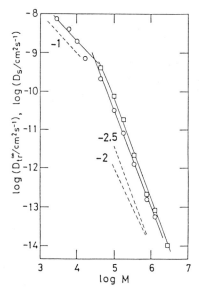

Fig. 8-11. M dependence of D_{tr}^{∞} (circles) and D_s (squares) derived from the data of Fig. 8-10 [34].

3.3.3 Isorefractive Solutions

We can use QELS to measure D_{tr} of a tracer polymer A in a solution consisting of a chemically different polymer B and a solvent, provided that B is thermodynamically compatible with A and that the solvent has the same refractive index as B. Such an "isorefractive" binary solution behaves optically as a single "solvent" (B becomes "invisible"). Dynamic light scattering from A dissolved in it can therefore be analyzed by the theory established for two-component systems to determine D_{tr} of A. No labeling is needed for probing A. This is a great experimental advantage.

In recent years, D_{tr} of polystyrene has been measured by this "isorefractive scattering" method [37–44]. The isorefractive solutions chosen were poly (methyl methacrylate) (PMMA) + benzene, poly(vinylmethylether) (PVME) + ortho–fluorotoluene (oFT), and PVME + toluene. In this section, we refer to a detailed study of Wheeler et al. [43] with PVME + oFT.

Wheeler et al. fixed the molecular weight M_p of B, PVME in their study, at 1.3×10^6 and chose four values ($6.5 \times 10^4, 1.79 \times 10^5, 4.22 \times 10^5, 1.05 \times 10^6$) for the molecular weight M of A, polystyrene. For each M they varied the PVME concentration c_p from 0 to 0.100 g cm^{-3}. Figure 8-12 shows, on a double–logarithmic plot, D_{tr} at fixed c_p and 30°C as a function of M. The data can be fitted by a power law relation $D_{tr} \sim M^{\beta(c_p)}$ over the entire range of

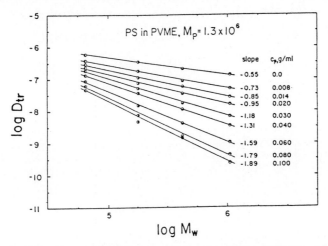

Fig. 8-12. M dependence of D_{tr} for PS in PVME + oFT solutions of different PVME concentrations c_p, with PVME molecular weight M_p fixed at 1.3×10^6 [43].

M studied. In Figure 8-13, β is plotted against reduced PVME concentration c_p/c_p^*, where c_p^* is the overlap concentration for PVME in oFT (actually, 0.00334 g cm^{-3}). This graph also shows the β values obtained by Hanley et al. [42] and Martin [38], though Martin's are concerned with toluene solutions. All the plotted points fall on a single continuous curve. We see that, with increasing c_p, β decreases continuously from -0.5 and reaches the reptation value -2 at a c_p more than 30 times c_p^*. Furthermore, it appears certain from the trend of the plotted points that β does not converge to but decreases below -2 at c_p higher than $30c_p^*$. Apparently, the reptation idea cannot explain this interesting finding of Wheeler et al.

When Figure 8-12 is compared to Figure 8-9, we find that both are consistent in that D_{tr} depends more strongly on M at higher c_p. However, the dependence follows a power law in the former but not in the latter. To explain this difference we may note the following points.

Firstly, we take note of the fact that Figure 8-12 concerns D_{tr} at a finite $M_p(=1.3 \times 10^6)$, while Figure 8-9 concerns D_{tr}^∞. As mentioned above, Kim et al. (and Nemoto et al. as well) showed that D_{tr} did not decrease to D_{tr}^∞ until M_p became 3 to 5 times larger than M. Since PVME with $M_p = 1.3 \times 10^6$ is similar in polymerization degree to polystyrene with $M \sim 2.2 \times 10^6$ [43], it is likely that the D_{tr} values for the two highest molecular weight polystyrenes in Figure 8-12 were somewhat higher than the corresponding D_{tr}^∞. Thus, Wheeler et al. might have observed non-linear $\log D_{tr}$ vs. $\log M$ relations similar to those reported by Kim et al. if they had done experiments using

259

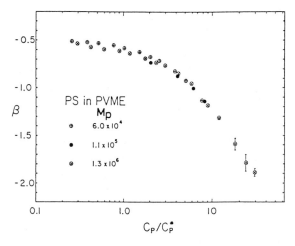

Fig. 8–13. Dependence of β on PVME concentration (β is defined by $D_{tr} \sim M^{\beta}$) at different PVME molecular weights M_p. $\oplus$, Hanley et al. [42](oFT); $\bullet$, Martin [38] (toluene); $\otimes$, Wheeler et al. [43] (oFT). $c_p{}^{\star}$ is the overlap concentration of the maxtrix solution.

PVME of a much higher M_p.

Secondly, as can be seen from Figure 8–10, D_{tr} increases sharply as M_p decreases below M, while it essentially stays constant for M_p above M. Thus, D_{tr} should be higher in a polydisperse isorefractive solution than in a monodisperse one when compared at the same average M_p. The difference is greater for a higher tracer molecular weight and for a larger polydispersity index M_w/M_n of the matrix polymer. The PVME sample used by Wheeler et al. was not sufficiently close to monodispersity (its polydispersity index was about 1.3). Hence it is reasonable to suspect that they obtained D_{tr} more or less higher than those for the monodisperse matrix in the region of higher M. Thus their observation of linear $\log D_{tr}$ vs. $\log M$ relations might have been due in part to a polydispersity of the PVME sample they used. This conjecture can be tested if a monodisperse PVME becomes available.

Thirdly, the solution chosen by Wheeler et al., containing chemically dissimilar polymers, was thermodynamically more complex than quasi–binary solutions as investigated by Kim et al. and Nemoto et al. In the former, thermodynamic interactions with the solvent should be different for the tracer and matrix polymers, while, in the latter, there is no such difference.

3.4 Remarks

3.4.1 Dynamic Structure Factor

At present, for one reason or another, the reptation model is overwhelmingly preferred. However, it seems to the author that the preference is too strong for the amount and range of reported experimental evidence. Earlier data which "confirmed" the reptation exponent −2 do not always cover sufficiently wide ranges of molecular weight and concentration. In fact, as illustrated above, some of the recent experiments which extended these ranges have revealed distinct departures from the reptation predictions. There are a non-negligible number of workers who advocate that such findings should be taken seriously as a warning for the rush to interpret dynamic behavior of polymer concentrates solely in terms of this idea.

Obviously, if found experimentally, the exponent −2 is only a sign of reptation motion but not a sufficient condition for it. Thus, in order to prove that reptation is a really dominant mode of chain motion in polymer concentrates, we have to test it not only with self–diffusion but also with other physical properties which reflect the local motion of polymer chains. One such property is the (coherent) dynamic structure factor $S(k, \tau)$ (see Section 3.2 of Chapter 4 for its definition). In fact, it was predicted theoretically [45–47] that the k dependence of its decay with τ in the range of k defined by

$$1/d_t > k > 1/R \tag{3.3}$$

distinctly differs depending on whether chain segments move isotropically or reptatively. Here, d_t is the diameter of the tube surrounding a chain and R a length of the order of the chain dimensions. In passing, we note that the incoherent dynamic structure factor $S_{ich}(k, \tau)$ for a spring–bead chain consisting of $N + 1$ beads is defined by

$$S_{ich}(k, \tau) = \sum_{j=0}^{N} \langle \exp[i\mathbf{k} \cdot (\mathbf{r}_j(\tau) - \mathbf{r}_j(0)]\rangle \tag{3.4}$$

Data on $S(k, \tau)$ can be obtained by quasi–elastic scattering of light or neutrons. The technique with light covers the range of small k up to about 2×10^{-2} nm^{-1}, while that with neutrons can measure $S(k, \tau)$ for large k down to about 3×10^{-1} nm^{-1} when the spin–echo method [48] is applied. Thus, the range $0.3 > k > 0.02$ (in nm^{-1}) remains to be filled by future techniques. For the melt of poly (dimethylsiloxane) studied by Richter et al. [49], Deutsch and Goldenfeld [50] estimated d_t to be 4 nm, for which eq 3.3 gives $0.25 > k$. This inequality implies that $S(k, \tau)$ for poly(dimethylsiloxane) melts at k defined by eq 3.3 cannot be measured by the techniques currently available. Richter et al. made spin–echo neutron scattering measurements at four k values ranging from 0.53 to 1.32 nm^{-1} and concluded (see also [51, 52]) that reptation cannot explain their experimental data (which are in favor of the Rouse–like chain). However, since these k values are too large to satisfy the condition $0.25 > k$, their experimental results do not always refute reptation, as pointed out by Deutsch and Goldenfeld [50].

Spin–echo neutron scattering measurements on molten poly (tetrahydrofuran) led Higgins et al. [53] to report that they found evidence for Rouse–like chain motion. The experiment was concerned with $k > 0.3$ nm^{-1}. However, such k does not satisfy the condition $1/d_t > k$, since d_t for molten poly (tetrahydrofuran) is estimated to be 3 nm [53]. Hence, it is doubtful if the finding of Higgins et al. can rule out the possibility of reptation.

Summarizing, we see that, because of the technical limitations, the previous studies on $S(k, \tau)$ failed to give a transparent answer to the question of whether reptation is a dominant chain motion or not. The achievement of a technical breakthrough is thus highly demanded. However, we have to remark that quasi-elastic neutron scattering experiments can at present be done only at a very few institutes in the world, because it requires huge and expensive instruments along with experienced technicians.

3.4.2   Scaling Equation of Phillies

Experimental data of $\log D_s$ plotted against $\log c$ usually exhibit no distinct break and can be fitted by a smoothly declining curve. According to the idea dominating at present, the change in the slope of this curve, after correction for the concentration dependence of ζ, may be interpreted as reflecting the gradual transition of chain motion from the Rouse–like to the reptative mode.

Recently, Phillies [54] challenged this prevailing view and concluded that probably no transition to the reptative mode occurs when a polymer solution is concentrated. This means that the self-diffusion of a polymer molecule is essentially isotropic regardless of the concentration of the solution.[3] His argument is based on the finding that non–reptative globular proteins in solution diffuse with $D_{tr}(c_p)$ which obeys the same scaling equation as does $D_{tr}(c_p)$ of flexible synthetic macromolecules over a broad concentration range. His scaling equation reads

$$D_{tr} = D_0 \exp(-\alpha c_p{}^\lambda) \qquad (3.5)$$

where D_0, α, and λ are adjustable parameters. The first should be treated as a function of the molecular weight M of the trace polymer, while the rest are

---

[3] He carefully reserves the possibility for reptation in polymer melts.

predicted to depend on the molecular weight M_p of the matrix polymer. For $M_p = M$ the left-hand side of eq 3.5 becomes D_s. Formally, D_0 is equal to D_s at infinite dilution, but since eq 3.5 is not analytical at $c = 0$ unless λ is a positive integer, D_0 is better treated as a floating parameter.

Phillies applied eq 3.5 to reported D_{tr} and D_s data on a number of polymer systems and showed that, for each species, a single set of the above three parameters suffices for all concentrations up to the highest studied (in one case, the melt). Figure 8-14 illustrates this fact with D_s data for polystyrene in CCl_4 covering a broad range of concentration. Table 8-2 lists the values of α and λ determined for each M by non-linear least squares fitting.

Table 8-2.   Parameters α and λ to fit the data of Figure 8-14 to eq 3.5, with D_0 taken as infinite-dilution D_s

$M \times 10^{-3}$	all data	data for $c < c^\star_0$
350	$\alpha = 0.211$	$\alpha = 0.221$
	$\lambda = 0.627$	$\lambda = 0.618$
	% RMS = 3.1	% RMS = 2.1
233	$\alpha = 0.192$	$\alpha = 0.186$
	$\lambda = 0.617$	$\lambda = 0.609$
	% RMS = 7.1	% RMS = 6.4
110	$\alpha = 0.0775$	$\alpha = 0.104$
	$\lambda = 0.75$	$\lambda = 0.67$
	% RMS = 6.2	% RMS = 2.8

The numerical values in Table 8-2 show that D_s for c above $c^\star_0$ can be predicted accurately by eq 3.5 with the parameter values evaluated from dilute solution data. Given this fact, it is not unreasonable to suppose that there is no substantial difference between the modes of chain motion in the dilute and semi-dilute regimes. It is from this supposition that Phillies concluded that reptation does not occur in semi-dilute solutions. In fact, he ends up his paper by stating 'Since there is no major difference between diffusion in "dilute" and "semidilute" regimes, D_s measurements with $c < c^\star$ and $c > c^\star$ must be considered on an equal footing.'

The basic weakness of Phillies' work is that eq 3.5 does not yet have a reasonable theoretical background. Thus, despite its apparent success, this equation is of little interest to those who want to know the mechanism responsible for polymer self-diffusion. The reader is advised to see Wheeler et al. [43] for more comments on it.

263

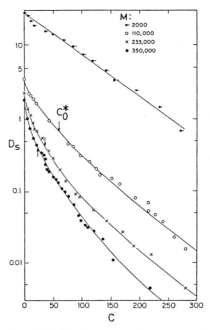

Fig. 8-14. Fit of Phillies' empirical equation 3.5 to D_s data of Callaghan and Pinder [31, 55] for PS in CCl_4. D_s is in units $10^{-7} cm^2 s^{-1}$ and c in g dm^{-3}.

4. Computer Simulations

4.1 Chain Motions on Various Timescales

The primitive chain in the Doi–Edwards theory is a smooth wire of constant length. Less coarse–grained and thus closer to actual polymer molecules is the spring–bead chain model. In this model, the beads wriggle individually under the constraint that they are connected by Gaussian springs. Their motions differ depending on the timescale on which we look at the chain. To discuss this problem we consider the discrete chain version of $g(t)$ defined by eq 2.16, i.e.,

$$g(t) = (1/N) \sum_{i=1}^{N} \langle [\mathbf{r}_i(t) - \mathbf{r}_i(0)]^2 \rangle \tag{4.1}$$

Here, $\mathbf{r}_i(t)$ $(i = 1, 2, \ldots, N)$ is the position of bead i at time t, and $\langle \ldots \rangle$ denotes the ensemble average. Timescales of observation are characterized by the variable t. The function $g(t)$ gives the average mean–square distance displaced by a bead for a time interval t. We may expect it to increase monotonically with t. Features of this increase can be approximately calculated under the simplifying assumption that neither excluded–volume repulsion nor hydrodynamic interaction acts between beads.

Reptating chains

If we assume the existence of a tube, we can proceed as follows [4]. For very small t the beads of each chain can move only very short distances and do not feel effects from other chains. Hence, the motion of the chain is essentially the same as that of a Rouse chain in dilute solution. This situation is sustained until t increases to τ_t at which $g(t)$ becomes comparable to $d_t{}^2$, i.e., the beads begin to feel the effect of other entangling chains. Here, as before, d_t denotes the diameter of the tube. Using the known expression for $g(t)$ of an isolated Rouse chain, we find

$$g(t) \sim d_t{}^2 (t/\tau_t)^{1/2} \quad (t < \tau_t) \tag{4.2}$$

where τ_t is given by

$$\tau_t \sim \frac{\tau_{\text{Rouse}}}{N^2} \left(\frac{d_t}{a} \right)^4 \tag{4.3}$$

and τ_{Rouse} is the Rouse relaxation time defined by

$$\tau_{\text{Rouse}} = \frac{\zeta a^2 N^2}{6\pi^2 k_B T} \tag{4.4}$$

Equation 4.2 presupposes $\tau_t \ll \tau_{\text{Rouse}}$. We may regard τ_t as a timescale for the onset of tube constraints.

For $t > \tau_t$, because of the constraint due to the tube wall, each bead is forced to wriggle parallel to the tube axis. This motion first tends to homogenize local densities of beads, and the chain reaches the state of uniform bead distribution on a timescale comparable to τ_{Rouse}. The accompanying change in $g(t)$ is represented by

$$g(t) \sim a d_t N^{1/2} (t/\tau_{\text{Rouse}})^{1/4} \quad (\tau_t < t < \tau_{\text{Rouse}}) \tag{4.5}$$

The primitive chain in the Doi–Edwards theory is a coarse–grained model of the chain in this state. The chain then begins to slither as a whole along the tube axis, i.e., to wriggle in the reptation mode. However, until t reaches the disengagement time τ_d (defined by eq 2.15), it is essentially trapped in the initial tube, and $g(t)$ follows the relation

$$g(t) \sim a^2 N (t/\tau_d)^{1/2} \quad (\tau_{\text{Rouse}} < t < \tau_d) \tag{4.6}$$

265

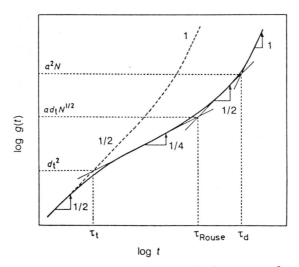

Fig. 8–15. Average mean–square displacement of a chain bead as a function of time. Solid curve, for reptating chains [4]. Dashed curve, for Rouse chains.

For $t > \tau_d$ the chain gradually escapes from the initial tube and $g(t)$ changes with time as

$$g(t) \sim D_s t \quad (\tau_d < t) \tag{4.7}$$

The solid curve in Figure 8–15 shows schematically the behavior of $g(t)$ (on a log–log plot) inferred from the above relations. The $t^{1/4}$ regime, first predicted by de Gennes [1, 45], is characteristic of the chain trapped in the tube. Checking its existence has thus become one of the central themes in recent computer simulations of the chain dynamics in concentrated systems.

Rouse chains

It can be shown that when the chain wriggles isotropically as a Rouse chain in dilute solution, $g(t) \sim t^{1/2}$ for $t \ll \tau_{\text{Rouse}}$ and $g(t) \sim D_s t$ for $t \gg \tau_{\text{Rouse}}$. Thus, its $g(t)$ shows a variation as sketched by the dashed curve in Figure 8–15. We note that this curve lacks the $t^{1/4}$ regime mentioned above, as should be expected.

4.2 Results from Typical Simulations

Since the correct picture of polymer dynamics in concentrated systems is hopelessly difficult to find analytically, there has evolved a great interest in approaching it by simulation experiments on a computer. Such work started only a decade ago, but it has already provided significant information, aided by

266

the remarkable progress of computer science and technology. Baumgärtner's review [56] summarizes and comments on the results reported by 1984. In this book, we focus on a recent study of Kolinski et al. [57], which is an extension of Kremer's earlier work [58] to higher concentrations and which the author believes has a considerable significance in clarifying the dynamics of chains in condensed systems.

The model system studied by Kolinski et al. is a collection of monodisperse random flight chains confined to a tetrahedral lattice. Each chain is made up of N beads connected by $N - 1$ bonds (rods) and occupies N lattice vertices. If there are n such chains per MC (Monte Carlo) box, subject to periodic boundary conditions, the volume fraction ϕ of the polymer is given by Nn/V, where $V = E^3/8$, the number of lattice sites per MC box of edge length E. The factor 8 comes from an integer representation of the diamond lattice vertices. Thus, the length of each rod vector a is $3^{1/2}$, with a being of the form $[\pm 1, \pm 1, \pm 1]$. The initial configuration of each system is specified by a set of three parameters (n, N, E).

A dynamic Monte Carlo simulation method was used to determine the time evolution of $g(t)$ and g_{cm} by allowing all chains to move simultaneously. Here g_{cm} is the mean–square displacement of the center of mass, i.e.,

$$g_{\text{cm}} = \langle [\mathbf{r}_G(t) - \mathbf{r}_G(0)]^2 \rangle \tag{4.8}$$

with $\mathbf{r}_G(t)$ the position of the center–of–mass at time t. The elementary motions employed are the following three:
(1) Three–bond jumps with a priori probability $p(3b)$ resulting from the exchange of bonds $\mathbf{a}_i$ and $\mathbf{a}_{i+2}$; the two beads are moved if $\mathbf{a}_i \neq \mathbf{a}_{i+2}$;
(2) Random orientation of chain ends involving one ($\mathbf{a}_1$ or $\mathbf{a}_{N-1}$) or two bonds ($\mathbf{a}_1$ and $\mathbf{a}_2$, or $\mathbf{a}_{N-1}$ and $\mathbf{a}_{N-2}$). The sum of a priori probabilities of the one and two bond motions, $p(1b)$ and $p(2b)$, respectively, is taken as equal to $p(3b)$, with the relative frequency of one–bond to two–bond flips equal to 1:2;
(3) Four–bond motions, with a priori probability $p(4b)$, which creates new orientations of bond vectors $\mathbf{a}_i$ and $\mathbf{a}_{i+3}$ within the chain; the necessary condition for this motion is $\mathbf{a}_i \neq \mathbf{a}_{i+3}$.

As Kolinski et al. mention, every conformational transition of a tetrahedral chain involving two or more beads can be decomposed into a succession of the above elementary motions. Thus, the only constraints on the dynamics of this chain come from the lattice and the excluded–volume restrictions. The choice of $p(3b)$ relative to $p(4b)$ was found to be immaterial for long chains, provided that both jump frequencies are of the same order. Thus, for actual computations $p(3b)$ was taken to be 0.25 (hence $p(4b) = 0.75$). Simulations were performed for eight values of N from 12 to 216 and at seven ϕ from 0 to 0.75. Here we do not refer to the algorithm and sampling procedure used.

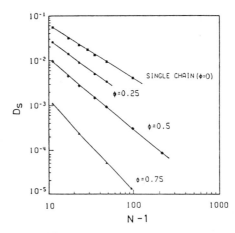

Fig. 8-16. Double–logarithmic plots of D_s vs. $N - 1$ at various volume fractions ϕ.

Figure 8-16 shows that the self–diffusion coefficient D_s scales for all concentrations as

$$D_s \sim (N - 1)^{-\alpha} \tag{4.9}$$

with the exponent α increasing monotonically from 1.15 at $\phi = 0$ to 2.06 at $\phi = 0.75$. It is apparent that α exceeds the reptation value 2 for $\phi > 0.75$. Kolinski et al. analyzed the data of Figure 8-16 to extract the topological factor F_s in the relation $D_s = k_B T/(\zeta N F_s)$ by assuming a free–volume expression [3] for ζ. However, their maneuver is dubious because what their computer simulations gave is actually not D_s but $1/F_s$. In another paper, Kolinski et al. [59] showed that D_s should diminish to zero at $\phi = 0.92$. This is due to the fact that as the concentration gets higher, the probability of having three–hole clusters becomes progressively smaller so that long distance chain motions tend to be shut down. In reality, however, such motions can take place even in the melt ($\phi = 1$) when the polymer is well above the glass transition temperature. The discrepancy indicates the limitation of simulation experiments with on–lattice chains. Thus, it is dangerous to extrapolate the α data of Kolinski et al. beyond the highest ϕ they studied. However, their conclusion that $D_s \sim (N - 1)^{-2.1}$ at high concentrations seems a bit too conservative.

Figure 8-17 illustrates $g(t)$ computed for $N = 216$ and $\phi = 0.50$. This $g(t)$ was not exactly the same as that defined by eq 4.1 but calculated for the 74 middle beads of each chain. However, its behavior would not differ much from that of the complete $g(t)$. If we compare Figure 8-17 with Figure 8-15, we see immediately that the plotted points do not exhibit the $t^{1/4}$ regime predicted for reptating chains, but follow a curve similar to that for Rouse chains. Kolinski

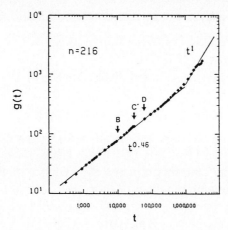

Fig. 8–17. Double–logarithmic plots of $g(t)$ vs. time t for $N = 216$ and $\phi = 0.5$. The arrows indicate the times at which the snapshopts shown in Fig. 8–19 are taken.

et al. observed the same behavior at $\phi = 0.75$, too. Interestingly, these findings were consistent with what Kremer [58] had already reported on a diamond lattice chain of $N = 200$ at a lower ϕ of 0.344. Furthermore, though for much shorter chains ($N \leq 48$), Crabb and Kovac [60] obtained similar results at ϕ as high as 0.900 from simulation experiments on a cubic lattice. Probably, it was only Deutsch [61] who ever observed the $t^{1/4}$ region by computer simulations in which all chains were allowed to move simultaneously. However, as pointed out by Crabb and Kovac [60], Deutsch's model probably builds in reptation–like motion.

These $g(t)$ data do not favor the reptation model, but are not definitive enough to rule out the possibility of reptation motion. Thus, Kolinski et al. [57] stepped further by calculating the longitudinal and lateral displacements of the primitive path used in the Doi–Edwards theory. Such calculations should give us direct information on local chain motions.

They divided the original N bead chain into blobs each consisting of $N_B + 1$ beads and defined a smooth random curve connecting the centers of mass of the blobs as the primitive path. Here, according to Doi and Edwards, N_B should be taken close to n_e, the number of beads between neighboring entanglements. Kolinski et al. computed the longitudinal displacement down the primitive path by projecting $r_i^\star(t_0 + \Delta t)$ on to the primitive path at time t_0 defined by $\{r_i^\star(t_0)\}$, where $r_i^\star$ denotes the position of the center–of–mass of blob i. If the projection of $r_i^\star(t_0 + \Delta t)$ is on to blob j in the original primitive path, then the displacement Δs of blob i down the original path is given by $a_B|i - j|$, with a_B being the average distance between the centers of the neighboring blobs. Thus, the mean–

269

square longitudinal displacement $g_\parallel(\Delta t)$ down the original primitive path for the time Δt is

$$g_\parallel(\Delta t) = a_B{}^2 \langle |i - j|^2 \rangle \tag{4.10}$$

The corresponding mean–square lateral displacement $g_\perp(\Delta t)$ can be calculated from

$$g_\perp(\Delta t) = \langle [\mathbf{r}_j^\star(t_0) - \mathbf{r}_i^\star(t_0 + \Delta t)]^2 \rangle \tag{4.11}$$

In these expressions, the average is over all beads in the middle third of the chain on the basis of the prediction that if the chain has to move through a tube, at least its middle part should retain the memory of the primitive path conformation for a substantial interval of time.

Figure 8–18 depicts the ratio $g_\perp(t)/g_\parallel(t)$ calculated with N_B and a_B taken to be 18 and 0.28, respectively. This ratio deviates from 2 when the chain motion is anisotropic, as expected by reptation theory. In the figure, the unfilled diamonds and circles refer to the case in which all chains are mobile, while the filled diamonds are concerned with the case in which matrix chains are partially frozen by pinning every 18 beads. Kolinski et al. [57] calculated the terminal relaxation time for the end–to–end vector $\mathbf{R}$ of the entire chain to be 5.2×10^5 for the unfilled diamond case and 1.52×10^6 for the unfilled circle case.

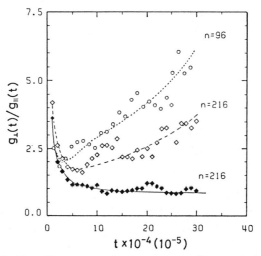

Fig. 8–18. Change in the anisotropy factor $g_\perp(t)/g_\parallel(t)$ with time [57]. $\circ$, $N = 96$, $\phi = 0.75$, t in 10^5; $\diamond$, $N = 216$, $\phi = 0.50$, t in 10^4; $\blacklozenge$, $N = 216$, $\phi = 0.50$, t in 10^4, the matrix is partially frozen.

If the lateral motion of a chain were largely suppressed by entangling chains, $g_\perp(t)$ could not grow beyond a certain limit and hence $g_\perp(t)/g_\parallel(t)$ should

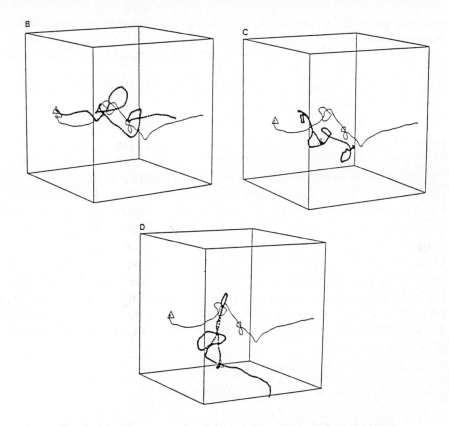

Fig. 8–19. Change in the shape and position of the primitive chain with time. Thin curves, at $t = 0$; thick curves, at $t = 10^4$(B), 3×10^4(C), 5×10^4 (D). $N = 216$, $\phi = 0.5$. The triangle locates the position of one of the chain ends.

decrease monotonically as t increases. The unfilled diamond and circle data do not at all conform to this prediction, yielding $g_\perp(t)/g_\parallel(t)$ which, except at small t, grows monotonically with increasing t. For $N = 96$ the increase is sustained even after the terminal relaxation time for **R**. On the other hand, as should be expected, $g_\perp(t)/g_\parallel(t)$ for the partially frozen matrix shows a monotonic decline. These findings are striking, since they indicate that dominant large scale motions of the chains in the model systems studied occur in the direction not longitudinal but lateral to the chain contour when all chains are mobile. In other words, they show no evidence for the tube model, but suggest that the chain motion is essentially Rouse-like.

Figure 8–19 illustrates how the primitive chain for $N = 216$ and $\phi = 0.5$ changes its shape and position with the elapse of time. In each box, the thin

271

curve shows a given chain at $t = 0$, and the thick curve its snapshot taken at t corresponding to one of the arrows in Figure 8–17. The pictures indicate that the chain migrates not reptatively but rather isotropically. This is a very interesting finding. The times at which these snapshots are taken are smaller than the relaxation time for R. Therefore, if the tube existed, a substantial part of the chain would stay in it at these times.

The simulation data of Kolinski et al. demonstrate that each chain in concentrated systems wriggles isotropically, thus casting doubt about the reptation hypothesis.[4] This isotropic motion cannot be the same as the motion of an isolated chain in dilute solution, because it goes under the influences of other chains, which include bead–bead frictions at the points of chain contact, the topological constraint due to chain entanglement, and the backflow described by Kolinski et al. The latter refers to the motion of other chains induced by a diffusing chain to conserve the local density of the system. In order to elucidate the mechanism of polymer self–diffusion in concentrated solutions and melts we have to solve the dynamics of many chains by taking all these interactions into account. No one can deny the great role played by the reptation theory in approaching this extremely difficult problem. However, it seems to the author that the reptation idea puts too strong an emphasis on the topological constraint of entanglements.

5. Viscosity

5.1 The 3.4 Power Law

Measurements of steady–state viscosity η (at zero shear rate) on many species of flexible linear polymers in concentrated solutions and melts have established that a simple empirical equation [3]

$$\eta \sim M_\text{w}^{3.4} \quad (M_\text{w} > M_\text{c}) \tag{5.1}$$

holds with accuracy over a wide range of M_w. Though sometimes exponents slightly different from 3.4 are reported, they seldom go outside the range 3.3 – 3.7. In eq 5.1, M_w is the weight average molecular weight of the polymer and M_c is a constant depending on the polymer and solvent species as well as the polymer concentration c. Typically, $M_\text{c} \sim 2 - 3M_\text{e}$, where M_e is the molecular

---

[4]One may criticize the work of Kolinski et al. by saying that even the longest chain ($N = 216$) studied is not long enough to refute the reptation model. However, Kolinski et al. [57] note that their new simulation experiments on cubic lattice chains up to $N = 800$ packed at $\phi = 0.5$ gave the same qualitative results as those on diamond lattice chains.

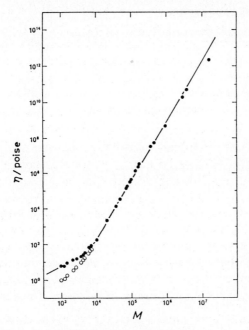

Fig. 8-20. Molecular weight dependence of η for PBD at
25°C [62]. •, corrected to the iso–free–volume state; ○,
original data. The lines have slopes of 1.0 and 3.4.

weight corresponding to n_e. Figure 8-20 illustrates accurate data of Colby et al.
[62] on narrow–distribution polybutadiene (PBD) melts. The data points spread
over the widest range of M ever covered by melt viscosity measurements.

The 3.4 power law for melt viscosity, eq 5.1, is of surprising generality, well
compared to the Houwink–Mark–Sakurada relation for intrinsic viscosity. Thus
its derivation by molecular theory has been one of the central themes in polymer
physics since it was proposed in 1951 by Fox and Flory [63]. Despite many
efforts no real success has as yet been achieved. The reader is advised to consult
Graessley's review [64] in 1974 concerning the earlier theories.

Dynamic properties of a polymer system reflect the process in which the polymer chains perturbed by an external force recover their equilibrium conformations by Brownian motion. Hence it is the key for their theoretical formulation
to clarify what mode of chain motions occurs in a given system. However, as
mentioned in the preceding sections, there is no established consensus about
this mode for entangled systems. This fact explains why the 3.4 power law for
viscosity still remains a mystery.

273

5.2 Theoretical Predictions

The phenomenological theory of linear viscoelasticity [3] relates η to τ_m and J_e by

$$\tau_m = J_e \eta \tag{5.2}$$

where τ_m is the maximum relaxation time for stress relaxation and J_e the recoverable shear compliance. Many published data show that J_e does not depend on M for M larger than about $4M_e$. Thus the 3.4 power law for η is actually equivalent to the 3.4 power law for τ_m, i.e.,

$$\tau_m \sim M^{3.4} \quad (M > 4M_e) \tag{5.3}$$

Stress relaxations at constant strain in entangled systems occur as the strained chains tend to recover equilibrium conformations by Brownian motion. Therefore it is reasonable to expect that τ_m has something to do with the rate at which the slowest mode of chain motion leading to steady self–diffusion takes place. An adequate measure of this rate is the average time τ_D in which the chain self–diffuses a distance equal to its radius of gyration. Thus, we have

$$\tau_D = \langle S^2 \rangle / (6D_s) \tag{5.4}$$

Since intrachain bead–bead repulsions are screened in entangled systems, $\langle S^2 \rangle$ may be equated to $\langle S^2 \rangle_\theta$ and hence vary linearly with M.

We ask what relation exists between τ_m and τ_D. Assuming that long–timescale motions of a chain occur dominantly in the reptation mode, Doi and Edwards [65] formulated an elegant theory of viscoelasticity of entangled polymer systems and showed that τ_m is related to τ_D by

$$\tau_m = \tau_d = (2/\pi^2)\tau_D \tag{5.5}$$

where τ_d is the disengagement time defined by eq 2.15 with eq 2.19 relating $D_p{}'$ to D_s. Equation 5.5 shows that τ_m is simply proportional to τ_D. Hence, a dimensionless parameter Γ defined by

$$\Gamma = \tau_m / \tau_D \tag{5.6}$$

is independent of M and equal to $2/\pi^2$ for systems of reptating chains. This parameter for systems of Rouse chains is also independent of M [64]. Thus, it is often postulated that the M independence of Γ holds for polymer concentrates regardless of the mechanism of long–timescale chain motion.

From this postulate it follows that if, as the reptation theory predicts, D_s varies linearly with M^{-2}, τ_m and hence η should increase in proportion to M^3.

274

In the early days, this consequence was highly praised as a shining success of the reptation model. However, many authors have shortly come to suspect that the difference between the predicted power 3 and the experimental one 3.4 is by no means minor.

Doi [66] was the first to attempt to explain this difference. He took thermal fluctuations of the length of the primitive chain into account and derived

$$\eta \sim M^3 \left[1 - \mu \left(\frac{M_e}{M} \right)^{1/2} \right]^3 \tag{5.7}$$

where μ is a parameter of order unity. The term multiplied by μ is the correction associated with chain length fluctuations. It was shown [66] that if an adequate value is chosen for μ, eq 5.7 gives a log η vs. log M relation which is closely approximated by the 3.4 power law in the range of M/M_e up to about 100, but eventually approaches the 3 power law at higher M/M_e. However, none of the previous authors have ever observed such asymptotic behavior. It is natural to consider that this might have been due to the limited ranges of M/M_e treated in their experiments. Thus, very recently, for a series of narrow–distribution polybutadiene samples with low vinyl content, Colby et al. [62] performed viscosity measurements over a wide range of M, from 1×10^3 to 1.65×10^7 (which corresponds to $0.5 < M/M_e < 8000$), and reported the data shown in Figure 8–20.

In the figure, we see that four data points in the high molecular weight region appear slightly below the indicated straight line of slope 3.4. To magnify the deviations Colby et al. [62] replotted the data as shown in Figure 8–21. The straight line represents the relation $\eta = 4.0 \times 10^{-12} M^{3.41}$ [62]. Several data points at M above about 10^6 apparently deviate from the 3.4 power law in the direction predicted by eq 5.7. However, the following remarks may be in order. The uppermost point in the graph is for a sample with $M = 3.74 \times 10^6$ and gives $\eta = 5.7 \times 10^{10}$ poises at 25°C. Experienced workers would not always agree that viscosities of this order and above in poises can be measured with an error less than 10%. Thus, the points showing definite departures from the indicated line are as uncertain as indicated by the error bars. Then it would be too strong to regard Figure 8–21 as evidence of the breakdown of the 3.4 power law. Colby et al. [62] showed that except for one at the highest M, the data points in this figure can be fitted by eq 5.7 with $\mu = 1.7$. However, this μ value is six times as large as the value of 0.29 estimated by Needs [67] from computer simulations. Needs showed that as M/M_e increases up to 60, chain length fluctuations raise the slope of log η vs. log M by only 0.06.

At present, there appears no definitive evidence against the general validity of the empirical 3.4 power law for τ_m. To reconcile this fact with the reptation

275

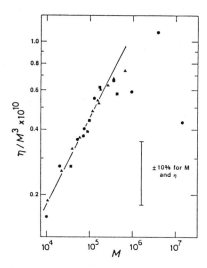

Fig. 8-21. Deviations of the melt viscosity of PBD from the 3 power law. •, data of Colby et al. [62]; ■, data of Struglinski and Graessley [68]; ▲, data of Roovers [69]. The line has a slope of 0.4.

prediction $D_s \sim M^{-2}$ it is necessary to abandon the postulate Γ = constant (see, for example, Ref. [70]). However, we then face a difficult problem of explaining why and how Γ depends on M. On the other hand, if we accept recent experimental data showing a stronger M dependence of D_s than predicted by the reptation theory, there is room for compromise between the 3.4 power law and constant Γ.

5.3 Remarks

1. The Doi–Edwards theory of linear viscoelasticity predicts 1.2 for $J_e G_N^0$, where G_N^0 is the plateau shear modulus. This value is significantly lower than typical experimental values found in the range 2–4 [3, 71]. This defect of the Doi–Edwards theory, along with its failure of predicting the 3.4 power law for viscosity, has been pointed out by Osaki and Doi [72]. It is associated with the fact that the Doi–Edwards theory yields a relaxation time distribution which is too narrow compared with observed ones [69]. Modifications of the Doi–Edwards theory have been made so as to bring $J_e G_N^0$ closer to measured values, but no remarkable success has as yet been achieved.

2. In many branches of science, technology, and engineering it is well known that relaxation processes in condensed systems including glasses and polymeric

276

materials are often described accurately by a stretched exponential form:

$$\phi(t) \sim \exp\left(-at^{\alpha}\right) \tag{5.8}$$

where a and α are constants and typically the latter is found in the range 0.3–0.7. This form is usually called the Williams–Watts equation [73]. In recent years, there has emerged a considerable interest in its theoretical derivation. Typical attempts are summarized and discussed by Rendell and Ngai [74]. Here we refer to the coupling theory proposed by Ngai [75]. When applied to the viscoelasticity of polymer concentrates, this very general theory predicts [74, 76] that not only is the relaxation modulus $G(t)$ represented by

$$G(t) = G_0 \exp[-(t/\tau_e)^{1-n}] \quad (0 < n < 1) \tag{5.9}$$

but the parameter τ_e is related to the "primitive" relaxation time τ_0 by

$$\tau_e \sim [(1-n)\tau_0]^{1/(1-n)} \tag{5.10}$$

where n is a measure of the strength with which the chain couples to its surroundings, and "primitive" refers to the uncoupled chain. Ngai and Plazek [76] took the terminal relaxation time of an isolated Rouse chain (τ_{Rouse} given by eq 4.4) for τ_0. Then it can be shown that

$$\eta \sim M^{2/(1-n)} \tag{5.11}$$

and

$$J_e G_N^0 = (1-n)\Gamma(2/(1-n))/[\Gamma(1/(1-n))]^2 \tag{5.12}$$

where Γ is the gamma function. If n is chosen to be 0.412, these equations give $\eta \sim M^{3.4}$ and $J_e G_N^0 = 2.11$, in good agreement with experiment.

McKenna et al. [77] equated τ_D in eq 5.4 to τ_e. Then it follows (with the assumption $\langle S^2 \rangle \sim M$) that

$$D_s \sim M^{-(1+n)/(1-n)} \tag{5.13}$$

If n is taken to be 1/3, this predicts $D_s \sim M^{-2}$. Thus the reptation exponent is derived without invoking the reptation idea. For $n = 0.412$ eq 5.12 gives $D_s \sim M^{-2.40}$, which comes closer to the recent data mentioned in the previous section. Hence we have no special reason to choose different n values for η and D_s, as McKenna et al. did.

In connection with these remarkable results, the following notes may be in order. Firstly, the coupling theory of Ngai does not specify the interactions of chains with their surroundings. Thus, it does not say anything about the dynamics of chain entanglement. Secondly, the assumption of McKenna et al. who replaced τ_D by τ_e seems ad hoc to the author. Thirdly, Ngai's theory refers to a monodisperse polymer. Its generalization to multi–component systems, for example, quasi–binary solutions, remains unexplored. It is not immediately apparent how this can be done. Fourthly, although various methods have been proposed to derive the Williams–Watts equation, as Rendell and Ngai [74] emphasize, none of them except for Ngai's have paid attention to the structure of the parmeter a in eq 5.8 (or the factor τ_e in eq 5.9). The advantage of Ngai's theory is that eq 5.10 makes it possible to test the theory with experiment through eq 5.11 and 5.12.

References

1. P.-G. de Gennes, J. Chem. Phys. 55, 572 (1971).
2. M. Tirrell, Rubber Chem. Tech. 57, 52 (1984).
3. J. D. Ferry, "Viscoelastic Properties of Polymers," Wiley, New York, 1980.
4. M. Doi and S. F. Edwards, "The Theory of Polymer Dynamics," Oxford Univ. Press, London, 1986.
5. S. F. Edwards, Polymer 6 , 143 (1977).
6. M. Doi and S. F. Edwards, J. Chem. Soc. Farad. Trans. II 74, 1789 (1978) (this paper deals with self–diffusion. See Ref. 65 concerning rheological properties).
7. J. Klein, Macromolecules 11, 852 (1978); 19, 105 (1986).
8. M. Daoud and P.-G. de Gennes, J. Polym. Sci., Polym. Phys. Ed. 17, 1971 (1979).
9. W. W. Graessley, Adv. Polym. Sci. 47, 67 (1982).
10. M. Adam and M. Delsanti, Macromolecules 10, 1229 (1977).
11. F. Brochard and P.-G. de Gennes, Macromolecules 10, 1157 (1977).
12. P. E. Rouse, J. Chem. Phys. 21, 1272 (1953).
13. W. Hess. Macromolecules 19, 1385 (1986).
14. J. Skolnick, R. Yaris, and A. Kolinski, J. Chem. Phys. 88, 1407 (1988).
15. F. Bueche, W. M. Cashin, and P. Debye, J. Chem. Phys. 20, 1956 (1952).
16. Y. Kumagai, H. Watanabe, K. Miyasaka, and T. Hata, J. Chem. Eng. Jpn. 12, 1 (1979).
17. R. Buchus and R. Kimmich, Polymer 24, 964 (1983).
18. G. Fleischer, Polym. Bull. 9, 152 (1983).
19. M. Antonietti, J. Coutandin, and H. Sillescu, Macromolecules 19, 793 (1986).
20. E. J. Kramer, cited in Ref. 2 as private communication.
21. F. Bueche, J. Chem. Phys. 48, 1410 (1968).
22. D. W. McCall, D. C. Douglas, and E. W. Anderson, J. Chem. Phys. 30, 771 (1959).
23. G. Fleischer, Polym. Bull. 11, 75 (1984); see G. Fleischer, Colloid Polym. Sci. 265, 89 (1987).
24. J. Klein and B. J. Briscoe, Proc. Roy. Soc. (London) A 365, 53 (1979).
25. J. Klein, D. Fletcher, and L. J. Fetters, Nature 304, 526 (1983).
26. C. R. Bartels, B. Crist, and W. W. Graessley, Macromolecules 17, 2707 (1984).
27. J. E. Tanner, Macromolecules 4, 748 (1971).
28. J. Klein, Phil. Mag. A 43, 771 (1981); Macromolecules 14, 460 (1981).
29. P. F. Green and E. J. Kramer, Macromolecules 19, 1108 (1986).

30. N. Nemoto, M. R. Landry, I. Noh, T. Kitano, J. A. Wesson, and H. Yu, Macromolecules **18**, 308 (1985).

31. P. T. Callaghan and E. W. Pinder, Macromolecules **17**, 431 (1984).

32. H. Kim, T. Chang, J. M. Yohanan, L. Wang, and H. Yu, Macromolecules **19**, 2937 (1987).

33. M. Adam and M. Delsanti, J. Phys. (Paris) **44**, 1185 (1983).

34. N. Nemoto, T. Kojima, T. Inoue, M. Kishine, T. Hirayama, and M. Kurata, Macromolecules **22**, 3793 (1989).

35. H. Yu, in "Molecular Conformation and Dynamics of Macromolecules in Condensed Systems," Ed. M. Nagasawa, Elsevier, Amsterdam, 1988.

36. M. R. Landry, Ph. D. Thesis, Univ. Wisconsin–Madison, 1985.

37. J. Hadgraft, A. J. Hyde, and R. W. Richards, J. Chem. Soc. Farad. Trans. II **75**, 1495 (1979). (1986).

38. J. E. Martin, Macromolecules **17**, 1279 (1984); **19**, 922 (1986).

39. N. Numazawa, K. Kuwamoto, and T. Nose, Macromolecules **19**, 2593 (1986).

40. N. Numazawa, T. Hamada, and T. Nose, J. Polym. Sci., Polym. Phys. Ed. **24**, 19 (1986).

41. T. P. Lodge, Macromolecules **16**, 1393 (1983).

42. B. Hanley, M. Tirrell, and T. P. Lodge, Polym. Bull. **14**, 137 (1985).

43. L. M. Wheeler, T. P. Lodge, B. Hanley, and M. Tirrell, Macromolecules **20**, 1120 (1987).

44. T. P. Lodge and L. M. Wheeler, Macromolecules **19**, 2983 (1986).

45. P.-G. de Gennes, Physics (USA) **3**, 37 (1967); the corrected result is given in Z. Akcasu, M. Benmouna, and C. C. Han, Polymer **21**, 866 (1980).

46. E. Doubois-Violette and P.-G. de Gennes, Physics (USA) **3**, 181 (1967).

47. P.-G. de Gennes, J. Phys. (Paris) **42**, 735 (1981).

48. F. Mozei, Z. Phys. **255**, 146 (1972).

49. D. Richter, A. Baumgärtner, K. Binder, B. Ewen, and J. B. Hayter, Phys. Rev. Lett. **47**, 109 (1981); Phys. Lett. **48**, 1695 (1982).

50. J. M. Deutsch and N. D. Goldenfeld, Phys. Rev. Lett. **48**, 1694 (1982).

51. D. Richter, A. Baumgärtner, K. Binder, B. Ewen, and J. B. Hayter, Phys. Rev. Lett. **48**, 1695 (1982).

52. B. Ewen, Pure & Appl. Chem. **56**, 1407 (1984).

53. J. S. Higgins, L. K. Nicholson, and J. B. Hayter, Polymer **22**, 163 (1981).

54. G. D. J. Phillies, Macromolecules **19**, 2367 (1986).

55. P. T. Callaghan and D. N. Pinder, Macromolecules **13**, 1085 (1980); **14**, 1334 (1981); **16**, 968 (1983).

56. A. Baumgärtner, Ann. Rev. Phys. Chem. **35**, 419 (1984).

57. A. Kolinski, J. Skolnik, and R. Yaris, J. Chem. Phys. **86**, 1567 (1987).

58. K. Kremer, Macromolecules **16**, 1632 (1983).

59. A. Kolinski, J. Skolnick, and R. Yaris, J. Chem. Phys. **84**, 1922 (1986).

60. C. C. Crabb and J. Kovac, Macromolecules **18**, 1430 (1985).

61. J. M. Deutsch, J. Phys. (Paris) **48**, 141 (1987).

62. R. H. Colby, L. J. Fetters, and W. W. Graessley, Macromolecules **20**, 2226 (1987).

63. T. G Fox and P. J. Flory, J. Phys. Colloid Chem. **55**, 221 (1955).

64. W. W. Graessley, Adv. Polym. Sci. **47**, 67 (1982).

65. M. Doi and S. F. Edwards, J. Chem. Soc. Farad. Trans. II **74**, 1802, 1818 (1978); **75**, 38 (1978); see also M. Doi, J. Polym. Sci., Polym. Phys. Ed. **18**, 1005 (1980).

66. M. Doi, J. Polym. Sci., Polym. Lett. Ed. **19**, 265 (1981); J. Polym. Sci., Polym. Phys. Ed. **21**, 667 (1983).

67. R. J. Needs, Macromolecules **17**, 437 (1984).

68. M. J. Struglinski and W. W. Graessley, Macromolecules **18**, 2630 (1985).

69. J. Roovers, Polym. J. **18**, 153 (1986).

70. H. Scher and M. F. Shlesinger, J. Chem. Phys. **84**, 5922 (1986).

71. W. W. Graessley, Polymer **22**, 1329 (1981).

72. K. Osaki and M. Doi, Polym. Eng. Rev. **4**, 35 (1984).

73. G. Williams and D. C. Watts, Trans. Farad. Soc. **66**, 80 (1970).

74. R. W. Rendell and K. L. Ngai, in "Relaxations in Complex Systems," Ed. K. L. Ngai and G. B. Wright, Naval Res. Lab., Washington, DC, 1986, p. 309.

75. K. L. Ngai, Comments Solid State Phys. **9**, 127 (1979); **9**, 141 (1980).

76. K. L. Ngai and D. J. Plazek, J. Polym. Sci., Polym. Phys. Ed. **23**, 2159 (1985); **24**, 619 (1986).

77. G. M. McKenna, L. K. Ngai, and D. J. Plazek, Polymer **26**, 1651 (1985); G. B. McKenna, Faraday Symp. Chem. Soc. **18**, 201 (1984).

Chapter 9   Phase Equilibria

1.   Thermodynamic Relations

1.1   Introduction

Though classic, the phase equilibrium of polymer solutions is not a closed subject of research. Academically, the most basic problem in its study is to find analytical expressions for $\triangle G$ which allow quantitative prediction of the phase relationships of various polymer solutions. Here $\triangle G$ denotes the Gibbs free energy of mixing for the system. Attempts to solve this problem date back to the famous Flory–Huggins theory. Yet, no fully satisfactory $\triangle G$ is found even for the simplest of polymer solutions, i.e., a binary system made up of a monodisperse (homo) polymer and a pure solvent. As defined in the previous chapter, quasi–binary solutions consist of two or more homologous monodisperse polymers and a single solvent. They are the simplest of multi-component polymer solutions, and have been the object of extensive research by those who were interested in seeing the effects of component multiplicity on the phase equilibrium behavior of polymer solutions. Despite a great many efforts, the expressions derived so far, either theoretically or empirically, for $\triangle G$ of quasi–binary solutions remain semi–quantitative.

Any actual synthetic homopolymer consists of a virtually infinite number of homologous species differing in chain length. Its solution in a single solvent thus gives a quasi–binary solution characterized by a continuous distribution of chain length. We may call such a solution truly polydisperse. Being formulated for discrete systems, the traditional thermodynamics is of no use for it. Surprisingly, it was only about a decade ago that the formulation of the thermodynamics of truly polydisperse solutions started, and the novel theory, usually called continuous thermodynamics, is still in the development stage. This means that at present we have no established procedure for thermodynamic analysis of phase equilibrium data on truly polydisperse solutions.

Other multicomponent polymer solutions of interest consist of chemically different polymers dissolved in a solvent or a single polymer in a mixture of different solvents. Their phase equilibrium behavior is important in relation to many practical problems, but it is much more complex and difficult to analyze and predict than that of quasi–binary systems.

In this chapter, we describe the development of typical thermodynamic studies that have been made for quantitative predictions of phase equilibria in binary and "paucidisperse" quasi–binary polymer solutions. In so doing, we are concerned only with systems which phase–separate when cooled, i.e., ones which have an upper critical point.

1.2 Thermodynamic Approach

In principle, it is possible to calculate ΔG of a given system by applying statistical mechanics to its relevant physical model. However, this theoretical method encounters many difficulties in complex systems such as polymer solutions. In fact, none of the available mathematical expressions for ΔG for polymer solutions are yet fully adequate for predicting phase equilibrium behavior quantitatively.

Another route to find ΔG is what is usually called the thermodynamic or phenomenological approach. In this, we assume an appropriate expression for ΔG on the basis of experiments or some theoretical considerations and evaluate the parameters involved from a set of relevant equilibrium data. We then apply the expression of ΔG so determined to calculate other equilibrium properties and compare the results with the corresponding experimental data. The comparison will show us how adequate the chosen ΔG is and how it should be modified for better agreement with experiment. Thus, in principle, it is possible to approach step by step a more correct ΔG of the system under study. Being purely empirical, this method may not be appealing for those who are primarily interested in events at the molecular level. However, it often allows us to know what thermodynamic factors play a role in controlling the phenomena concerned. The availablity of such information is indeed essential for theoreticians who want to make a more accurate formulation of ΔG by statistical mechanical theory.

1.3 Basic Variables

We consider a quasi–binary solution of $q + 1$ components. When $q = 1$, the system will be referred to as binary. We designate the solvent as component 0 and q monodisperse homologous homopolymers as components $1, 2, \ldots, q$ in the order of increasing relative chain length. Here, the relative chain length P_i of component i is defined by

$$P_i = V_i/V_0 \qquad (1.1)$$

with V_i being the partial molar volume of component i. We assume that V_i is independent of solution composition as well as temperature T and pressure p. This assumption is to neglect volume changes of the system which occur with changes in T and p and upon mixing different components.[1] With it, P_i becomes a constant characteristic of component i (we note $P_0 = 1$), and the variable p can be omitted from all the expressions presented below.

---

[1] This assumption is inadequate in discussing systems which phase–separate when T is raised.

We denote the amount of component i in moles by n_i. Then the volume V of the solution can be expressed by

$$V = V_0(n_0 + \sum_{i=1}^{q} P_i n_i) \tag{1.2}$$

The composition of the solution can be represented by a set of q independent variables $\phi_i (i = 1, 2, \ldots, q)$, where

$$\phi_i = V_0 n_i / V \tag{1.3}$$

Subject to the above-mentioned assumption that no volume change occurs on mixing, ϕ may be referred to as the volume fraction of component i. For simplicity, we denote the q independent ϕ_i by a symbol $\{\phi_i\}$. With eq 1.1 and 1.2, we may write eq 1.3 as

$$\phi_i = P_i n_i / (n_0 + \sum_{j=1}^{q} P_j n_j) \tag{1.4}$$

The total volume fraction ϕ of the polymer mixture in the solution is the sum of ϕ_i over all polymer components, i.e.,

$$\phi = \sum_{i=1}^{q} \phi_i \tag{1.5}$$

We introduce the weight fraction w_i of polymer component i in the polymer mixture, i.e.,

$$w_i = \Omega_i / \sum_{k=1}^{q} \Omega_k \quad (i = 1, 2, \ldots, q) \tag{1.6}$$

where Ω_i is the weight of polymer component i. Since the sum of w_i over all polymer components is unity, we have $q - 1$ independent w_i, which are hereafter denoted by $[w_i]$. The weight-basis chain length distribution $f(P)$ in the polymer mixture is then determined by $[w_i]$ and $\{P_i\}$, where $\{P_i\}$ stands for a set of $P_1, P_2, \ldots, P_q$. As is well known experimentally, the specific volumes of homologous polymers do not depend on chain length except for oligomers. Hence we hereafter equate $[w_i]$ to $[\xi_i]$, where $[\xi_i]$ denotes a set of $q - 1$ independent ξ_i, with ξ_i defined by

$$\xi_i = \phi_i / \phi \quad (i = 1, 2, \ldots, q) \tag{1.7}$$

284

We may then express $f(P)$ in terms of $[\xi_i]$ and $\{P_i\}$. In what follows, we choose $\xi_2, \xi_3, \ldots, \xi_q$ for $[\xi_i]$.

The composition of a quasi–binary solution can be specified by $\{\phi_i\}$, but, with eq 1.7, it also can be represented by a set of ϕ and $[\xi_i]$. Practically, the latter representation is sometimes more useful, since the total polymer concentration ϕ is convenient as an experimental variable.

We denote ΔG per unit volume of solution by $\Delta\overline{G}$. Thermodynamics tells us that $\Delta\overline{G}$ of a quasi–binary solution is a function of T, p, and $\{\phi_i\}$. For a given polymer species it also may depend on the distribution of relative chain length in the polymer mixture. Thus, when the pressure effect is not considered, the basic variables for $\Delta\overline{G}$ of quasi–binary solutions are T, ϕ, $[\xi_i]$, and $\{P_i\}$. Since the combination of $[\xi_i]$ and $\{P_i\}$ defines $f(P)$, we may take T, ϕ, and $f(P)$ as the basic variables. For a paucidisperse polymer, q is finite and $f(P)$ is represented by q delta functions with different strength, while for a truly poly-disperse polymer, q is infinitely large and $f(P)$ becomes a continuous function of P. Thus, $\Delta\overline{G}$ for quasi–binary solutions of a truly polydisperse polymer is a functional with respect to $f(P)$, and requires sophisticated mathematics for its treatment.

1.4 Phase Relationships

1.4.1 Spinodal and Critical State

The derivation of the phase relationships for quasi–binary solutions is given in textbooks of polymer thermodynamics [1, 2]. Here, the results necessary for the subsequent discussion are summarized.

The boundary (line, surface, etc.) demarcating a metastable region from an unstable region in the phase space[2] for a system is called the **spinodal** and is given by the equation

$$
J_{\text{sp}} \equiv
\begin{vmatrix}
\Delta\overline{G}_{11} & \Delta\overline{G}_{12} & \cdots & \Delta\overline{G}_{1q} \\
\Delta\overline{G}_{21} & \Delta\overline{G}_{22} & \cdots & \Delta\overline{G}_{2q} \\
\vdots & \vdots & \ddots & \vdots \\
\Delta\overline{G}_{q1} & \Delta\overline{G}_{q2} & \cdots & \Delta\overline{G}_{qq}
\end{vmatrix}
= 0
\tag{1.8}
$$

where

$$
\Delta\overline{G}_{ij} = \partial^2 \Delta\overline{G}/\partial\phi_i\partial\phi_j
\tag{1.9}
$$

---

[2] A $q + 1$–dimensional space spun by $q + 1$ variables consisting of T and q composition variables if p is precluded.

The relation determined by combining eq 1.8 with the equation

$$
\begin{vmatrix}
\partial J_{\mathrm{sp}}/\partial \phi_1 & \partial J_{\mathrm{sp}}/\partial \phi_2 & \cdots & \partial J_{\mathrm{sp}}/\partial \phi_q \\
\Delta \overline{G}_{12} & \Delta \overline{G}_{22} & \cdots & \Delta \overline{G}_{q2} \\
\vdots & \vdots & \ddots & \vdots \\
\Delta \overline{G}_{1q} & \Delta \overline{G}_{2q} & \cdots & \Delta \overline{G}_{qq}
\end{vmatrix} = 0 \qquad (1.10)
$$

defines a point, line, etc. on the spinodal line, surface, etc., respectively. Each point, line, etc. thus defined is called the **critical state**. Recently, Šolc and Koningsveld [3] pointed out an important precaution to be taken when eq 1.8 and 1.10 are combined.

1.4.2   Binodal and Coexistence Curves

When the solution under consideration separates into two or more equilibrium phases at temperature T, the following phase equilibrium relations must hold:

$$
\Delta \mu_i(T, \phi', f(P)') = \Delta \mu_i(T, \phi'', f(P)'') \quad (i = 0, 1, \ldots, q) \qquad (1.11)
$$

Here $\Delta \mu_i$ is the chemical potential of component i in the solution minus that in the pure state (this definition of $\Delta \mu_i$ implies that we have defined ΔG by choosing the pure state of each component as the reference state), and single and double primes signify the pair of separated phases.

The law of mass conservation combined with the condition of no volume change on mixing allows us to obtain

$$
\frac{\phi_i - \phi_i'}{\phi_1 - \phi_1'} = \frac{\phi_i'' - \phi_i'}{\phi_1'' - \phi_1'} \quad (i = 2, 3, \ldots, q) \qquad (1.12)
$$

where the quantity with no prime refers to the solution before phase separation, which is hereafter referred to as the "mother" solution.

Equation 1.11 along with eq 1.12 gives us $2q$ relations for $2q$ variables consisting of ϕ', ϕ'', $f(P)'$, and $f(P)''$. Hence, in principle, it is possible to calculate from these equations the compositions of the two phases separated at temperature T from a mother solution specified by ϕ and $f(P)$. In practice, the calculation becomes exponentially difficult as the number of polymer components increases. The composition of each phase is geometrically represented by a point in a q–dimensional space defined by q composition variables. The point describes a curve when the composition of the mother solution is varied. The resulting curve is called the **binodal**. It varies with changing T, giving a surface which is called the binodal surface. Though the proof is omitted here, the binodals for different phases separated at T meet a point, and this point defines the critical state at the temperature concerned.

The values of ϕ' and ϕ'' for a given mother solution vary with T. The curves obtained when T is plotted against ϕ' and ϕ'' are called **conjugate coexistence curves**. They change as the composition of the mother solution is varied.

1.4.3 Cloud-point and Shadow Curves

When a uniform mother solution is cooled very slowly, a second phase begins to separate at a definite temperature T_p, which is called the **cloud-point temperature** or the cloud point. The **cloud-point curve** refers to T_p plotted against the composition ϕ of the mother solution. Since the second phase is infinitesimal in volume at the cloud point, we see from eq 1.11 that the cloud-point curve must satisfy the conditions:

$$\Delta \mu_i(T_p, \phi, f(P)) = \Delta \mu_i(T_p, \phi'', f(P)'') \quad (i = 0, 1, \dots, q) \tag{1.13}$$

Given $f(P)$, these $q + 1$ equations can, in principle, be solved for T_p, ϕ'', and $f(P)''$ as functions of ϕ. The resulting relation between T_p and ϕ gives the cloud-point curve, while that between T_p and ϕ'' gives a line called the **shadow curve**. The latter cannot be determined experimentally, since the second phase is too small in volume to be analyzed for the composition. It can be shown that the cloud-point and shadow curves coincide with the conjugate coexistence curves when and only when the solution is strictly binary. This fact is important, because some authors make no distinction between cloud-point curve and coexistence curve in describing phase equilibria of polydisperse solutions.

1.4.4 Experimental

Experimentally, T_p can be determined by finding the temperature at which a given mother solution begins to be turbid when cooled very slowly. Coexistence curves can be determined by allowing a given solution to phase-separate at a series of T and measuring the total polymer volume fractions in the two phases. The binodal at given T can be obtained by measuring the compositions of the phases separated at that temperature from mother solutions with different ϕ. The composition analysis is effectively made by use of high-resolution gel permeation chromatography (GPC). The binodal can also be constructed from a set of cloud-point curves for mother solutions with different $f(P)$. The spinodal projected on the $T - \phi$ plane can be obtained by using the principle that the intensity of scattered light diverges as T approaches the spinodal temperature T_s. The pulse-induced critical scattering (PICS) technique [4] is known to be very useful for the routine determination of T_s.

1.5 Empirical Expressions for $\Delta\overline{G}$ and Chemical Potentials

To approach $\Delta\overline{G}$ of quasi-binary polymer solutions in an empirical way it is usual to invoke the Koningsveld–Staverman formalism [5], which chooses the familiar Flory–Huggins (FH) athermal solution as the reference system and expresses $\Delta\overline{G}$ as

$$\Delta\overline{G} = RT[(1 - \phi)\ln(1 - \phi) + \sum_{j=1}^{q}(\xi_j\phi/P_j)\ln(\xi_j\phi)$$
$$+ g(T, \phi, f(P))\phi(1 - \phi)] \tag{1.14}$$

where RT has the usual meaning. The factor g combines all the departures of a given quasi-binary solution from the reference FH solution, and may contain both entropic and enthalpic contributions. We note that there is no guarantee that the mixing entropy for quasi-binary solutions is given exactly by the Flory–Huggins term. Thermodynamics tells us nothing about g, but we can say that as ϕ approaches unity, this factor should diverge to infinity in such a way that $g(1 - \phi)$ remains finite. This is because otherwise $\Delta\overline{G}$ of an undiluted mixture of polymers $1, 2, \ldots, q$ is given by $RT\sum_{j=1}^{q}(\xi_j/P_j)\ln\xi_j$, which is not correct in general.

It follows from eq 1.14 that

$$\Delta\mu_0 = RT[\ln(1 - \phi) + (1 - P_\mathrm{n}^{-1})\phi + \chi(T, \phi, f(P))\phi^2] \tag{1.15}$$

where P_n is the number–average relative chain length of the polymer mixture in the solution, expressed as

$$P_\mathrm{n}^{-1} = \sum_{j=1}^{q}\xi_j/P_j \tag{1.16}$$

and χ is related to g by

$$\chi = g - (1 - \phi)(\partial g/\partial\phi)_{T,f(P)} \tag{1.17}$$

Equation 1.15 may be regarded as the defining equation for the parameter χ, since $\Delta\mu_0$ can be evaluated by use of such standard techniques as osmotic pressure, light scattering, and sedimentation equilibrium. In the classic FH theory, χ is taken as the strength of the polymer–solvent interaction. However, in the Koningsveld–Staverman formalism, it has no such meaning but is an empirical function which absorbs all the deviations of $\Delta\mu_0$ in an actual quasi-binary solution from that in the reference FH solution.

Since, differing from $\Delta\mu_0$, $\Delta\overline{G}$ cannot be evaluated by experiment, g is not a measurable quantity, but it can be expressed in terms of χ as

$$g = g(T, 0, f(P)) - \frac{1}{1-\phi} \int_0^\phi \chi(T, u, f(P))\, du \qquad (1.18)$$

which follows from eq 1.17. The integration constant $g(T, 0, f(P))$ has to be left as an unknown. However, this fact does not matter for effecting thermodynamic approaches to phase equilibrium problems.

To solve the phase equilibrium relations we have to derive the expressions for $\Delta\mu_i$ for all the polymer components ($i = 1, 2, \ldots, q$). When $\Delta\overline{G}$ is given by eq 1.14, they are as follows:

$$\begin{aligned}
\Delta\mu_i = RT[&\ln(\xi_i\phi) - (P_i - 1) + P_i(1 - P_n^{-1})\phi \\
&+ P_i(1 - \phi)\{(1 - \phi)[g + (\partial g/\partial\phi)\phi] \\
&+ \phi[m_i(\partial g/\partial\xi_i) - \sum_{j=1}^{q-1} \xi_j(\partial g/\partial\xi_j)]\}]
\end{aligned} \qquad (1.19)$$

where $m_i = 1$ for $i \neq q$ and $= 0$ for $i = q$.

1.6 Special Cases

When g or χ does not depend on $f(P)$, it follows from eq 1.8 that the spinodal projected on the $T - \phi$ plane is given by

$$(1 - \phi)^{-1} + (P_w\phi)^{-1} - 2\chi(T, \phi) - \phi(\partial\chi/\partial\phi)_T = 0 \qquad (1.20)$$

where P_w is the weight–average relative chain length of the polymer mixture, defined by

$$P_w = \sum_{j=1}^q \xi_j P_j \qquad (1.21)$$

Moreover, it can be shown that eq 1.10 for such g reduces to

$$(1 - \phi)^{-2} - P_z(P_w)^{-2} - 3(\partial\chi/\partial\phi)_T - \phi(\partial^2\chi/\partial\phi^2)_T = 0 \qquad (1.22)$$

where

$$P_z = \sum_{j=1}^q \xi_j P_j^2 / P_w \qquad (1.23)$$

Equations 1.20 and 1.22 were first derived by Koningsveld and Staverman [6], and allow the critical point (T_c, ϕ_c) on the $T - \phi$ plane to be determined when

289

g or χ is given as a function of T and ϕ. If g or χ is independent of ϕ as well as of $f(P)$, these equations give

$$\phi_c = 1/(1 + P_w P_z^{-1/2}) \qquad (1.24)$$

$$\chi(T_c) = (1/2)(1 + P_z^{1/2} P_w^{-1})(1 + P_z^{-1/2}) \qquad (1.25)$$

which were derived by Stockmayer [7] as early as 1949.

The original FH theory corresponds to the case in which g depends only on T. Being relatively easy to treat analytically, the phase relationships of this special case have been theoretically investigated in detail. Notably, Šolc [8] clarified how appreciably they are affected by the shape of $f(P)$ and how complex they become as the number of polymer components increases. His theoretical work provides useful guidelines in analyzing phase separation data on actual polymer solutions, but g of actual systems seldom depends only on T.

2. Binary Solutions

2.1 Pioneering Work of Shultz and Flory

Whatever property we may be concerned with, detailed work on binary so-lutions is the prerequisite for exploring more complex solutions. Thus, phase equilibria in binary solutions of a monodisperse homopolymer and a pure sol-vent have been extensively investigated by many authors. However, progress toward more accurate g or χ has taken place rather slowly. It consisted of step by step determination of the factors which control these empirical functions. The results are sketched in this section 2.

For a binary solution the function $f(P)$ reduces to a single parameter P, the relative chain length of the polymer, so that g or χ depends on T, ϕ, and P. The basic task in the thermodynamic study of binary solutions is to find this dependence by appropriate experiments.

Modern work on phase equilibrium in polymer solutions was initiated by Shultz and Flory [9] in 1952. They measured accurately the cloud–point curves for well–fractionated polystyrene and poly(isobutylene) with different P in cy-clohexane for the former and in diisobutyl ketone for the latter, and examined how closely the observed data can be described by the theory with g assumed to depend only on T. In Figure 9–1, the solid curves are drawn to fit the data points, while the dashed curves are calculated results with $\chi = -0.556 + 324.3/T$ (T is in kelvin units).

Theoretically, the critical point (T_c, ϕ_c) for any binary system appears at the top (sometimes at the bottom) of the cloud–point curve, regardless of whether g or χ depends on ϕ or not [2]. Thus, Shultz and Flory identified the maxima

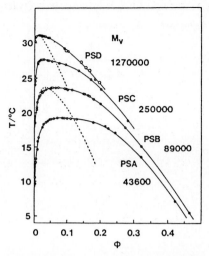

Fig. 9–1. Cloud–point curves for PS fractions with different M (as indicated) in CH. Solid lines, fitting data points. Dashed lines, calculated with χ depending only on temperature [9].

of the solid curves in Figure 9–1 as the critical points for respective molecular weights, and found that the experimental T_c can be predicted accurately if the above expression for $\chi(T)$ is used. For a binary solution with χ dependent only on temperature, eq 1.24 reduces to

$$\phi_c = 1/(1 + P^{1/2}) \tag{2.1}$$

The values of ϕ_c calculated from this equation give the concentrations at the tops of the dashed curves in Figure 9–1. Though not apparent from the figure, these concentrations are significantly lower than those at the tops of the corresponding solid curves. Another discrepancy is that the dashed curves give much narrower miscibility gaps than the solid ones. In this connection, we note that if judged from the data obtained later on narrow–distribution polystyrene, the polymer samples used by Shultz and Flory were not always well–fractionated.

The data obtained for poly(isobutylene) in diisobutyl ketone [9] were similar to those displayed in Figure 9–1. However, the discrepancies between theory and experiment concerning ϕ_c and the miscibility gap were not as marked as seen in this figure.

2.2   Concentration Dependence

The findings of Shultz and Flory showed that g must be treated as a function of ϕ and P as well as T in order to quantify the description of phase equilibrium

291

behavior. Already in 1952 Tompa [10] found ϕ_c to increase appreciably if χ is allowed to increase linearly with ϕ. Later in 1961 Maron and Nakajima [11] showed that calculated miscibility gaps are markedly widened when a factor essentially corresponding to g increases linearly with ϕ. Thus, the importance of the concentration dependence of g was clear in the early 1960s.

More generally, g as a function of ϕ may be expressed by a polynomial of ϕ. Koningsveld et al. [12] in 1970 truncated it at the quadratic term and tried to evaluate the coefficients from their own critical–point data for polystyrene in cyclohexane. Actually, when only critical–point data are available, this trunca-tion is mandatory even if the coefficients are assumed to be independent of T. Their result is

$$g = 0.4099 + 90.65/T + 0.2064\phi + 0.0518\phi^2 \qquad (2.2)$$

which yields for χ

$$\chi = 0.2035 + 90.65/T + 0.3092\phi + 0.1554\phi^2 \qquad (2.3)$$

This equation predicts the enthalpy of mixing to be concentration–independent, which is not valid in general [2]. To be free from this inconsistency it is necessary to make the coefficients of the polynomial temperature–dependent. However, we then have to truncate the polynomial at the term linear in ϕ, provided that no phase data other than critical–point ones are available. By applying such an alternative maneuver to the critical–point data of Koningsveld et al. Kurata [2] obtained

$$\chi = 0.3015 + 61.0/T + (-0.300 + 190/T)\phi \qquad (2.4)$$

However, eq 2.3 predicts more accurately χ derived from osmotic pressure, light scattering, and sedimentation equilibrium measurements on one–phase solutions [12].

Koningsveld and Kleintjens [13] proposed to express g in a closed form as

$$g = \alpha + \frac{\beta_0 + \beta_1/T}{1 - \gamma\phi} \qquad (2.5)$$

where α, β_0, β_1, and γ are parameters independent of T and P. Some theoretical support for this form exists [13]. It was shown that eq 2.5 also describes the critical–point data of Koningsveld et al. [12] accurately if the values of the parameters are suitably chosen. But, when eq 2.5 with the same parameter values was used to calculate spinodals for different P, the results deviated significantly from the experimental ones reported by Scholte [14], as can be

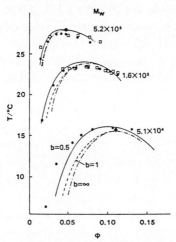

Fig. 9-2. Spinodals for PS in CH, Points, measured [4, 14]. Dot-dash lines, calculated with eq 2.5 for g. Solid and dashed lines, calculated with the hybrid theory [15].

seen in Figure 9-2 (the dot-dash lines for $b = \infty$ give the calculated values based on eq 2.5). We note that the discrepancy is more marked at at lower ϕ and P.

2.3 Chain Length Dependence

The finding of Koningsveld and Kleintjens showed that treating g as a function of T and ϕ did not suffice to attain an accurate description of phase equilibrium behavior of binary solutions. It appeared that Koningsveld and his group had recognized by the early 1970s the need for taking the chain length effect into account. In an important paper of 1974, Koningsveld et al. [15] offered three mechanisms which would make g dependent on P: (i) the chain-end effect, (ii) excluded-volume effect, and (iii) dilute-solution effect. They considered (i) to be unimportant, and formulated a theory of g incorporating (ii) and (iii), which is called the hybrid theory.

The basic concept of Koningsveld et al. was that the thermodynamic behavior of a polymer solution should differ in the dilute and concentrated regimes, because the spatial distribution of chain segments is heterogeneous in the former and virtually uniform in the latter. Thus, they proposed expressing g at a given concentration by an interpolation formula as

$$g = Qg_{\mathrm{dil}} + (1 - Q)g_{\mathrm{conc}} \tag{2.6}$$

Here, g_{dil} and g_{conc} are the g factors characterizing dilute and concentrated

293

solutions, respectively, and Q is the probability that a microscopic volume in the solution contains no polymer segments. By a simple argument Koningsveld et al. showed that Q can be approximately represented by $\exp\left(-\phi/\phi^\star\right)$, with the overlap concentration $\phi^\star$ relating to P by $\phi^\star = 1/(bP^{1/2})$, where b is a proportionality constant. After approximating eq 2.6 by

$$g = g_{\text{conc}} + Qg^\star \tag{2.7}$$

with

$$g^\star \equiv \lim_{\phi \to 0}\left(g_{\text{dil}} - g_{\text{conc}}\right) \tag{2.8}$$

they used eq 2.5 for g_{conc} and estimated g_{dil} at infinite dilution by a two-parameter theory for the second virial coefficient A_2. This maneuver allows g to vary with P through Q and $g^\star$. The P dependence of $g^\star$ is associated with mechanism (ii), and that of Q with mechanism (iii). However, as we have noted in Section 2.8 of Chapter 2, it is unlikely that A_2 below θ obeys the two-parameter approximation. This indicates that the estimation of $g^\star$ by Koningsveld et al. needs a reconsideration. Furthermore, Koningsveld et al. did not anticipate that g_{conc} would depend on P.

Figure 9–2 includes the spinodals calculated by the hybrid theory. If we compare the curves for $b = 0.5$ with those for $b = \infty$, we see that the new theory substantially improves agreement with experiment, especially at low ϕ and P. This improvement owes much to the P dependence of Q, i.e., the fact that the dilute regime persists up to a higher concentration as P is lowered. In what follows, all that is associated with this dependence is referred to as the dilute–solution effect.

Although the hybrid theory of Koningsveld et al. was a milestone in the development of polymer solution thermodynamics, it had missed an important effect. This fact became apparent when χ was directly estimated by light scattering measurements in the studies described below.

2.4 Light Scattering Determination of χ

2.4.1 Work of Scholte

In the determinations of g described above, the parameters in the assumed forms of g were all evaluated by use of critical–point data. This maneuver has the shortcoming that we have to use in advance part of the phase equilibrium data to be predicted. A more preferable method is to determine g in the one-phase region of the system as a function of temperature and composition and to extrapolate the result to the region in which phase separation takes place. The necessary experiment has to be done at temperatures below θ, so that it needs skill and care as remarked in Section 2 of Chapter 4.

Scholte [14, 16] attempted it by light scattering for polystyrene in cyclohexane and showed that the χ data obtained as a function of temperature, concentration, and molecular weight were fitted closely by an expression similar to eq 2.2, except at low concentrations. In fact, the temperature dependence of χ at low concentrations significantly deviated from the form linear in $1/T$. It is likely that this could be accounted for by the dilute–solution effect mentioned above. However, we have to remark that Scholte's work appeared prior to the hybrid theory of Koningsveld et al. Another finding of Scholte was that the concentration–independent terms in eq 2.2 varied with P.[3] When extrapolated to the two–phase region his equations for χ yielded spinodals which came fairly close to the measured ones [14]. However, the agreement still left something to be desired.

Because of their failure at low concentrations, Scholte's equations for χ were essentially concerned with the behavior at high concentrations. Then his finding that their concentration–independent terms varied with P implied that g_{conc} should depend on P. This implication, however, conflicted with the prevailing notion that the segment–solvent interaction in concentrated solutions has nothing to do with the chain–connectivity of polymer segments so that g should become independent of chain length as the concentration increases. It is probably for this reason that Scholte's finding received little attention from other workers.

2.4.2   Work of Einaga et al.

Because of the significant implication of Scholte's work Einaga and coworkers considered it worthwhile to repeat similar experiments with care. Thus, they carried out light scattering measurements first on polystyrene in cyclohexane [17] and then on poly(isoprene) in dioxane. They obtained essentially similar results for the two systems. Here, the data for polystyrene in cyclohexane are illustrated.

The χ data were found to be represented accurately by an empirical equation

$$\chi = \chi_{conc}^{0} + (1/3)\phi + (A/B)[\phi^2/2 - 1/B + \ln(1 + B\phi^2)/B^2\phi^2]$$
$$+ (\chi_{dil}^{0} - \chi_{conc}^{0})Q(P^{1/2}\phi) \tag{2.9}$$

---

[3] Scholte's results are as follows:

$$M = 51000 : \chi = 0.2975 + 62/T + f(w)$$
$$M = 163000 : \chi = 0.3310 + 52/T + f(w)$$
$$M = 520000 : \chi = 0.3438 + 48/T + f(w)$$

where $f(w) = 0.306w + 0.30w^2$, with w the weight fraction of the polymer component.

with

$$Q(x) = (0.754/x^2)[1 - (1 + 1.875x + 0.864x^2 + 0.525x^3)$$
$$\times \exp(-1.875x - 0.432x^2 - 0.175x^2)] \qquad (2.10)$$

Here

$$\chi^0_{\text{dil}} = 0.5 + 0.26(\Theta/T - 1) + 4.6(\Theta/T - 1)^2 \qquad (2.11)$$

$$\chi^0_{\text{conc}} = 0.4930 + 0.345(\Theta/T - 1) + 0.0029 \exp[-30(\Theta/T - 1)] \qquad (2.12)$$

$$A = 1.4P^{1/3} \qquad (2.13)$$

$$B = 7P^{1/3} \exp[-18(\Theta/T - 1)] \qquad (2.14)$$

with Θ being the theta temperature in kelvin units. In eq 2.9, the first and second rows represent the contributions from the concentrated–solution and dilute–solution parts of the solution, respectively, and the superscript 0 indicates the state of infinite dilution.

The cloud–point curves, spinodals, and critical temperatures and concentrations calculated with eq 2.9 are compared with typical experimental data [12, 19–25] in Figures 9-3 to 9-5. Agreement between the calculated and measured results is fairly good. In this connection, the following remarks are relevant.

First, the concentrated–solution part of eq 2.9 depends on P through B in the brackets (the term A/B is independent of P), but this P dependence vanishes as P increases. Second, the calculated T_c^{-1} vs. $P^{-1/2}$ relation deviates slightly downward in the region near $P^{-1/2} = 0$, whereas the data points appear to follow a straight line over the entire range of $P^{-1/2}$ indicated. However, the ordinate intercept of this line gives a θ temperature about 1°C lower than 34.5°C, that chosen by Einaga et al. for their calculations. Third, ϕ_c varies almost linearly with $P^{-1/3}$, while eq 2.1 predicts ϕ_c to depend linearly on $P^{-1/2}$ for large P. This difference indicates that the critical concentration is sensitive to the dependence of χ on concentration and chain length. Recently, Perzynski et al. [26] derived the relation $\phi_c \sim P^{-0.38}$ for polystyrene in cyclohexane.

2.4.3 Approach of Nies et al.

Nies et al. [27] modified eq 2.5 of Koningsveld and Kleintjens by adding a term representing the dilute–solution effect and moreover by allowing α to vary with T and P. Their proposed expression for g was

$$g = \frac{\beta_0 + \beta_1/T}{1 - \gamma\phi} + [\alpha_{01} + \alpha_{11}/P + (\alpha_{20} + \alpha_{21}/P)(T - \Theta)]$$
$$+ (1/P)[g_1 + g_2(T - \Theta)] \exp(-bP^{1/2}\phi) \qquad (2.15)$$

296

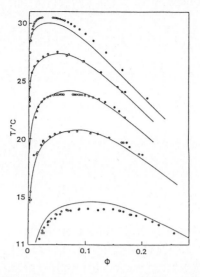

Fig. 9–3. Cloud–point curves for PS in CH. Points, experimental ($M_w \times 10^{-4}$ = 156 [19], 49.8 [20], 20.0 [20], 10.3 [20], 4.53 [22] from top to bottom). Lines, calculated with χ given by eq 2.9.

The parameters were evaluated by a simultaneous optimization of the agreement of well–documented data for χ, cloud point, spinodal, and critical point for polystyrene + cyclohexane with the relevant equations derived from eq 2.15. Figure 9–6 illustrates the results for cloud points. It was crucial to make the term α in eq 2.5 P dependent in order to get good agreement between calculated and experimental values. Because this term along with the first term in eq 2.15 governs g at high concentration, we see that Nies et al. also came to notice a sizable P dependence of g_{conc}.

2.5 Approach using the Apparent Second Virial Coefficient

In Figure 9–7, the solid curves show $-\Delta\mu_0/RT$ for polystyrene with $M = 20 \times 10^4$ in cyclohexane at several temperatures above and below T_c as a function of ϕ, calculated by using eq 2.9 for χ. The dashed line indicates the corresponding quantity for the Flory–Huggins athermal solution. The term $\chi\phi^2$ in eq 1.15 equals the remainder of the former subtracted by the latter. As can be seen from the figure, in the ranges of T and ϕ shown, the magnitude of $\Delta\mu_0/RT$ for the actual solution is far smaller than that for the FH solution. This means that $\chi\phi^2$ is comparable in magnitude to the latter. Therefore, unless χ is determined with a high absolute accuracy, we will obtain $\Delta\mu_0/RT$ which contains gross relative

297

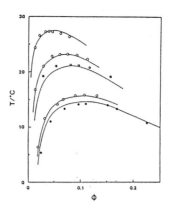

Fig. 9-4. Spinodals for PS in CH. Points, experimental
($M_w \times 10^{-4} = 52$ [14], 16.3 [14], 5.1 [14], 4.36 [25] from
top to bottom). Lines, calculated with χ given by eq 2.9

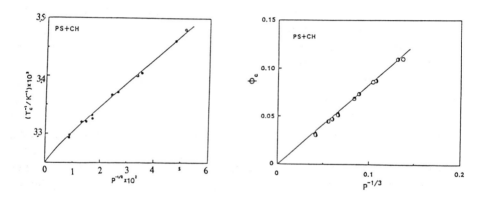

Fig. 9-5. Chain length dependence of critical temperature
T_c (left) and critical composition ϕ_c (right). Points, experimental [12, 19, 21, 22, 23]. Lines, calculated with χ given
by eq 2.9. Note that if χ depends only on temperature, ϕ_c
varies with $P^{-1/2}$ for $P^{1/2}$ much larger than unity (see eq
1.24).

errors. This fact implies that the use of χ, i.e., the choice of the FH solution
as the reference, is not advantageous for describing thermodynamic behavior of

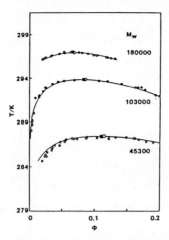

Fig. 9–6. Cloud–point curves for PS in CH. Points, experimental. Lines, calculated with g given by eq 2.15 [27].

polymer solutions in the region where phase separation takes place.

The appreciable departure of the dashed line from the solid curves is due to the gross overestimation of the FH theory for the entropy of mixing. This is understandable, because as the temperature is lowered in poor solvents, the polymer chain tends to collapse and thus to behaves more like a compact particle than an interpenetrable open domain. Thus, we may expect that $\Delta \overline{G}$ for polymer solutions near or below the θ temperature does not differ much from that for the regular solution. Furthernore, if the solutions are relatively dilute, the latter may be replaced by its limiting for very low concentrations. This limit defines a hypothetical solution whose $\Delta \mu_0$ is given by $-(RT/P)\phi$ over the entire range of ϕ (P should read P_n for quasi–binary solutions). We name it the van't Hoff (VH) solution. The dot–dash line in Figure 9–7 shows the van't Hoff solution. As expected, it is located much closer to the solid curves than the dashed curve. Therefore, if we choose the van't Hoff solution as a new reference system and express $\Delta \mu_0$ for an actual solution by

$$\Delta \mu_0 = RT[-(\phi/P) + \Gamma(T, \phi, P)\phi^2] \qquad (2.16)$$

the second term in the brackets should remain small in poor solvent systems in comparison with the first one. We hereafter call Γ the apparent second virial coefficient. This coefficient absorbs all the thermodynamic deviation of the actual solution from the van't Hoff solution, and hence it corresponds to χ in the Koningsveld–Staverman formalism (see eq 2.23).

299

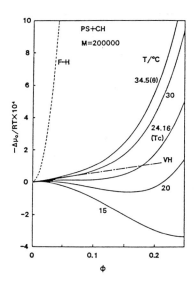

Fig. 9–7. Solid lines, calculated with χ given by eq 2.9.
Dashed line, FH solution. Dot–dash line, VH.

The apparent second virial coefficient was used by Vink [28] in 1976. He
expressed Γ by a polynomial in ϕ, but Einaga et al. [29] left it as an arbitrary
function to be determined by measurement of quantities related to $\Delta\mu_0$. For
example, according to the theory of light scattering, Γ can be determined by use
of the formula:

$$\Gamma = \frac{2}{\phi^2} \int_0^\phi J(T, u, P) u \, du \qquad (2.17)$$

where J is defined by

$$J = \tfrac{1}{2}[(P\phi)^{-1} - K(\Delta R_0)^{-1}] \qquad (2.18)$$

with ΔR_0 the excess Rayleigh ratio at zero scattering angle and K an optical
parameter.

Einaga et al. [29] applied eq 2.17 to the light scattering data reported in Ref.
[17] and obtained

$$\Gamma = J_{\text{conc}}^0 + \tfrac{1}{2} J_{\text{conc}}^1 \phi^2$$
$$+ 2(J_{\text{dil}}^0 - J_{\text{conc}}^0) P^{-1} \phi^{-2} [1 - (1 + \phi P^{1/2}) \exp(-\phi P^{1/2})]$$
$$\qquad (2.19)$$

300

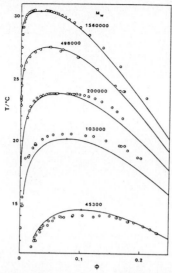

Fig. 9–8. Cloud–point curves for PS in CH. Points, experimental data (the same as those shown in Fig. 9–3). Lines, calculated with Γ given by eq 2.19 [29].

where

$$J^0_{\text{conc}} = 0.036 P^{-1/3} - 0.23[(\Theta/T) - 1] \tag{2.20}$$

$$J^1_{\text{conc}} = 0.47 - 3.5[(\Theta/T) - 1] \tag{2.21}$$

$$J^0_{\text{dil}} = -0.26[(\Theta/T) - 1] - 4.6[(\Theta/T) - 1]^2 \tag{2.22}$$

The first and second rows in eq 2.19 represent the concentrated– and dilute–solution effects, respectively. Thus we see that J depends on P in the concentrated regime, as χ does. This is expected, because Γ is related to χ by

$$\Gamma = -\chi - [\ln(1 - \phi) + \phi]/\phi^2 \tag{2.23}$$

The cloud–point curves calculated with eq 2.19 are compared with experimental data in Figure 9–8. The agreement is not as good as that seen in Figure 9–3 or 9–6, but we should note that eq 2.19 is much simpler than eq 2.9.

Tong and Einaga [30] found that eq 2.19 with different coefficients can describe the Γ data of Takano et al. [18] for poly(isoprene) (PIP) in dioxane. Figure 9–9 shows fairly good agreement between the calculated and measured cloud–point curves and spinodals for this system.

301

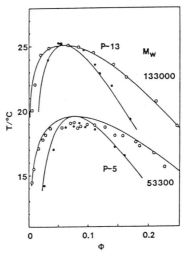

Fig. 9-9. Cloud-point curves and spinodals for PIP in dioxane. Points, experimental [18]. Lines, calculated by the Γ method [30].

2.6 Remarks

We have described some recent studies which have led to the conclusion that the factor g (or χ or Γ) has to be treated as a function of T, ϕ, and P in order to achieve quantitative prediction of the phase relationships of binary solutions. Among many others, the most significant was the finding of the P dependence of g in concentrated solutions. We can indeed show that the high concentration branches of measured cloud-point curves and spinodals for different P cannot be consistently and quantitatively described unless g is allowed to vary with P at high concentration.

According to the original FH theory, χ is interpreted as the strength of segment-solvent interactions. Since, in concentrated solutions, this strength is expected to depend essentially on the local segment density, χ should become independent of P as the concentration increases. This prediction contradicts the sizable P dependence of g found at high concentration. However, the discrepancy does not matter, since, within the Koningsveld-Staverman formalism, g is a free empirical parameter. Nonetheless, it is frustrating that we cannot yet explain why g shows a significant P dependence in concentrated solutions. Anyway we have to check whether this phenomenon is general or not by further experiments.

3. Quasi–binary Solutions

3.1 Classic Data of Shultz and Flory

Now that the P dependence of g for binary solutions has become apparent, it is clear that this factor for quasi–binary solutions has to be treated as a function of the chain length distribution $f(P)$. Historically, the $f(P)$ dependence of g is not a recent finding but has become gradually recognized through analyses of observed phase relationships. In what follows, we outline some typical contributions which played a role in this process.

Again we start with the pioneering work of Shultz and Flory [9] in 1952. In that, they measured cloud–point curves for mixtures of two well–fractionated polystyrene samples, PSB (P_{PSB} = 768) and PSD (P_{PSD} = 10950), in cyclohexane. Their data are reproduced in Figure 9–10, where the curves PSD, 1, 2, 3, 4, and PSB correspond to ξ_{PSB} (the weight fraction of PSB) = 0, 0.50, 0.75, 0.90, 0.98, 1.0, respectively. Among the features seen in Figure 9–10, the most interesting is the marked difference between curves 4 and PSB. We see that only 2% of the high–molecular–weight PSD mixed in the low–molecular–weight PSB drastically suppresses the solubility of the latter in dilute solutions. This phenomenon suggests that great attention should be paid to the presence of high–molecular–weight fractions in measuring dilute solution properties at temperatures below θ. As Tompa [1] showed, it can be explained, though qualitatively, with g dependent only on T and hence is not a characteristic effect due to the dependence of g on $f(P)$.

Figure 9–11 shows the triangular (or triaxial) phase diagram for the ternary system studied by Shultz and Flory [9]. Here, the solid curve is the binodal at 25°C constructed by cutting the cloud–point curves with a horizontal line at this temperature, and the solid straight lines show the experimental tie lines. On the other hand, the dashed curve is the corresponding binodal calculated by assuming χ to depend only on T (actually given by the expression that Shultz and Flory established for binary polystyrene + cyclohexane). The two curves are similar, but the theoretically predicted two–phase region is much smaller than the observed. This discrepancy may be compared to the big difference observed in Figure 9–1 between the solid and dashed cloud–point curves for the system polystyrene + cyclohexane.

3.2 Critical Point

The top of a cloud–point curve is called the threshold point. The critical point on the $T - \phi$ plane for a binary solution is situated at the threshold point, but this is no longer the case for quasi–binary solutions. The critical point on the $T - \phi$ plane for a quasi–binary solution should appear somewhere on the

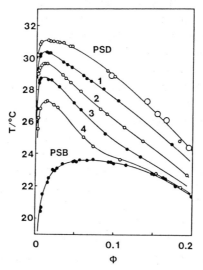

Fig. 9–10. Cloud–point curves for mixtures of PS fractions PSB and PSD in CH [9]. ξ_{PSB} = 0, 0.50, 0.75, 0.90, 0.98, 1.0 for curves PSD, 1, 2, 3, 4, PSB, respectively.

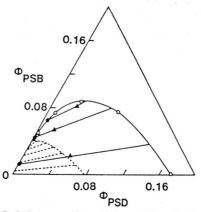

Fig. 9–11. Solid curve, binodal at 25°C derived from the cloud–point curves in Fig. 9–10. Dashed curve, calculated with χ depending only on temperature [9].

high–concentration branch of the cloud–point curve [2]. Probably, Rehage et al. [31] were the first to confirm this prediction. Figure 9–12 reproduces their data for a polydisperse polystyrene in cyclohexane (note that the concentration here is expressed in terms of the polystyrene weight fraction w). The star on the cloud–point curve indicates the location of the critical point. As can be seen, the critical concentration distinctly differs from the threshold one. The dashed lines show the coexistence curves for mother solutions with different w. One particular pair of conjugate coexistence curves (for $w = 10\%$) meet horizontally at the critical point. This property is often utilized to determine the position of the critical point on the $T - \phi$ plane. The critical point may also be found by use of the phase volume method devised by Koningsveld and Staverman [32]. Whichever method is used, the determination of critical points needs considerable experimental effort and patience.

Cloud–point curves of quasi–binary solutions often exhibit a dent as seen in the data of Rehage et al. (see also curve 4 in Figure 9-1), and the critical point is usually located somewhere in the dent. In some cases, the dent has a break, as sketched in Figure 9-13 (actual examples are seen, for instance, in Koningsveld's data [33] on the system polyethylene + diphenyl ether). When this happens the critical point either appears very close to the break point or disappears in the immiscibility region below the observable cloud–point curve. In the latter case, the cloud–point curve has an invisible loop with cusps as shown by the dashed line sketched in Figure 9-13.

These features of the critical point exemplify a great variety of phase equilibria of quasi–binary solutions, which are associated with the multiplicity of polymer components and the dependence of g on T, ϕ, and $f(P)$. Among many others, Šolc [8] made a most detailed theoretical study on the effects arising from the many–component nature of quasi–binary solutions. Since he assumed for simplicity that g depends only on temperature, his findings are of limited use for quantitative purposes. A good summary of Šolc's theory can be seen in Kurata's monograph [2].

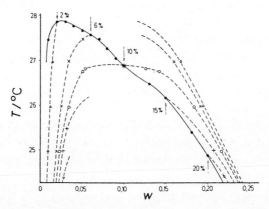

Fig. 9-12. Solid line, cloud–point curve for PS in CH. Dashed lines, coexistence curves for PS in CH corresponding to the indicated mother solution concentrations. Star, critical point [31].

305

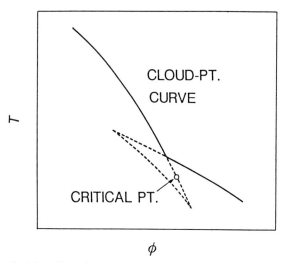

Fig. 9–13. Cloud–point curve with a break point. The critical point appears on the dashed part which is not experimentally measurable.

3.3 Concentration Dependence

The work of Shultz and Flory showed clearly that g cannot be a function of T only, and it was natural that attempts toward a more accurate description of phase equilibria in quasi–binary solutions began by choosing ϕ as a second variable of this factor. Koningsveld and Staverman [6] showed for the first time that if g depends on both T and ϕ, the spinodal and critical state for quasi-binary solutions are given by eq 1.20 and 1.22. These general equations make the following important predictions.

Firstly, eq 1.20 predicts that the spinodals for homologous polymers different in polydispersity but identical in P_w form a single curve. Scholte [14] in 1971 and Derham et al. [4] in 1974 demonstrated that this prediction did not hold for polystyrene in cyclohexane. Figure 9–14 displays the data of Derham et al. Here, the spinodal labeled PS166 is for a narrow–distribution sample with $M_w = 166 \times 10^3$ and one labeled PSM5 is for a binary mixture with $M_w = 165 \times 10^3$ and $M_z/M_w = 3.68$. The two curves do not superimpose but intersect at $\phi = 0.075$.

Secondly, it follows from eq 1.22 combined with eq 1.20 that homologous polymers different in polydispersity but having the same P_w and P_z have critical points at the same point in the $T - \phi$ plane. As far as the author is aware, no experimental test of this prediction is as yet reported. It is probably not a simple task to prepare polymer samples needed for such a test.

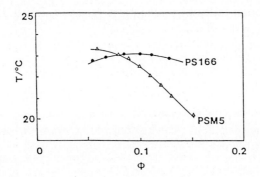

Fig. 9–14. Spinodals for PS in CH [4]. Line PS166, for near–monodisperse PS with $M_w = 166 \times 10^3$. Line PSM5, for binary PS mixture with $M_w = 165 \times 10^3$ and $M_z/M_w = 3.68$. The two lines should form a composite curve if g is independent of $f(P)$.

3.4 Dependence on Chain Length Distribution

3.4.1 Ad Hoc Maneuvers

The findings of Scholte and Derham et al. revealed that g for quasi–binary solutions ought to depend not only on T and ϕ, but on at least one more variable. It is not evident whether, in the early 1970s, one had a clear recognition that this third variable should be the chain length distribution $f(P)$ or, to be equivalent, that g should depend on the concentrations of the individual polymer components $(\phi_1, \phi_2, \ldots, \phi_q)$, not the sum of them ϕ. Undoubtedly, when this is the case, the thermodynamic approach to $\Delta \overline{G}$ for quasi–binary solutions becomes more complex as the number of polymer components increases, and we cannot help invoking some assumptions or approximations.

The earliest example is due to Koningsveld and Kleintjens [13] in 1971, who proposed to choose P_n given by eq 1.16 as the third variable. In actuality, they allowed β_0 and β_1 in eq 2.5 to vary in proportion to P_n^{-1}, and this maneuver was taken over by Kennedy et al. [34] in 1972. Later, Derham et al. [35] assumed that the coefficients g_k ($k \geq 2$) in the expansion

$$g = g_0 + g_1 \phi + g_2 \phi^2 + \ldots$$

are arbitrary functions of P_w defined by eq 1.21. The underlying idea of these ad hoc maneuvers was to describe effects of $f(P)$ in terms of a limited number of average chain lengths of low order.

307

3.4.2   Separation Factor

One of the important quantities relating to phase equilibria is the separation factor (or distribution coefficient) σ_i defined by

$$\sigma_i = P_i^{-1} \ln(\phi_i''/\phi_i') \quad (i = 1, 2, \ldots, q) \tag{3.1}$$

where ϕ_i' and ϕ_i'' are the volume fractions of polymer i in the dilute and concentrated phases, respectively, separated from a mother quasi–binary solution. If g does not depend on $f(P)$, it can be deduced [2] from the phase equilibrium relations $\Delta\mu_i' = \Delta\mu_i''$ that

$$\sigma_i = \Delta[\ln(1 - \phi) + (2\phi - 1)g - \phi(1 - \phi)(\partial g/\partial\phi)] \tag{3.2}$$

where ΔX stands for the difference $X'' - X'$. Since g is assumed to depend only on T and ϕ, the right–hand side of eq 3.2 has nothing to do with the index i, giving the same σ_i for all i. Hence, a straight line is obtained if $\ln(\phi_i''/\phi_i')$ is plotted against P_i as was done by Breitenbach and Wolf [36].

Figure 9–15 shows the Breitenbach–Wolf plots at 25.7°C for three mother solutions of a very polydisperse polystyrene ($M_n = 82 \times 10^3$, $M_w = 410 \times 10^3$, $M_z = 1220 \times 10^3$) in cyclohexane [37]. The three curves are distinctly non-linear, evidencing that g for the system depends on $f(P)$. If this dependence is taken into account, eq 3.2 is replaced by

$$\sigma_i = \Delta\{\ln(1 - \phi) + 2(\phi - 1) - \phi(1 - \phi)(\partial g/\partial\phi)$$
$$- (1 - \phi)[m_i(\partial g/\partial\xi_i) - \sum_{j=1}^{q-1} \xi_j(\partial g/\partial\xi_j)]\} \tag{3.3}$$

with $m_i = 1$ for $i \neq q$ and = 0 for $i = q$. Since $f(P)$ is different in the two separated phases, the derivative $\partial g/\partial\xi_i$ should vary differently with the index i in the two phases, thus making σ_i dependent on P_i. In other words, we should obtain a non–linear Breitenbach–Wolf plot.

3.5   Two Polystyrenes in Cyclohexane

In the late 1970s when they looked over the existing literature, Hashizume et al. noticed that even for the simplest of quasi–binary solutions, i.e., ternary solutions containing two homologous polymers in a single solvent, systematic phase equilibrium data were virtually lacking. Since such data seemed essential for testing any proposed expression for g for quasi–binary solutions, they [22] undertook extensive measurements of cloud–point curves, binodals, and critical points on mixtures of two narrow–distribution polystyrenes f4 and f10 dissolved

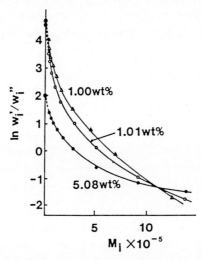

Fig. 9-15. Breitenbach–Wolf plots for a polydisperse PS in CH at 25°C determined by GPC [37]. Each curve refers to different initial concentration.

in cyclohexane. The work was continued by Tsuyumoto et al. [20], who replaced f10 with a higher molecular weight narrow–distribution polystyrene f40. The weight–average molecular weights of f4, f10, and f40 were 4.52×10^4, 10.3×10^4, and 49.8×10^4, respectively. These samples were nearly monodisperse when checked by GPC. Some of the typical results are illustrated below and compared with the predictions of an empirical theory described in the next section.

Figure 9-16 shows the cloud–point curves obtained for a series of the weight fraction ξ_4 of f4 from 0 to 1. The overall pattern of the curves in the lower panel is similar to that reported earlier by Shultz and Flory [9] (see Figure 9-1). Again we note that addition of a very small amount of a high–molecular–weight polymer appreciably raises cloud points of the solutions of a low–molecular–weight polymer at low concentrations.

Figures 9-17 and 9-18 depict some binodal data in the triangular phase diagram. The data points were determined by composition analysis using GPC and also from the cloud–point curves.

Finally, in Figure 9-19, the composition dependence of T_c and ϕ_c for the two systems is shown. For either system the critical point was determined over the entire range of ξ_4. We note that, in a ternary solution in which the chain length ratio of the polymer components, P_2/P_1, is much larger than unity, the critical point for ξ_1 very close to unity happens to hide below the cloud–point curve as in Figure 9-13 and thus becomes unobservable.

309

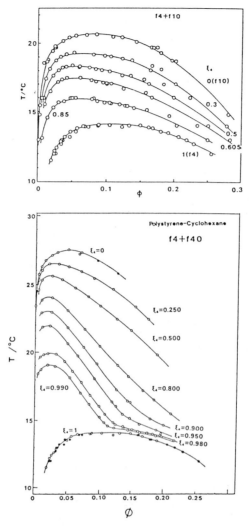

Fig. 9–16. Cloud–point curves for PS f4 + PS f10 (upper panel) [22] and PS f4 + PS f40 (lower panel) [20] in CH. ξ_4 denotes the weight fraction of PS f4.

3.6 Empirical Form of the Apparent Second Virial coefficient

For a quasi-binary solution eq 2.16 should read

$$\Delta\mu_0 = RT[-(\phi/P_n) + \Gamma(T, \phi, f(P))\phi^2] \tag{3.4}$$

At present, we have no established information allowing us to infer the $f(P)$ dependence of Γ. Hence, we have to use some ad hoc maneuver to proceed

310

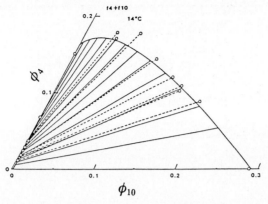

Fig. 9–17. Binodal data for PS f4 + PS f10 in CH at 14°C [22]. Points, experimental. Solid lines, calculated [29]. ϕ_4 and ϕ_{10} are the volume fractions of PS f4 and PS f10, respectively.

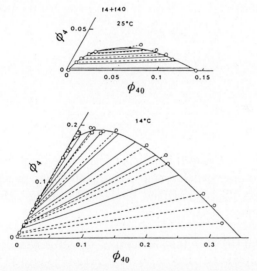

Fig. 9–18. Binodal data for PS f4 + PS f40 in CH at 25 and 14°C [20]. Points, experimental. Solid lines, calculated [29].

with eq 3.4. Here we refer to the maneuver proposed by Einaga et al. [29], who thereby came to interesting findings.

Their proposal is to replace P involved in $\Gamma(T, \phi, P)$ for binary solutions by certain average chain lengths. Thus, it is basically the same as those invoked by Koningsveld and others in the 1970s. Actually, (i) P in J^0_{conc} in eq 2.19 was

311

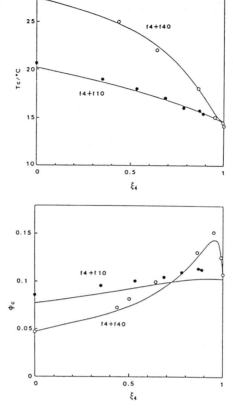

Fig. 9-19. Composition dependence of critical temperature (upper panel) and concentration (lowe panel) for PS f4 + PS f10 [22] and PS f4 + PS f40 in CH [20]. Solid lines, calculated [29].

replaced by P° and (ii) P in the last term of the same equation by $P^\bullet$, where

$$(P^\circ)^{-1/3} = \sum_{j=1}^{q} \xi_j P_j^{-1/3} \tag{3.5}$$

$$(P^\bullet)^{-1/2} = \sum_{j=1}^{q} \xi_j P_j^{-1/2} \tag{3.6}$$

The substitution (ii) is equivalent to approximating the overlap concentration of a quasi-binary solution by the average $\sum_{j=1}^{q} \phi_j^\star \xi_j$, where $\phi_j^\star$ is the overlap concentration for the binary solution of polymer i. These substitutions introduce

312

neither adjustable parameters nor limitations to q. Hence, it is possible to predict any phase relationship of cyclohexane solutions containing any number of polystyrene homologues by using eq 2.19 for Γ combined with eq 3.5 and 3.6 (see the Appendix).

Figure 9–20 depicts the cloud–point curves, shadow curves, and loci of critical points calculated in this way for the ternary solutions studied by Hashizume et al. and Tsuyumoto et al. The cloud–point curves reproduce well the features of the experimental data in Figure 9–16, though quantitative agreement leaves something to be desired. The solid lines in Figures 9–17 to 9–19 are not for eye guiding but the results from the same calculation. Their fit to the data points is quite good.

For a further test of the ad hoc substitutions (i) and (ii), Tong et al. [38] prepared three polymer mixtures, each containing four narrow–distribution polystyrenes ($M_w \times 10^{-3}$ = 11, 44.5, 195, 807) at a prescribed weight ratio, allowed each of them to separate into two phases in cyclohexane at various T and ϕ, and measured the compositions of the two phases by GPC. They found that the separation factors for the four components computed from the experimental data agreed almost quantitatively with the values calculated by the above method of Einaga et al. This result was significant in that the system contained as many polymer components as four.

3.7 Three–phase Separation

According to Gibbs' phase rule, a ternary solution may separate into three liquid phases in equilibrium under certain conditions of T and ϕ. As early as 1949 Tompa [1] investigated this phenomenon theoretically with g assumed to depend on T only. In 1967, Koningsveld and Staverman [39] verified Tompa's prediction with the system in which two polyethylene samples with $M_w = 540 \times 10^3$ and 12×10^3 or 25×10^3 were dissolved in diphenyl ether. Since these samples were considerably polydisperse, the solutions were not thermodynamically well–defined three–component systems. Later, Koningsveld et al. [12] made a similar study with a mixture of narrow–distribution polystyrenes with $M_w = 51 \times 10^3$ and 1500×10^3 and observed three–phase separation to occur in the temperature range predicted by Tompa's theory.

Tompa [1] and later Šolc [8] showed that at the limit of very large chain length the ratio P_2/P_1 must be larger than about 10 for three–phase separation to occur in ternary solutions if g depends only on temperature. Fujita and Einaga [40] extended the theory to g varying linearly with ϕ. However, now that, in general, g has to be regarded as a function of $f(P)$, the conclusions from these calculations cannot have much quantitative significance.

Recently, Dobashi and Nakata [41] measured precisely the coexistence curves for a ternary system consisting of two polystyrenes ($M_w = 17.3 \times 10^3$ and

313

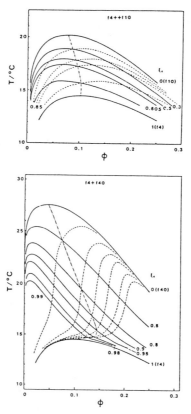

Fig. 9-20. Calculated cloud-point curves (solid lines), shadow curves (dashed lines), and loci of critical points (dot-dashed lines) for PS f4 + PS f10 (upper panel) and PS f4 + f40 (lower panel) in CH [29].

719×10^3) and methylcyclohexane and found that three-phase equilibria took place in the very narrow temperature range $24.34 - 24.70°$C. Using a g similar to that in eq 2.5, they calculated the three-phase region for their experimental system, but the predicted region was considerably narrower than the observed. This failure is understandable, because their g contained no term relating to the chain length distribution. It is desirable for a more detailed study of three-phase separation to measure the composition of each phase as well as the coexistence curve. Such a measurement on quasi-binary polymer solutions, however, was not reported until a recent paper of Einaga et al. [42], who studied mixtures of polystyrenes f4 and f128 ($M_w = 43.6 \times 10^3$ and 1260×10^3) in cyclohexane.

Figure 9-21 depicts the triangular phase diagram obtained at $13.8°$C. The triangle formed by thick solid lines defines the three-phase region which may be

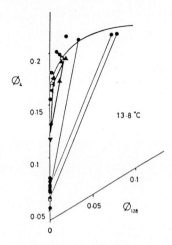

Fig. 9–21. Triangular phase diagram for PS f4 + PS f128
in CH at 13.8°C [42]. Triangle formed by thick solid lines
defines the experimental three–phase region. Unfilled triangles, calculated compositions of three phases. Thin solid
curve, calculated binodal.

inferred from the phase compositions indicated by filled triangles. The unfilled
triangles show the three phases predicted by the empirical method of Einaga et
al. mentioned in Section 3.6. Except for the dilute phase, the calculated phase
compositions come near the measured ones. Figure 9–22 displays the phase
relations projected on the $T - \phi$ plane for the mixture with $\xi_2 = 0.05$, where ξ_2
is the weight fraction of polystyrene f128. The lines were all calculated by the
method of Einaga et al. The three–phase coexistence curve indicated by the thin
solid line shows the narrowness of the temperature range in which the system
separates into three equilibrium phases. It passes close by the triangular data
points representing the polymer volume fractions in the three phases, except the
two for the dilute phase.

Multi–phase equilibria reflect very sensitively the dependence of $\Delta \overline{G}$ on its
variables. Precise data on them, especially those on the phase compositions,
therefore, should be useful for strict testing of any proposed expressions for
$\Delta \overline{G}$ of multicomponent solutions. It is clear now that whatever approach may
be used, it cannot be self-contained unless taking into account the dependence
of g (or the corresponding factors) on $f(P)$ in the concentrated regime as well
as in the dilute one. However, this fact does not seem to be well recognized by
current workers who are interested in the phase equilibrium of multicomponent
polymer solutions.

315

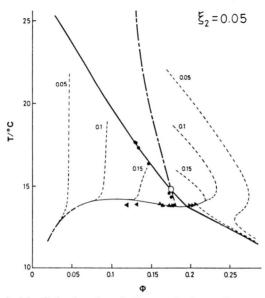

Fig. 9–22. Calculated and observed phase diagrams for PS f4 + PS f128 (the weight fraction of the latter is 0.05) + CH on the $T-\phi$ plane. Thick solid line, calculated cloud–point curve. Dot–dashed line, calculated shadow curve. Dashed lines, calculated two–phase conjugate coexistence curves for the indicated polymer volume fractions. Thin solid line, three–phase coexistence curve. Unfilled circle, calculated critical point. Filled circles, measured cloud points. Filled triangles, measured polymer volume fractions in three separated phases.

4. Remarks

4.1 Critical Exponents

We return to binary solutions. In approaching their $\Delta\overline{G}$ by the empirical method, we usually express the dependence of g on ϕ and T by some analytical function. This operation is an assumption, and is sometimes called the mean–field or classical approximation. It leads to the general prediction that the cloud–point curve in the vicinity of the critical point $(T_c,\ \phi_c)$ is described by a universal relation, called the critical scaling law:

$$|\phi - \phi_c|/\phi_c \sim \epsilon^{0.5} \qquad (4.1)$$

or

$$(\phi'' - \phi')/\phi_c \sim \epsilon^{0.5} \qquad (4.2)$$

316

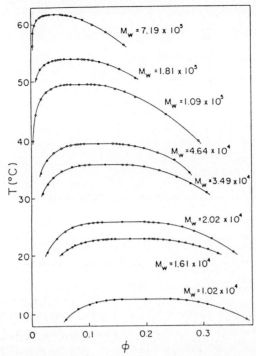

Fig. 9–23. Cloud–point curves for PS in MCH at a series of different M_w [43].

Here, ϵ is a dimensionless temperature defined by

$$\epsilon = |T - T_c|/T_c \qquad (\ll 1) \tag{4.3}$$

and the single and double primes indicate the dilute and concentrated phases, respectively.

Having been calculated with an analytical g, the cloud–point curves shown in Figure 9–3 obey eq 4.1. Interestingly, as seen from the figure, some of them clearly do not fit the data points in the region near the maxima. The latter follow a flatter curve than the former. The discrepancy is not an exception but a rule. This means that the data showing no such discrepancy are probably in error. In fact, the high–precision data of Dobashi et al. [43] for polystyrene in methylcyclohexane (MCH), reproduced in Figure 9–23, show that the cloud–point curve for any molecular weight studied is distinctly flat over a range of ϕ encompassing its maximum.

The mean–field approximation is expected to break down in the region near the critical point, because the fluctuations of polymer segment densities tend

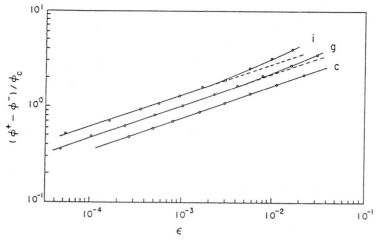

Fig. 9–24. Tests of eq 4.4 with the data shown in Fig. 9–23 [44]. $M_w \times 10^{-3}$ = 17.3, 109, and 719 for c, g, and i, respectively.

to correlate over a large distance as the solution approaches this characteristic point. Many experiments have shown that cloud–point curves of binary mixtures of simple liquids at temperatures close to T_c are represented by an empirical relation

$$(\phi'' - \phi')/\phi_c = A\epsilon^\beta \quad (\epsilon \ll 1) \tag{4.4}$$

with the "critical exponent" β essentially universal. Systems with the same β are said to belong to a universality class.

Dobashi et al. [44] found that the data of Figure 9–23 also follow eq 4.4 with $\beta = 0.332 \pm 0.001$ independent of molecular weight. Figure 9–24 illustrates this fact for three different M_w. We see that the range of ϵ in which eq 4.4 holds diminishes with increasing molecular weight. Thus, for the case i, it is restricted to $T_c - T < 0.7°$C. Dobashi et al.'s β is consistent with 0.327 ± 0.004 previously obtained by Nakata et al. [45] for a polystyrene with $M_w = 1560 \times 10^3$ in cyclohexane and also with 0.326 ± 0.003 for methanol + cyclohexane ($\epsilon < 0.06$), 0.328 ± 0.007 for aniline + cyclohexane ($\epsilon < 0.03$), and 0.328 ± 0.004 for isobutyric acid + water ($\epsilon < 0.006$). Dobashi et al. thus concluded that the system polystyrene + methylcyclohexane belongs to the same universality class as binary mixtures of ordinary simple liquids.

Another critical exponent relating to the phase equilibrium in binary solutions is μ defined by

$$[(\phi' + \phi'')/2\phi_c] - 1 = B\epsilon^\mu \quad (\epsilon \ll 1) \tag{4.5}$$

In the mean–field approximation, this exponent equals unity. Dobashi et al.

[44] analyzed the data of Figure 9–23 by eq 4.5 and obtained $\mu = 0.858 \pm 0.005$ independent of molecular weight. This μ is in excellent agreement with 0.860 ± 0.006 reported in the above–cited paper of Nakata et. al. [45]. Dobashi et al. [44] found that, differing from β and μ, the factors A and B depended on molecular weight. The former decreased appreciably $(A\phi_c \sim M_w^{-0.23})$ and the latter increased slowly with increasing M_w. As shown in Figure 9–5, ϕ_c for the system polystyrene + cyclohexane varies in proportion to $M^{-1/3}$. If we assume that the same relation holds for the system polystyrene + methylcyclohexane too, the finding of Dobashi et al. gives $A \sim M^{0.10}$, which is in good agreement with Sanchez's deduction [46] $A \sim M^{0.102 \pm 0.002}$. Calculations of β and μ by statistical mechanics still remain to be done.

Nakata et al. [45] showed that the Wegner expansion [47]

$$\phi'' - \phi' = \epsilon^\beta (A_0 + A_1 \epsilon^{1/2} + A_2 \epsilon + \cdots) \tag{4.6}$$

can be fitted accurately to their data over an extended range of ϵ $(T_c - T < 7.0\mathrm{K})$ when truncated at the third term. This type of cloud–point data analysis tacitly assumes that phase separation is a kind of critical phenomenon so that it cannot be adequately treated by the mean–field approximation. On the other hand, the Koningsveld–Staverman approach and its modifications as discussed in the previous sections expect that except for a region very near the critical point, the mean–field approximation is good enough to describe phase equilibrium behavior of polymer solutions.

4.2 Continuous Thermodynamics

The phase relations for quasi–binary solutions outlined in Section 1 are general and exact under the basic assumptions made. However, the computational work with them becomes exponentially difficult as the number of components increases. In fact, it is virtually impossible to solve the phase equilibrium equations for solutions of actual synthetic polymers, which contain an almost infinite number of components. We thus need a novel approach to analyze phase equilibrium data on such systems. The discipline called continuous thermodynamics has emerged to meet this requirement. It deals with mixtures of molecules whose physical properties such boiling point, molecular weight, and so forth vary continuously, and is the correct method for treating solutions of a truly polydisperse polymer (see Section 1.1 of this chapter for its definition).

Continuous thermodynamics was first formulated by Aris and Gavalas [48] in 1966, but its progress has taken place in the last decade [49–54]. Rätzsch and coworkers [54] have developed it in relation to polymer solutions. Yet their theory is limited to the simple case in which g does not depend on $f(P)$. When g is allowed to depend on $f(P)$, the phase equilibrium relations become functional equations, which are very difficult to solve. It would be relevant to conclude that the thermodynamics of phase equilibria of polydisperse solutions is yet in an infancy state of exploration.

Appendix

We consider a quasi–binary solution containing polymers 1 and 2 in a solvent 0 and derive the expressions for $\Delta\mu_1$ and $\Delta\mu_2$ when $\Delta\mu_0$ is given by eq 3.4. The Gibbs–Duhem relation at constant T and p gives

$$\frac{(1-\phi)}{V_0}\left(\frac{\partial\mu_0}{\partial\phi_1}\right) + \frac{\phi_1}{V_1}\left(\frac{\partial\mu_1}{\partial\phi_1}\right) + \frac{\phi_2}{V_2}\left(\frac{\partial\mu_2}{\partial\phi_1}\right) = 0 \qquad (9A\text{-}1)$$

$$\frac{(1-\phi)}{V_0}\left(\frac{\partial\mu_0}{\partial\phi_2}\right) + \frac{\phi_1}{V_1}\left(\frac{\partial\mu_1}{\partial\phi_2}\right) + \frac{\phi_2}{V_2}\left(\frac{\partial\mu_2}{\partial\phi_2}\right) = 0 \qquad (9A\text{-}2)$$

After rewriting these in terms of ϕ and ξ_2 $(=\phi_2/\phi)$ we can derive the following relations:

$$(1-\phi)\left(\frac{\partial\mu_0}{\partial\phi}\right) + \frac{\phi(1-\xi_2)}{P_1}\left(\frac{\partial\mu_1}{\partial\phi}\right) + \frac{\phi\xi_2}{P_2}\left(\frac{\partial\mu_2}{\partial\phi}\right) = 0 \qquad (9A\text{-}3)$$

$$(1-\phi)\left(\frac{\partial\mu_0}{\partial\xi_2}\right) + \frac{\phi(1-\xi_2)}{P_1}\left(\frac{\partial\mu_1}{\partial\xi_2}\right) + \frac{\phi\xi_2}{P_2}\left(\frac{\partial\mu_2}{\partial\xi_2}\right) = 0 \qquad (9A\text{-}4)$$

After introduction of eq 3.4, these two differential equations can be integrated to give

$$\Delta\mu_1 = \mu_1^\infty - \mu_1^0 + RT\{\ln[\phi(1-\xi_2)] - \phi + [1 - (P_1/P_2)]\phi\xi_2$$

$$+ \Gamma P_1\phi(1-\phi) + P_1\int_0^\phi\left[\Gamma - \xi_2\left(\frac{\partial\Gamma}{\partial\xi_2}\right)\right]du\} \qquad (9A\text{-}5)$$

and

$$\Delta\mu_2 = \mu_2^\infty - \mu_2^0 + RT\{\ln(\phi\xi_2) - \phi - [(P_2/P_1) - 1]\phi(1-\xi_2)$$

$$+ \Gamma P_2\phi(1-\phi) + P_2\int_0^\phi\left[\Gamma + (1-\xi_2)\left(\frac{\partial\Gamma}{\partial\xi_2}\right)\right]du\} \qquad (9A\text{-}6)$$

where μ_i^0 denotes the chemical potential of component i in the pure state, and μ_i^∞ is defined by

$$\mu_i^\infty = \lim_{\phi\to 0}(\mu_i - RT\ln\phi_i) \qquad (9A\text{-}7)$$

Both μ_i^0 and μ_i^∞ are functions of T and p that thermodynamics cannot evaluate. With eq 9A-5 and 9A-6 introduced into eq 1.11 we obtain the phase equilibrium equations for polymers 1 and 2. Since μ_i^0 and μ_i^∞ drop out of both sides, these equations contain only Γ as an unspecified function. It can be shown that the same holds for quasi–binary solutions containing more than two polymer components. In conclusion, the phase equilibrium behavior of any quasi–binary solution is completely determined if Γ for the chemical potential of the solvent component is specified as a function of T, ϕ, and $f(P)$.

References

1. H. Tompa, "Polymer Solutions," Butterworths, London, 1956.
2. M. Kurata, "Thermodynamics of Polymer Solutions," (translated by H. Fujita from the Japanese), Harwood Academic, Chur, Switzerland, 1982.
3. K. Šolc and R. Koningsveld, J. Phys. Chem. **89**, 2237 (1985).
4. K. E. Derham, J. Goldsbrough, and M. Gordon, Pure & Appl. Chem. **38**, 97 (1974).
5. R. Koningsveld and A. J. Staverman, J. Polym. Sci. A-2 **6**, 305 (1968).
6. R. Koningsveld and A. J. Staverman, J. Polym. Sci. A-2 **6**, 325 (1968).
7. W. H. Stockmayer, J. Chem. Phys. **17**, 588 (1949).
8. K. Šolc, Macromolecules **3**, 665 (1970); **8**, 819 (1975); **10**, 1101 (1977); **16**, 236 (1983); **17**, 573 (1984); **19**, 1166 (1986); J. Polym. Sci., Polym. Phys. Ed. **12**, 1865 (1974); **20**, 1974 (1982). K. Šolc, A. Kleintjens, and R. Koningsveld, Macromolecules **17**, 573 (1984).
9. A. R. Shultz and P. J. Flory, J. Am. Chem. Soc. **74**, 4760 (1952).
10. H. Tompa, C. R. 2^e Reunion Soc. Chim. Phys. Paris 163 (1952); quoted in Ref. 2.
11. S. H. Maron and N. Nakajima, J. Polym. Sci. **54**, 587 (1961).
12. R. Koningsveld, L. A. Kleintjens, and A. R. Shultz, J. Polym. Sci. A-2 **8**, 1261 (1970).
13. R. Koningsveld and L. A. Kleintjens, Macromolecules **4**, 637 (1971).
14. Th. G. Scholte, J. Polym. Sci. A-2 **9**, 1553 (1971).
15. R. Koningsveld, W. H. Stockmayer, J. W. Kennedy, and L. A. Kleintjens, Macromolecules **7**, 73 (1974).
16. Th. G. Scholte, Eur. Polym. J. **6**, 1063 (1970).
17. Y. Einaga, S. Ohashi, Z. Tong, and H. Fujita, Macromolecules **17**, 527 (1984).
18. N. Takano, Y. Einaga, and H. Fujita, Polym. J. **17**, 1123 (1985).
19. M. Nakata, T. Dobashi, N. Kuwahara, and M. Kaneko, Phys. Rev. A **18**, 2683 (1978).
20. M. Tsuyumoto, Y. Einaga, and H. Fujita, Polym. J. **16**, 229 (1984).
21. M. Nakata, N. Kuwahara, and M. Kaneko, J. Chem. Phys. **62**, 4278 (1975).
22. J. Hashizume, A. Teramoto, and H. Fujita, J. Polym. Sci., Polym. Phys. Ed. **19**, 1405 (1981).
23. N. Kuwahara and M. Nakata, Polymer **14**, 415 (1973).
24. J. Kojima, N. Kuwahara, and M. Kaneko, J. Chem. Phys. **63**, 333 (1975).
25. S. Ohashi, B. S. Thesis, Osaka University, 1983.
26. R. Perzynski, M. Delsanti, and M. Adam, J. Phys. (Paris) **48**, 115 (1987).
27. E. Nies, R. Koningsveld, and L. A. Kleintjens, Progr. Colloid & Polymer Sci. **71**, 2 (1985).

28.   H. Vink, Eur. Polym. J. **12**, 77 (1976).
29.   Y. Einaga, Z. Tong, and H. Fujita, Macromolecules **18**, 2258 (1985).
30.   Z. Tong and Y. Einaga, unpublished paper.
31.   G. Rehage, D. Möller, and O. Ernst, Makromol. Chem. **88**, 232 (1965); G. Rehage and D. Möller, J. Polym. Sci. C **16**, 1787 (1967).
32.   R. Koningsveld and A. J. Staverman, J. Polym. Sci. C **16**, 1775 (1967).
33.   R. Koningsveld and A. J. Staverman, J. Polym. Sci. A-2 **6** , 349 (1968).
34.   J. W. Kennedy, M. Gordon, and R. Koningsveld, J. Polym. Sci. C **39**, 71 (1972).
35.   K. W. Derham, J. Goldbrough, M. Gordon, R. Koningsveld, and L. A. Kleintjens, Makromol. Chem. Suppl. **1**, 401 (1975).
36.   J. W. Breitenbach and B. Wolf, Makromol. Chem. **108**, 263 (1967).
37.   L. A. Kleintjens, R. Koningsveld, and W. H. Stockmayer, Brit. Polym. J. **8**, 144 (1978).
38.   Z. Tong, Y. Einaga, and H. Fujita, Macromolecules **18**, 2264 (1985).
39.   R. Koningsveld and A. J. Staverman, Kolloid-Z. Z. Polymere **229**, 31 (1967).
40.   H. Fujita and Y. Einaga, J. Polym. Sci., Polym. Phys. Ed. **21**, 1187 (1983).
41.   T. Dobashi and M. Nakata, J. Chem. Phys. **84**, 5775 (1986).
42.   Y. Einaga, Y. Nakamura, and H. Fujita, Macromolecules **20**, 1083 (1987).
43.   T. Dobashi, M. Nakata, and M. Kaneko, J. Chem. Phys. **72**, 6692 (1980).
44.   T. Dobashi, M. Nakata, and M. Kaneko, J. Chem. Phys. **72**, 6685 (1980).
45.   M. Nakata, T. Dobashi, N. Kuwahara, M. Kaneko, and B. Chu, Phys. Rev. A **18**, 2683 (1978).
46.   I. C. Sanchez, J. Appl. Phys. **58**, 2871 (1985). See also S. Stepanow, J. Phys. (Paris) **48**, 2037 (1987).
47.   F. J. Wegner, Phys. Rev. B **5**, 4529 (1972).
48.   R. Aris and G. R. Gavalas, Trans. Roy. Soc. (London) A **260**, 351 (1966).
49.   E. Dickinson, J. Chem. Soc. Farad. Trans. II **76**, 1458 (1980).
50.   E. R. Smith and J. S. Rowlinson, J. Chem. Soc. Farad. Trans. II **76**, 1468 (1980).
51.   J. A. Gulatien, J. M. Kincaid, and G. Morrison, J. Chem. Phys. **77**, 521 (1982).
52.   J. J. Salacuse and G. Stell, J. Chem. Phys. **77**, 3714 (1982).
53.   J. G. Briano and E. D. Glandt, Fluid Phase Equil. **14**, 91 (1983)
54.   M. T. Rätzsch, H. Kehlen, and J. Bergmann, Z. phys. Chem. (Leipzig) **264**, 318 (1983); H. Kehlen and M. T. Rätzsch, Z. phys. Chem. (Leipzig) **264**, 1153 (1983); H. Kehlen, M. T. Rätzsch, and J. Bergmann, Z. phys. Chem. (Leipzig) **265**, 318 (1984); M. T. Rätzschand H. Kehlen, J. Macromol. Sci. – Chem. A **22**, 323 (1985); M. T. Rätzsch, H. Kehlen, and D. Browarzik, J. Macromol. Sci. – Chem. A **22**, 1679 (1985); J. Bergmann, H. Kehlen, and

M. T. Rätzsch, Z. angew. Math. Mech. **65**, 343 (1985); D. Browarzik, H. Kehlen, M. T. Rätzsch, and Th. Schlezel, Z. phys. Chem. (Leipzig) **266**, 177 (1985); M. T. Rätzsch, H. Kehlen, and D. Thieme, J. Macromol. Sci. – Chem. A **23**, 811 (1986).

Chapter 10   Some Other Topics

1.   Ring Polymers

1.1   Matrix Effects on Tracer Diffusion of Ring Polymers

We consider a primitive chain trapped in a curvilinear tube fixed in space and allowed only to reptate along the centroid of the tube. If the chain is linear, it can change its conformation by creeping out of the mouths of the tube and can thus self–diffuse. On the other hand, if the chain is cyclic, it cannot disengage from the tube and hence undergoes no self–diffusion. Thus, we may expect that if only reptation is possible for it, a ring polymer in an entangled system can self–diffuse only when its surroundings relax by tube renewal. This prediction can be tested by examining whether the tracer diffusion coefficient D_{tr} of a ring polymer in the matrix of a linear polymer diminishes to zero, regardless of the length of the ring, as the linear becomes infinitely long.

Recently, by using forward recoil spectroscopy, Mills et al. [1] measured D_{tr} of ring polystyrene of molecular weight M in the range $10 \times 10^3 - 180 \times 10^3$ in linear deuteropolystyrene of molecular weight M_{p} in the range $55 \times 10^3 - 915 \times 10^3$. Their data are shown in Figure 10–1 (a), where $\log D_{\mathrm{tr}}$ at five fixed M_{p} is plotted against $\log M$. We see that the plots asymptotically converge to a straight line as M_{p} increases. Thus, if D_{tr} is expressed as

$$D_{\mathrm{tr}} \sim f(M_{\mathrm{p}})M^{-n} \qquad (1.1)$$

the index n approaches about 3.2 and $f(M_{\mathrm{p}})$ levels off to a non–zero value as M_{p} increases. This behavior of $f(M_{\mathrm{p}})$ indicates that the rings of any length can self–diffuse even in the matrix of the linears in which tube renewal is not likely to take place, thus contradicting the above–mentioned prediction. However, the recent idea of Klein [2] explains within the framework of reptation theory why $f(P)$ remains non–zero in the matrix of very long linear chains.

Figure 10–2 sketches the possible configurations that Klein imagined for a ring molecule in a linear chain matrix. Here, the dots show the cross–sections of uncrossable linear chains fixed in space. The ring is threaded by linears in (a), and not threaded in (b) and (c). The unthreaded ring in (b) has large loops and is assumed to behave as a star–branched chain, while that in (c) contains only small extrusions and is assumed to behave as a linear chain whose length is half that of the ring. If no motion other than reptation occurs, the ring in (a) simply rotates around a mean position and undergoes no translational diffusion, but the rings in (b) and (c) can migrate by wriggling through the space between the linears. Klein estimated D_{tr} associated with these rings by applying the known theories for reptating star and linear chains.[1]

---

[1] For star chains Klein extended the previous formulation due to de Gennes [3].

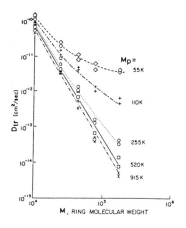

Fig. 10-1. M dependence of tracer diffusion coefficient D_{tr} for ring PS in linear PS matrices of indicated M_p [1].

As shown by Mills et al. [1], however, Klein's formulas do not fit their experimental results. For example, they fail to predict that eq 1.1 with $n = 3.2$ holds for large M_p. The probability of finding unthreaded rings in a linear chain matrix should be very low, except for relatively short rings. Therefore, it seems unreasonable to interpret the D_{tr} data of Mills et al. only in terms of the unthreaded rings. Mills et al. considered the possibility that a once–threaded ring diffuses like a smoke ring along the threading linear chain. This motion of the ring is no longer reptative but more or less Rouse–like. Mills et al. formulated it to obtain an approximate expression $D_{tr} \sim P_1 M^{-1} M_p^{-1}$, where P_1 is the probability that a ring in the melt is once–threaded. However, this predicts D_{tr} to vanish at the limit of large M_p and hence does not conform to the experimental finding. Nonetheless, the smoke–ring motion through a linear chain seems to be a promising idea worthy of further pursuit.

At present, no reported data on ring self–diffusion in polymer concentrates are available other than those of Mills et al. and no theory of this subject exists other than Klein's. Thus we see a virgin field of research open before us. What seems most needed is experimental data for self–diffusion in the melt and concentrated solutions of rings. Diffusion of linear chains in ring chain matrices should also be instructive, as pointed out by Mills et al. The reptation idea now dominating the study of polymer self–diffusion will face crucial tests when accurate and systematic diffusion data on these systems become available.

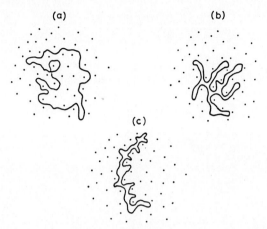

Fig. 10-2. Possible configurations of a ring chain in a matrix of linear chains. Dots show the cross-sections of linears. (a), general, threaded by linears; (b), unthreaded, behaving like a star chain; (c), unthreaded, behaving like a linear.

1.2 Melt Viscosity

Probably, Dodgson et al. [4] were the first to obtain data on the viscosity $\eta(r)$ for a ring polymer in bulk. They made measurements (25°C) on sharp fractions of linear and ring poly (dimethylsiloxane) over the range $500 < M_w < 25000$. The results are displayed in Figure 10-3. Interestingly, the values of $\eta(r)$ appear above those of viscosity $\eta(\ell)$ for the corresponding linears at $M_w < 4000$, and the reverse is the case at $M_w > 4000$. As Dodgson et al. showed, this crossover phenomenon is attributed to the fact that the activation energy for viscous flow, E_η, is larger for the rings than for the linears, where E_η is defined by

$$\eta = A \exp(E_\eta/RT) \tag{1.2}$$

In fact, the factor A, i.e., the viscosity corrected for this difference in E_η,[2] was found to be larger for the linear species than for the ring one over the entire range of molecular weight studied.

The work of Dodgson et al. was limited to relatively low molecular weights. For polystyrene in bulk Roovers [5] measured $\eta(r)$ and $\eta(\ell)$ up to $M_w = 3.34 \times 10^5$ for the former and $M_w = 8.6 \times 10^5$ for the latter. Figure 10-4 shows the reported results. We see that the behavior of $\eta(r)$ relative to $\eta(\ell)$ in the

---

[2] This correction is to reduce η to an iso–free volume state. The average free volume per repeat unit of a linear chain is larger than that of the corresponding ring, because the repeat units near the chain ends are more mobile than those inside the chain.

327

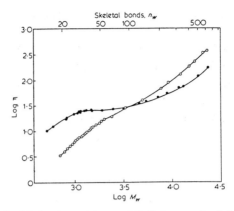

Fig. 10–3. Melt viscosities at 25°C for cyclic (•) and linear (○) PDMS as functions of weight–average molecular weight M_w [4].

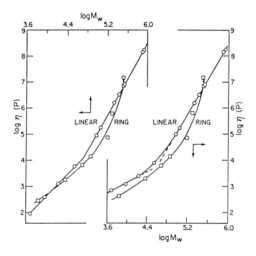

Fig. 10–4. Left panel, raw viscosity data for ring and linear PS in bulk at 169.5°C. Right panel, free–volume corrected data at the same temperature. Both are due to Roovers [5].

region of low M_w is essentially similar to that observed by Dodgson et al. for poly(dimethylsiloxane). Up to $M_w \sim 1 \times 10^5$ the free–volume corrected values of $\eta(\ell)$ are about twice as higher as those of $\eta(r)$. Then, the latter increases more sharply with M_w than the former, leading to their crossover at an M_w of about 3.4×10^5. As expected, the data points for $\eta(\ell)$ in this region are fitted by the familiar 3.4 power law [6](see Section 5.1 of Chapter 8). Roovers tried to justify the crossover, but a more recent study of McKenna et al. [7]

on polystyrene melts revealed no crossover phenomenon at least in the range of molecular weight studied.

McKenna et al. carried out sample preparation and viscosity measurements with admirable care and precision. Here in Figure 10–5, we quote as the most important of their findings the data of free–volume corrected $\eta(r)$ for the ring samples (presumably less knotted and less contaminated with linears), which were the "best" selected from Roovers' as well as their own. We may note the following.

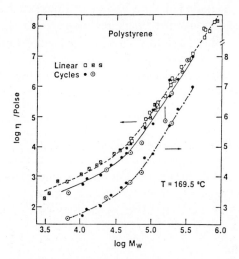

Fig. 10–5. Solid line, fitting (free–volume corrected) $\eta(r)$ for cyclic PS in the melt at 169.5°C (•, samples of McKenna et al.; ⊙, Roovers' samples). Dashed line, fitting the corresponding $\eta(\ell)$ for linear PS.

(i)   The solid broken line fits the data points closely. This shows that for the best selected rings the data of McKenna et al. and of Roovers are consistent with one another;

(ii)   When compared at the same M_w, $\eta(r)$ is consistently smaller than $\eta(\ell)$ over the entire range indicated. Thus, no crossover phenomenon is seen. The difference remains quite small, giving $\eta(r)/\eta(\ell)$ equal to about 0.5 for $M_w < M_c (\sim 5 \times 10^4)$ and closer to unity for $M_w > M_c$. Here, M_c is the molecular weight at the break for either the solid or the dashed line;

(iii)   These lines have the same slope of 1.5 for M_w below M_c, while their slopes for M_w above M_c are 3.46 for the rings and 3.40 for the linears.

These features show striking similarities between ring and linear polystyrenes with respect to the molecular weight dependence of melt viscosity. Of especial interest is the virtual agreement of $\eta(r)$ and $\eta(\ell)$ in entangled systems, i.e., in

329

the region above M_c, suggesting that steadily sheared linear and ring polymer melts relax by essentially similar chain motions. Thus, the experimental results of McKenna et al. and Roovers put doubt on the reptation mechanism currently favored for linear chains in entangled systems, because it is hard to visualize reptating rings within rings.

As noted above, the solid and dashed lines below the breaks in Figure 10–5 have a slope of 1.5. This finding is not consistent with the usual notion that the melt viscosity below M_c depends linearly on molecular weight M so that polymer chains in unentangled systems behave as Rouse chains in dilute solution (for example, the data below the break in Figure 8–20 show this molecular weight dependence). Thus, McKenna et al. [7] proposed to interpret their data by an interpolation formula which gives η (reduced to an iso–free volume state) the linear M dependence at very low M and the 3.4 power M dependence at very high M. It reads

$$\eta = AM + CM^{\delta} \tag{2.1}$$

where A, C, and δ are adjustable constants (McKenna et al. left δ free instead of fixing at 3.4). Though not displayed here, continuous curves calculated by eq 2.1 were found to fit closely the data points for both linear and cyclic polystyrene fractions over the range of M_w from 10^4 to 10^6. However, this data analysis is of no special interest, because eq 2.1 lacks the theoretical foundation.

1.3   Remarks

1.   McKenna et al. [7] conjectured that Roovers' finding of a very strong M dependence of $\eta(r)$ (Figure 10–4) in the entangling region was an artifact due to the self–knotting of some ring samples and also to the contamination of others with linear chains. In this connection, the experiment done by McKenna and Plazek [8] is worth referring to. These authors measured η of blends consisting of linear and cyclic polystyrene fractions of similar molecular weights. For weight fractions $w(r)$ of the rings from 0.85 to 1.0 it was found[3] that η varied approximately as $w(r)^{-5.6}$. This result indicates that the melt viscosity of a ring polymer in bulk is appreciably enhanced with slight contamination of linear counterparts of similar molecular weights. Thus, if ring samples of higher molecular weights were more contaminated with linears, they would give $\eta(r)$ depending more strongly on M than the pure ring samples. Anyway, it is crucial for accurate work on ring polymers to prepare samples which have minimal contamination by linear species, self–knotted rings, etc. When working at high temperature, attention has to be paid to the degradation of rings to linears.

---

[3] More recently, Roovers [9] found a stronger effect of the linear species from a study on polybutadiene.

2. From dynamic mechanical measurements on ring and linear polystyrenes with comparable M_w ($\sim 3.4 \times 10^5$) Roovers [5] obtained $G_N^0(r)/G_N^0(\ell) \sim 0.5$, where G_N^0 denotes the plateau shear modulus. Since the molecular weight between entanglements, M_e, is given by $\rho RT/G_N^0$ [6] (where ρ is the density of the system), this finding implies that M_e for the ring is about twice that for the linear, and hence one may imagine that a ring within rings assumes the configuration of a double–stranded linear chain as illustrated in Figure 10–2 (c). However, one will feel hard pressed to accept such a postulate. Very recently, Roovers [9] found from a similar study on polybutadiene that $G_N^0(r)$ was about 1/5 of $G_N^0(\ell)$ at the same molecular weight. Thus, the number of entanglements in the ring–ring polybutadiene system is significantly smaller than that in the ring–polystyrene system. We do not know what is associated with this difference.

We have just begun to understand the dynamic behavior of ring polymers in the condensed state. This situation will motivate theoretical and experimental work needed for clarification of the mechanism in which rings within rings can relax and diffuse.

2. Spinodal Decomposition

2.1 Introduction

Figure 10–6 (a) sketches the phase diagram for a two–component system on the $T - \phi$ plane, where, as in Chapter 9, T is the temperature and ϕ the volume fraction of one component. For the convenience of later discussions, we are concerned here with a system having a lower critical point. The solid and dashed curves in the figure represent the binodal and the spinodal, respectively. The curve in Figure 10-6 (b) shows the equilibrium Gibbs free energy of mixing, ΔG_e, at $T = T_0$ as a function of ϕ. Thermodynamics tells us that the system is unstable in the region inside the spinodal and metastable in the region between the spinodal and the binodal. Thus, when a stable mother solution outside the binodal is brought to a state between points B and B', it cannot stay uniform at its initial concentration but eventually separates into two distinct phases having the compositions at points A and A'.

Until recently, the kinetics of this first–order transition in polymer systems has received relatively little attention, in contrast to the numerous studies made on systems of small molecules and atoms as reviewed by many authors (for example, Gunton et al. [10]). Probably, Nishi et al. [11] in 1975 were the first to study it experimentally. Greater research activity has emerged in the 1980s among physicists interested in polymer dynamics and the engineers wishing to design composite materials sometimes called polymer alloys. The results are

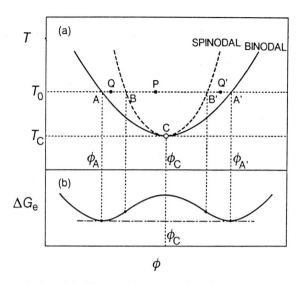

Fig. 10-6. (a) Phase diagram of a binary system with a lower critical point; solid line, binodal (or cloud–point curve); dashed line, spinodal; C, Critical point. (b) Composition dependence of ΔG_e at $T = T_0$.

adequately summarized in the review articles cited in Section 2.6. In this chapter, confining ourselves to the kinetic process called spinodal decomposition, we briefly discuss some typical theories of it along with related experimental data on polymer systems. The aim is to help the reader understand what problems arouse current interest in the dynamics of polymer phase separation.

2.2 Spinodal Decomposition

In a stable solution, whenever local concentration fluctuations occur, matter is spontaneously transported down the concentration gradient (downhill diffusion) generated to damp them, and the solution stays uniform macroscopically. When by a sudden change in temperature (this operation is sometimes called quenching, even though it is heating) the solution is brought to the region inside the spinodal (say, point P in Figure 10-6 (a)), some of the concentration fluctuations present in it grow by spontaneous transport of matter against the concentration gradient (uphill diffusion). This occurs because then the Gibbs free energy of mixing ΔG decreases. Thus, there develops in the solution a composition distribution ϕ which, with increasing time, will deviate more from $\phi(0)$, where $\phi(0)$ is the composition distribution present in the solution at the instant of quenching. In what follows, we simply refer to the difference

$\phi - \phi(0)$ as the composition distribution. The change in this distribution continues until the solution eventually reaches the equilibrium state consisting of two distinct phases of macroscopic size, each having the composition ϕ_A or $\phi_{A'}$ in Figure 10-6 (a). This transition process starting from small composition fluctuations to equilibrium phase separation is called **spinodal decomposition**. Roughly classified, it goes through three steps, called early, intermediate, and late stages. The early stage is the period in which the amplitude of the composition distribution remains sufficiently small compared with ϕ_0, where ϕ_0 is the average of $\phi(0)$ over the solution and hereafter called the initial composition of the system. In the intermediate stage, while its amplitude keeps increasing toward the equilibrium value $\phi_{A'} - \phi_A$, the composition distribution is gradually coarsened, i.e., its small irregularities are smeared out. Sooner or later, the system is coarse-grained to a state which may be viewed as consisting of vaguely separated sub-macroscopic domains whose compositions are close to either ϕ_A or $\phi_{A'}$. The late stage is the period in which these domains grow in size and their boundaries are sharpened.

2.3 Nucleus Growth

When a solution is quenched to the region between the spinodal and the binodal (say, point Q in Figure 10-6 (a)), small concentration fluctuations are damped, because their growth leads to an increase in ΔG. Non-decaying local phases can appear (i.e., ΔG diminishes) only when there occur concentration fluctuations which are so large in amplitude that their composition comes between the opposite branches of the spinodal and binodal (say, point Q' in Figure 10-6 (a)). Such concentration fluctuations are called nuclei or droplets. The nuclei are not in equilibrium but tend to settle down to the equilibrium composition at point A' in Figure 10-6 (a). This induces uphill diffusion of matter from their outlying regions on to their surface. As a result, the phase containing nuclei grows in size and is concentrated from Q' to at A', whereas the remaining phase is diluted from Q to A. This mechanism leading to equilibrium phase separation is called **nucleus growth**. In this book, we will not concern ourselves with it.

2.4 Cahn–Hilliard Theory

2.4.1 Basic Relations

We start from a brief account of the phenomenological theory of spinodal decomposition due essentially to Cahn and Hilliard [12, 13]. We consider a binary system consisting of chemically different, but mutually compatible components 1 and 2, with the volume fraction ϕ assigned to component 2. For simplicity,

333

the system is assumed to be one-dimensional, i.e., it has a cross-section A and extends in the x direction. Its volume is denoted by V.

The basis of the Cahn–Hilliard theory [12] is the assumption that ΔG per unit volume of the system, $\Delta \overline{G}$, is given by

$$\Delta \overline{G} = (A/V) \int [\Delta \overline{G}_e(\phi) + \kappa (\partial \phi / \partial x)^2] \, dx \qquad (2.1)$$

which is called the Ginzburg–Landau form in statistical physics. Here, $\Delta \overline{G}_e(\phi)$ is the equilibrium free energy of mixing per unit volume, and κ a constant related to the interfacial tension arising between different concentrations. In the following, $\Delta \overline{G}_e$ is designated by f, as is done in many papers concerning this subject.

From the conservation of mass we obtain the continuity equation:

$$\frac{\partial \phi(x, t)}{\partial t} = -\frac{\partial J(x, t)}{\partial x} \qquad (2.2)$$

where $J(x, t)$ is the flux of component 2 at position x and time t. We assume that J is given by

$$J = -M(\phi) \left(\frac{\delta \Delta \overline{G}}{\delta \phi} \right) \qquad (2.3)$$

where $M(\phi)$ is the mobility of polymer 2 assumed to depend only on ϕ, and δ denotes the functional differential. With eq 2.1 and 2.3 it follows from eq 2.2 that $\phi(x, t)$ is governed by

$$\frac{\partial \phi}{\partial t} = \frac{\partial}{\partial x} \left\{ M(\phi) \left[\left(\frac{\partial^2 f}{\partial \phi^2} \right) \left(\frac{\partial \phi}{\partial x} \right) - 2\kappa \left(\frac{\partial^3 \phi}{\partial x^3} \right) \right] \right\} \qquad (2.4)$$

This non-linear partial differential equation is usually called the Cahn–Hilliard equation.

2.4.2   Early Stage

In the early stage of spinodal decomposition, ϕ remains close to the average concentration ϕ_0 of the initial solution. If we neglect the terms higher than first order in concentration fluctuation u defined by

$$u = \phi - \phi_0 \qquad (2.5)$$

we can linearize eq 2.4 to the form

$$\frac{\partial u}{\partial t} = M_0 \left[\left(\frac{\partial^2 f}{\partial u^2} \right)_0 \left(\frac{\partial^2 u}{\partial x^2} \right) - 2\kappa \left(\frac{\partial^4 u}{\partial x^4} \right) \right] \qquad (2.6)$$

where the subscript 0 signifies the evaluation at $u = 0$, i.e., $\phi = \phi_0$. The relevant solution of this differential equation is

$$u(x,t) = \int_{-\infty}^{\infty} U(k) \exp\left[R(k)t\right] \exp\left(ikx\right) dk \qquad (2.7)$$

where $i = \sqrt{-1}$, $R(k)$ is the rate–determining factor given by

$$R(k) = -M_0 k^2 [(\partial^2 f / \partial u^2)_0 + 2\kappa k^2] \qquad (2.8)$$

and $U(k)$ is related to $u(x, 0)$ by

$$u(x,0) = \int_{-\infty}^{\infty} U(k) \exp\left(ikx\right) dk \qquad (2.9)$$

Here $u(x, 0)$ denotes the distribution of concentration fluctuations in the system straight after quenching. Equation 2.9 shows its Fourier–integral representation, with $U(k)$ as the amplitude of the Fourier component for a wavenumber k.

For a pair of ϕ_0 and T located inside the spinodal, the derivative $(\partial^2 f / \partial u^2)_0$ is negative, so that $R(k)$ can be positive for $k < k^\star$, where $k^\star$ is the value of k for which $(\partial^2 f / \partial u^2)_0 + 2\kappa k^2$ vanishes (we note that M_0 is a positive quantity). Thus, as time elapses, the Fourier components in eq 2.7 for $k < k^\star$ grow exponentially and those for $k > k^\star$ decay. It follows from eq 2.8 that $R(k)$ has a maximum at $k = k_m$, where

$$k_m = \frac{1}{2} \left[-\frac{(\partial^2 f / \partial u^2)_0}{\kappa} \right]^{1/2} \qquad (2.10)$$

This means that the Fourier component for $k = k_m$ in the initial concentration fluctuations dominates the growth of u in the early stage of spinodal decomposition. In passing, we note that

$$k_m = 2^{-1/2} k^\star \qquad (2.11)$$

2.4.3 Thermal Noise Effect

The Cahn–Hilliard theory is deterministic, ignoring the random generation of concentration fluctuations by thermal agitation. This thermal noise effect, first pointed out by Cook [14], adds a new term to the right–hand side of eq 2.4. Since the timescale of thermal agitation is much shorter than that of the slow uphill diffusive process concerned here, the added term may play a role

335

in the early stage of spinodal decomposition. This should be especially the case for the phase separations occurring in the vicinity of the critical point (as realized in experiments using what is called VSCQ in Section 2.6.2), because the term $\partial^2 f/\partial\phi^2$ in eq 2.4 is kept very small during such processes. Cook [14] developed a theory accounting for the thermal noise effect. His theory is included as a special case of a more general theory described in the next section.

2.5 LBM Theory

2.5.1 Structure Factor

In general, concentration fluctuations at different points in a system do not occur independently but are correlated. The correlation of the lowest level is represented by a pair correlation function $h(x, t)$ defined as

$$h(x, t) = \langle u(0, t)u(x, t)\rangle$$

$$= \iint u(0, t)u(x, t)\rho_2(u(0), u(x), t)\delta u(0)\delta u(x) \tag{2.12}$$

where ρ_2 denotes the two–point distribution functional. The change in the concentration distribution accompanying a spinodal decomposition causes $h(x, t)$ to vary with time. Hence, information about the kinetics of spinodal decomposition ought to be obtained by studying this function theoretically and experimentally. Though $h(x, t)$ cannot be measured directly, its Fourier transform $S(k, t)$, i.e.,

$$S(k, t) = A \int_{-\infty}^{\infty} h(x, t)\exp(ikx)\,dx \tag{2.13}$$

can be measured as a function of time t and wavenumber k (the magnitude of the scattering vector) by using various scattering techniques (light, x–ray, and neutrons). Usually, $S(k, t)$ is called the structure factor. If desired, $h(x, t)$ can be calculated by Fourier inversion of the data for $S(k, t)$. However, in many cases, $S(k, t)$ is more useful and convenient for discussing phase separation kinetics. Thus, theoreticians interested in this type of kinetics pay more attention to formulating and solving the equation governing the structure factor.

We define a reduced pair correlation function $H(x, t)$ by

$$H(x, t) \equiv h(x, t)/h(0, t) \tag{2.14}$$

and note $h(0, t) = \langle u(0, t)^2\rangle \equiv \overline{u(t)^2}$ by eq 2.12. Then, eq 2.13 can be rewritten as

$$S(k, t) = A\overline{u(t)^2} \int_{-\infty}^{\infty} H(x, t)\exp(ikx)\,dx \tag{2.15}$$

where $\overline{u(t)^2}$ is the mean–square composition fluctuation at any point in the mixture at time t and hereafter called the amplitude of the correlation function.

A general equation for $S(k,t)$ was derived by Langer et al. [15] in 1975 (see also Langer [16]) from statistical considerations of molecular flow processes. Their formulation invokes eq 2.1 for $\Delta\overline{G}$ and incorporates the thermal noise effect. Omitting its details, we can write the basic equation of Langer et al. as

$$\frac{\partial S(k,t)}{\partial t} = -2M_0 k^2 \left\{ \left[\left(\frac{\partial^2 f}{\partial u^2} \right)_0 + 2\kappa k^2 \right] S(k,t) \right.$$
$$\left. + \sum_{n=3}^{\infty} \frac{1}{(n-1)!} \left(\frac{\partial^n f}{\partial u^n} \right)_0 S_n(k,t) \right\} + 2M_0 k_{\mathrm{B}} T k^2 \qquad (2.16)$$

Here, $S_n(k,t)$ (with $S_2(k,t) \equiv S(k,t)$) is defined by

$$S_n(k,t) = A \int_{-\infty}^{\infty} h_n(x,t) \exp(ikx)\, dx \qquad (2.17)$$

with
$$h_n(x,t) = \langle u(x,t)^{n-1} u(0,t) \rangle \qquad (2.18)$$

and k_{B} is the Boltzmann constant. The angle brackets have the same meaning as in eq 2.12. We note that Langer et al. treated M as concentration–independent and that eq 2.16 corresponds to the Cahn–Hilliard equation in which $\partial^2 f/\partial\phi^2$ is expanded in powers of u and all higher terms are retained. Thus, eq 2.16 is non–linear in u, though this is not apparent explicitly.

On the basis of eq 2.16 Langer et al. developed a sophisticated theory of phase separation in binary alloys. It gave a starting point for recent statistical theories of phase separation dynamics [10], and called the LBM (Langer–Baron–Miller) theory. Here we touch upon it.

2.5.2 Early Stage

The early stage of spinodal decomposition corresponds to the special case of eq 2.16 in which the non–linear sum term is negelcted, i.e.,

$$\frac{\partial S(k,t)}{\partial t} = 2R(k)S(k,t) + 2M_0 k_{\mathrm{B}} T k^2 \qquad (2.19)$$

where $R(k)$ is given by eq 2.8. This linear equation for $S(k,t)$ had already been given by Cook [15]. It can be integrated to give [14, 15]

$$S(k,t) = S_{\mathrm{T}}(k) + [S_0(k) - S_{\mathrm{T}}(k)] \exp(2R(k)t) \qquad (2.20)$$

337

where

$$S_0(k) = S(k,0) \tag{2.21}$$

$$S_T(k) = k_B T/[(\partial^2 f/\partial u^2)_0 + 2\kappa k^2] \tag{2.22}$$

Equation 2.20 predicts for the early stage of spinodal decomposition that $S(k,t)$ at any fixed k smaller than $k^\star$ increases exponentially, where $k^\star$ has been defined in Section 2.4.2, and that $S(k,t)$ grows with time most rapidly at $k = k_m$, where k_m is given by eq 2.11. The $S(k,t)$ vs. k relation may show a maximum, but the position of the maximum depends on $S_0(k)$ and S_T, i.e., it is not always found at $k = k_m$ and moves with time. Finally, substitution of eq 2.20 leads to

$$h(x,t) = \frac{1}{\sqrt{2\pi}A} \int_{-\infty}^{\infty} \{[S_T(k') + [S_0(k') - S_T(k')]$$
$$\times \exp(2R(k')t)\} \exp(-ik'x)\,dk' \tag{2.23}$$

2.5.3 Intermediate Stage

After the early stage has passed, the sum term in eq 2.16 cannot be neglected, and the equation becomes non-linear. Langer et al. [15] tried to circumvent the arising difficulty by a decoupling approximation to ρ_2, which allows $S_n(k,t)$ to be replaced by $(\langle u^n \rangle / \langle u^2 \rangle)S(k,t)$. Then, eq 2.16 is transformed into

$$\frac{\partial S(k,t)}{\partial t} = [2R(k) - 2M_0 k^2 E(t)]S(k,t) + 2M_0 k_B T k^2 \tag{2.24}$$

where

$$E(t) = \sum_{n=3}^{\infty} \frac{1}{(n-1)!} \left(\frac{\partial^n f}{\partial u^n}\right)_0 \frac{\langle u^n \rangle}{\langle u^2 \rangle} \tag{2.25}$$

and

$$\langle u^n \rangle = \int u(x,t)^n \rho_1(u(x),t)\delta u(x) \tag{2.26}$$

with $\rho_1(u(x),t)$ being the one–point distribution functional. The problem is thus reduced to evaluating ρ_1, but it is still difficult to solve. Langer et al. [15] devised a numerical method of solution. Here, in Figure 10–7, we display their calculated time evolutions of $S(k,t)$ and $\rho_1(u,t)$ for a critical quenching, i.e., quenching of a binary mixture whose initial composition is the critical value ϕ_c.

With increasing time, the bell–shaped curve for the reduced structure factor $\hat{S}(\hat{k},\hat{t})$ is markedly sharpened and its peak shifts to lower k. The latter behavior is one of the non–linear effects incorporated in eq 2.16. The overall features of

338

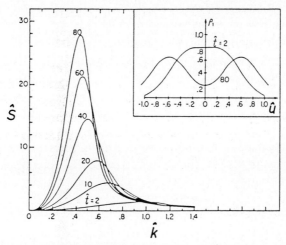

Fig. 10–7. Reduced structure factor $\hat{S}(\hat{k}, \hat{t})$ for a critical spinodal decomposition at various reduced times $\hat{t}$, where $\hat{k}$ is a reduced wavenumber. The insert shows ρ_1 at two values of $\hat{t}$ [15, 10].

Figure 10–7 are in qualitative agreement with Monte Carlo simulations and experimental results [10] in the intermediate stage, suggesting that the decoupling approximation is reasonable for this stage.

The LBM theory and the Cahn–Hilliard theory as well approximate the composition distribution in a spinodally decomposing system by a smooth continuous function. This continuum approximation may be good enough for mathematical description of the early and intermediate stages of spinodal decomposition. However, it should become inadequate as the system approaches or enters the late stage, in which the composition distribution undergoes considerable coarsegraining. A better approximation to this stage is to view the system as consisting of discrete grains (phases) separated by unstable interfaces. Various theories have been proposed for the dynamic growth of such grains [10]. As far as the author is aware, there exists no theory designed to apply over the enitre period of a spinodal decomposition.

2.5.4   Interpretation in Terms of the Correlation Function

Langer et al. [15] showed by numerical calculation that the term $E(t)$ for critical quenching increases sharply from zero to a certain limit with increasing t (the change is more complex for off–critical quenching). This effect is qualitatively equivalent to decreasing $|(\partial^2 f / \partial u^2)_0|$, and hence to decreasing $k_{\rm m}$ (see eq 2.10). Thus, the shift of the maximum position to the left as seen

339

in Figure 10–7 is a manifestation of the non–linear effect incorporated in $E(t)$.

The monotonic rise of the maximum of $S(k,t)$ is not a simple reflection of the increase in the amplitude $\overline{u(t)^2}$ toward the equilibrium value $\overline{u(\infty)^2}$. As seen from the double–peak curve of ρ_1 in the insert of Figure 10–7, $\overline{u(t)^2}$ comes close to its equilibrium value at $\hat{t} = 80$. However, the calculated $S(k,t)$ curves do not show the trend that the maximum stops rising shortly after this reduced time. Hence, the rise of the peak in the later stage has to be attributed to the time dependence of $H(x,t)$. The function $H(x,t)$ represents how strongly the composition fluctuations at two points separated by a distance x are correlated on the ensemble average. The correlation length ξ characterizes the range of x for which $H(x,t)$ remains effectively non–zero. Immediately after quenching, ξ is very small, and $H(x,t)$ may look like a delta function (this can be deduced from eq 2.23). As uphill diffusion proceeds, smaller composition fluctuations are smeared out and the composition distribution is coarsened to ordered local domains. This change implies that, as spinodal decomposition goes on, $H(x,t)$ spreads to larger x so that ξ increases. It is reasonable to consider that $H(x,t)$ has a damped oscillating shape as sketched in Figure 10–8 (the regions of plus and minus H represent individual ordered local domains). As a simple form of such $H(x,t)$ we assume

$$H(x,t) = (b-1)^{-1}[b\exp(-b|x|/\xi) - \exp(-|x|/\xi)] \qquad (2.27)$$

where b is a parameter which may depend on time. Substituting this into eq 2.15, we obtain

$$S(k,t) = A\overline{u(t)^2} \frac{2\xi(1+b)(k\xi)^2}{[(k\xi)^2 + 1][(k\xi)^2 + b^2]} \qquad (2.28)$$

This equation gives a bell–shaped S vs. k curve which rises in proportion to k^2 for small k, declines in proportion to k^{-2} for large k, and has a maximum at $k = k_{\mathrm{m}} = b/\xi$. Hence, the position of the peak of $S(k,t)$ shifts to small k as the correlation length increases, i.e., the correlation function spreads, with increasing time. We note that this shift depends on the time change in b, but has nothing to do with the time dependence of $\overline{u(t)^2}$. With $k_{\mathrm{m}}\xi = b$, eq 2.28 gives for $S(k_{\mathrm{m}},t)$, the maximum of $S(k,t)$,

$$S(k_{\mathrm{m}},t) = A\overline{u(t)^2}\xi(1+b)/(1+b^2) \qquad (2.29)$$

Thus, the peak of S becomes higher with the increases in both amplitude and spread of the correlation function, and this behavior is affected by the time dependence of b.

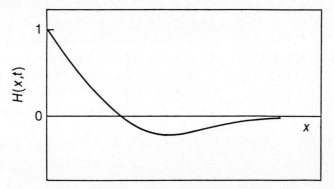

Fig. 10–8. Sketch of the reduced correlation function $H(x,t)$ given by eq 2.27. Roughly speaking, the positive and negative parts determine the behavior of Z in the regions of large and small k, respectively.

2.6 Scaling Laws

2.6.1 Scaling for $S(k,t)$

It has been postulated that after some initial transient time following a quench, $H(x,t)$ should change its shape self–similarly with time. In this section, we examine what comes out of this postulate. The self–similarity of $H(x,t)$ is mathematically equivalent to the condition that there exists a single time-dependent correlation length $\xi(t)$ which satisfies the relation

$$H(r,t) = H(r/\xi(t)) \tag{2.30}$$

Here, we have generalized the discussion to a binary mixture extending isotropically in three dimensions and changed x to the polar radius r. Accompanying this generalization, we have to replace eq 2.15 by

$$S(k,t) = \overline{u(t)^2} \int H(r,t) \exp{(i\mathbf{k} \cdot \mathbf{r})}\, d\mathbf{r} \tag{2.31}$$

Substitution of eq 2.30 into eq 2.31 gives

$$S(k,t) = \overline{u(t)^2}[\bar{k}(t)]^{-3} F(k/\bar{k}(t)) \tag{2.32}$$

where

$$\bar{k}(t) = 1/\xi(t) \tag{2.33}$$

341

and F is an unknown function. What is actually measured is the scattering intensity $I(k,t)$ which, if the background scattering is neglected, is related to $S(k,t)$ by

$$I(k,t) = KS(k,t) \qquad (2.34)$$

where K may be considered a constant if multiple scattering is ignored. We define a normalized scattering intensity $\bar{I}(k,t)$ by

$$\bar{I}(k,t) = I(k,t) / \int_{k'}^{k''} I(k,t)k^2 \, dk \qquad (2.35)$$

where k' and k'' denote the k values at which $I(k,t)$ virtually vanishes. With eq 2.32 and 2.34, this equation yields

$$\bar{I}(k,t) = [\bar{k}(t)]^{-3} \overline{F}(k/\bar{k}(t)) \qquad (2.36)$$

where $\overline{F}(\eta)$ is defined by

$$\overline{F}(\eta) = \frac{F(\eta)}{\int_{\eta'}^{\eta''} F(\eta)\eta^2 \, d\eta} \qquad (2.37)$$

with

$$\eta = k/\bar{k}(t) \qquad (2.38)$$

and η' and η'' are the η values corresponding to k' and k'', respectively. Equation 2.36 gives a dynamic scaling law which states that plots of $\bar{I}(k,t)[\bar{k}(t)]^3$ vs. k for different t are scaled by a dimensionless variable η. We note that this scaling law invokes only the self–similarity postulate for $H(k,t)$.

To test eq 2.36 by experiment we have to relate $\bar{k}(t)$ to a measurable quantity. Many authors have equated it to $k_m(t)$. With this choice, eq 2.36 has been confirmed for many binary systems of atoms and small molecules [10]. Tests on polymer mixtures have begun in recent years, as will be illustrated later.

2.6.2   Scaling for $k_m(t)$ and $\bar{I}_m(t)$

When a binary mixture with the critical composition ϕ_c is suddenly brought to a temperature T inside the spinodal but sufficiently close to the critical temperature T_c, we call the operation a very shallow critical quench (VSCQ). During the spinodal decomposition after a VSCQ the mixture stays in the vicinity of the critical state so that we may expect its dynamic properties at different T to be scaled by a T–dependent correlation length $\xi_c(T)$ and a T–dependent time constant $t_c(T)$ which characterize near–critical mixtures. The expected scaling laws for $k_m(t,T)$ and $\bar{I}_m(t,T)$ $(= \bar{I}(t,k)$ at its maximum) are

$$k_m(t,T)\xi_c(T) = q_m(\tau) \qquad (2.39)$$

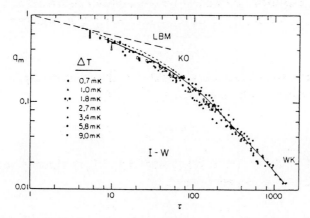

Fig. 10-9. Reduced peak position q_m as a function of the reduced time τ for VSCQ of the system isobutyric acid + water [17]. KO stands for the theory of Kawasaki and Ohta [18]. Data points are for the indicated values of $\Delta T = T - T_c$.

$$\bar{I}_m(t,\tau)/[\xi_c(\tau)]^3 = \tilde{I}_m(\tau) \tag{2.40}$$

Here, τ is a dimensionless time defined by

$$\tau = t/t_c(T) \tag{2.41}$$

with the time constant $t_c(T)$ calculated from

$$t_c(T) = [\xi_c(T)]^2/D(T) \tag{2.42}$$

where D is the diffusion coefficient.

Figures 10-9 and 10-10 illustrate a famous verification of the scaling laws for q_m and $\tilde{I}_m$ by Chou and Goldburg [17] with data on a binary fluid. Figure 10-9 shows that the exponent α in $q_m(\tau) \sim \tau^{-\alpha}$ changes continuously from about 1/3 to 1 with increasing τ from about 7 to 1000. The LKB lines in the figures indicate the slopes predicted by Langer et al.[15] and are significantly off the experimental data, suggesting the LBM theory to be of a semi-quantitative nature. On the other hand, the KO lines due to Kawasaki and Ohta [18], who incorporated hydrodynamic interactions into the basic equation of the LBM theory, agree more closely with experiment than do the LBM lines. Like the LBM theory, the KO theory is not expected to hold up to the late stage of spinodal decomposition.

343

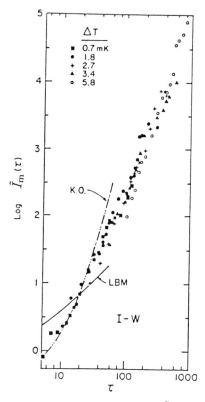

Fig. 10–10. Reduced peak intensity $\tilde{I}_m$ as a function of τ for the same system as in Fig. 10–9.

2.7   Experimental Studies on Polymer Systems

2.7.1   Introduction

Research on the spinodal decomposition of polymer solutions and mixtures began in relatively recent years, but a considerable body of experimental data has already been accumulated by several groups of workers. Major findings have been reviewed by Hashimoto [19, 20] and Nose [21], along with a relevant summary of the related theories. In this section, we describe some typical studies on polymer spinodal decomposition performed mainly to test the theoretical predictions mentioned in the preceding sections. We do not intend an extensive or complete survey of related papers.

Polystyrene (PS) and poly(vinylmethylether)(PVME) are mutually compatible over the entire range of composition, and their blends have thus often been chosen for experimental studies on the phase separation dynamics of binary polymer mixtures. The subsequent discussion depends heavily on the data re-

344

reported on them. In this connection, the following remarks may be in order. Differing from PS, no narrow–distribution sample of PVME is as yet available, and no effective fractionation method is known for this polymer. Thus, all the PVME samples used in previous studies were quite polydisperse. Hence, the polystyrene/PVME data cited below (we follow the recent fashion using a slant to denote a polymer blend) are not for strictly binary systems. The same is more or less true of other polymer blends investigated so far. Furthermore, PVME is so hygroscopic that its use is not always advantageous for experimental work.

At present, the most orthodox method of studying phase separation kinetics is to measure the time evolution of the structure factor $S(k, t)$ as a function of the wavenumber k (actually, the scattering angle ψ) by use of a time–resolved light (or X–ray) scattering apparatus [22]. The total integrated scattering intensity I_{total} is proportional to $\int S(k, t)k^2 dk$. Substituting eq 2.31 and noting that $H(0, t) = 1$, we find that $I_{\text{total}} \sim \overline{u(t)^2}$, which allows us to see how the amplitude of the correlation function changes in the process of phase separation [23].

2.7.2 Early Stage

1. **Typical Data and Analysis** To begin with, we illustrate in Figure 10–11 the binodal and spinodal curves for a PS–1 ($M_w = 200 \times 10^3$ and $M_w/M_n = 1.05$)/ PVME–1 ($M_w = 47 \times 10^3$ and $M_w/M_n = 1.5$) blend [24]. These curves show that the system has a lower critical point at $\phi_c = 0.80$ and $T_c = 95.8°\text{C}$. In what follows, we designate the volume fraction of PVME by ϕ. Actually, the dashed curve is not a binodal but a cloud–point curve, since the system is not binary (see Chapter 9). Nonetheless it is called binodal in the ensuing description. We note that the gap between the binodal and the spinodal is quite narrow.

Snyder et al. [23] were among the earlier workers who studied the early stage of spinodal decomposition of polymer blends, with the distinct aim of testing the Cahn–Hilliard theory. They measured I_{total} as a function of time for three PS/PVME blends and found that $\ln(I_{\text{total}}^{1/2})$ increased linearly with time t over a certain range of early time. If $R(k)$ defined by eq 2.8 has a sharp maximum at $k = k_m$, eq 2.7 may be approximated by $u(x, t) \sim \exp(R(k_m)t)$, so that $\overline{u(t)^2} \sim \exp(2R(k_m)t)$. Therefore, the initial slope of $\ln(I_{\text{total}}^{1/2})$ vs. t can be equated to $R(k_m)$. Snyder et al. used this idea to analyze their data (the same idea had already been used by Nishi et al. [11]). However, we have to note that the peak of $R(k)$ given by eq 2.8 is not so sharp as to justify the approximation used.

In his review article, Hashimoto [19] cited a number of studies which examined the early stage of polymer phase separation. Here in Figure 10–12 we illustrate the data of Hashimoto et al. [25] for a critical PS–1/PVME–1 blend,

345

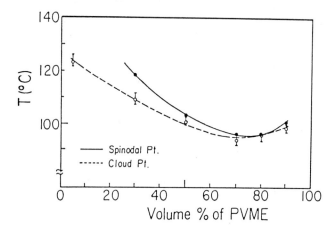

Fig. 10-11. Binodal and spinodal for a PS/PVME blend [24]. See the text for the component polymers.

shown as $\log I$ plotted against t at different fixed values of k. The data points for each k follow a straight line over a considerable range. If $I(k,t)$ is directly proportional to $S(k,t)$, this behavior is consistent with the relation

$$S(k,t) = S_0(k)\exp\left(2R(k)t\right) \qquad (2.43)$$

which can be derived from eq 2.20 by neglecting the thermal noise term $S_T(k)$. Thus, we would proceed as follows.

First, we equate the initial slope of each curve in Figure 10-12 to $2R(k)$ and evaluate $R(k)$. Then we plot $R(k)/k^2$ against k^2. The resulting plot should be linear, because eq 2.8 can be rewritten

$$R(k)/k^2 = D_{\mathrm{app}} - 2\kappa M_0 k^2 \qquad (2.44)$$

with D_{app} being an apparent diffusion coefficient defined by

$$D_{\mathrm{app}} \equiv -M_0(\partial^2 f/\partial u^2)_0 \qquad (2.45)$$

Figure 10-13 substantiates this prediction for three quench temperatures. Finally, using eq 2.44, we evaluate D_{app} and $2M_0\kappa$ from the intercept and slope of the straight line obtained. If separate information about $f(u)$ is available, we can calculate the more basic parameters M_0 and κ from them. Sometimes the Flory–Huggins theory is used for $f(u)$, but probably it is only of semi–quantitative use for the two–phase region.

346

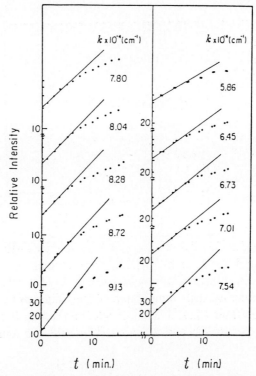

Fig. 10–12. Time evolution of scattering intensity at various k for a critical PS–1/PVME–1 blend quenched to $T_q = 96.8°C$ (1.0°C above T_c) [25].

The straight lines in Figure 10–13 have definite negative slopes. This means that κ is not negligible, i.e., the interfacial energy term in eq 2.1 plays a role in the early stage spinodal decomposition of the system studied. By extrapolation of the D_{app} values at the three T_q indicated in Figure 10–13 we find that D_{app} vanishes at 95.8°C. This is consistent with the fact that since the blend has the critical composition, its T_c is just the spinodal temperature at which $\partial^2 f/\partial u^2$ vanishes.

2. Improved Analysis

The above data analysis is based on two assumptions. One is that the thermal noise effect may be neglected, and the other is that the measured scattering intensity $I(k,t)$ is directly proportional to $S(k,t)$. The first assumption is considered valid for polymer mixtures, but the second one is not obtained in usual experimental setups, in which some stray light unavoidably enters into the detector for the measurement of scattering intensity. Thus, it is

347

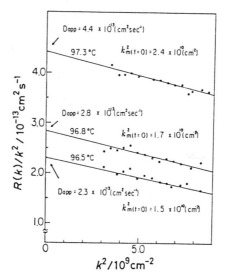

Fig. 10–13. Linear relations between $R(k)/k^2$ and k^2 for the indicated quench temperatures T_q. The system is the same as in Fig. 10–12 [25].

reasonable to relate $I(k,t)$ to $S(k,t)$ by

$$I(k,t) = KS(k,t) + I_b \tag{2.46}$$

with K an optical constant and I_b an unknown constant representing the background intensity of the stray light. With I_∞ and I_0 defined by

$$I_\infty = KS_T + I_b \quad \text{and} \quad I_0 = KS_0 + I_b \tag{2.47}$$

we obtain from eq 2.20 and 2.46

$$I(k,t) = I_\infty + (I_0 - I_\infty)\exp(2R(k)t) \tag{2.48}$$

It follows from 2.48 that the initial slope of $\ln I(k,t)$ plotted against t is no longer $2R(k)$ but $2\{1 - [I_\infty(k)/I_0(k)]\}R(k)$. Hence it turns out that I_∞ need to be known in order to determine $R(k)$, since I_0 may be obtained by extrapolating measured $I(k,t)$ to $t = 0$. Recently, Sato and Han [26] derived the following series from eq 2.48:

$$\left[\frac{t}{I(k,t) - I_0}\right]^{1/3} = \frac{1}{[2(I_0 - I_\infty)R(k)]^{1/3}}[1 - (1/3)R(k)t \\ + (1/81)(R(k)t)^3 + \cdots] \tag{2.49}$$

348

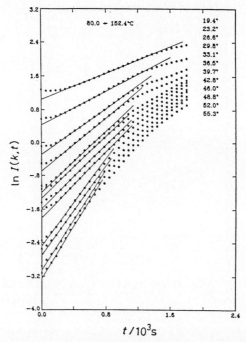

Fig. 10–14. Time evolution of scattering intensity $I(k,t)$ for a near–critical d–PS/PVME blend ($\phi = 0.81$) and $T_q = 152.4°C$) [26]. Scattering angles shown are refraction-corrected values and correspond to different k.

This equation shows that if its right–hand side calculated from experimental $I(k,t)$ and I_0 is plotted against t, the data points for $R(k)t < 1$ fall on a straight line. The slope and intercept of the line allow us to determine I_∞ and $R(k)$ at the same time. The success of this method depends on how accurately we can extrapolate I_0.

Sato and Han [26] studied blends of deuterated PS and PVME; M_w and M_w/M_n were 4.02×10^5 and 1.42 for the former and 2.10×10^5 and 1.32 for the latter. This system has a critical point at $T_c \cong 151°C$ and $\phi_c \cong 0.80$. Figure 10–14 shows their data for the quench from 80°C to $T_q = 152.4°C$ of a near–critical blend. The solid lines in the figure fit the data points, which appear to follow a linear relation over a broad range. Interestingly, the data show upward deviations from these lines in the region of small t, i.e., they pass a transient approach before entering into the linear region, while the data of Hashimoto et al. in Figure 10-12 show no such approach. Sato and Han showed that eq 2.49 closely fits the initial part of their $I(k,t)$ data for all T_q and ϕ studied, as should be expected. The resulting values of $R(k)$ for

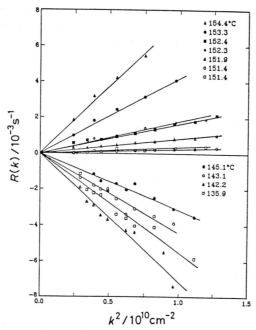

Fig. 10–15. $R(k)$ for a near–critical d-PS/PVME blend with $\phi = 0.81$. Upper part: for quenching from $80°$C to the indicated T_q. Lower part: for reverse quenching from $149.9°$C to the indicated T_r.

$\phi = 0.81$ appear plotted against k^2 in the upper part of Figure 10–15. To a good approximation, the set of data for each T_q falls on a straight line passing the origin. Sato and Han found the same to hold at $\phi = 0.90$ and 0.66, and thus concluded that $\kappa \cong 0$, i.e., the interfacial energy term in eq 2.1 makes no substantial contribution to the early stage spinodal decomposition. When extrapolated, the slopes of the lines in the upper part of Figure 10-15 predict D_{app} to vanish at $151.2°$C. This is the spinodal temperature T_s for $\phi = 0.81$. In the same way, Sato and Han found T_s to be 151.3 and $154.1°$C for $\phi = 0.90$ and 0.66, respectively. These values indicate that the spinodal of the d-PS/PVME blend they studied is quite flat over a range encompassing the critical point.

As noted above, the data of Sato and Han differ from those of Hashimoto et al. in having a short approach before entering the region in which $\ln I$ depends linearly on t. According to eq 2.48, this difference merely implies that I_∞ was significant in the experimental setup of Sato and Han, but negligible in that of Hashimoto et al. Surprisingly, it led to inconsistent findings that $\kappa \cong 0$ by Sato and Han, while $\kappa \neq 0$ by Hashimoto et al. In the following, we consider possible causes of this discrepancy.

3. On the Parameter κ The scattering due to thermal fluctuations of a one-phase equilibrium binary polymer blend is described in terms of S_T defined by eq 2.22. Using the random phase approximation (see Section of Chapter 6), de Gennes [27] derived

$$[S_T(k)]^{-1} = [\phi_1 P_1 H_1{}^0(k)]^{-1} + [\phi_2 P_2 H_2{}^0(k)]^{-1} - [2\chi + \phi_1(\partial\chi/\partial\phi_1)] \quad (2.50)$$

where ϕ_i, P_i, and $H_i{}^0(k)$ are the volume fraction, the relative chain length, and the Debye scattering function (see Chapter 6) of polymer i, respectively, and χ denotes the interaction parameter in the Flory–Huggins theory (see Chapter 9). Expanding $H_i{}^0(k)$ in powers of k and comparing this equation with eq 2.22, we arrive at the following expression for κ:

$$\kappa = (RT/6)[\langle S_1{}^2\rangle_\theta /(V_1\phi_1) + \langle S_2{}^2\rangle_\theta /(V_2\phi_2)] \quad (2.51)$$

where $\langle S_i{}^2\rangle_\theta$ and V_i are the unperturbed mean–square radius of gyration and the molar volume of polymer i, respectively.[4] Sato and Han [26] computed κ from this equation by using appropriate values for the parameters and showed that the second term $2\kappa k^2$ in the brackets of eq 2.8 is almost negligible in comparison with the first term over the k range treated in their experiment for $\phi = 0.81$. This result predicts that $R(k)$ virtually changes in direct proportion to k^2. In fact, the data in the upper part of Figure 10–15 conform to the prediction. Thus, it is reasonable to conclude that Sato and Han correctly measured and analyzed the early stage spinodal decomposition of their polymer blend. Then we wonder if the data analysis of Hashimoto et al. was really concerned with the initial behavior of this process.

When, as Hashimoto et al. did with the data of Figure 10–12, Sato and Han equated the slopes of the straight lines in Figure 10–14 to $2R(k)$, they obtained $R(k)$ whose k dependence is essentially similar to that found by Hashimoto et al. and thus conflicts with de Gennes' prediction. However, as mentioned above, when they analyzed the initial part of the same data by eq 2.48, they did obtain $R(k)$ consistent with de Gennes' theory. These findings lead us to wonder if the "linear region," i.e., where $\ln I$ shows a distinct linear dependence on t, really represents the early stage. If not, the linearity between $\ln I$ and t is a consequence of the non–linear effects controlling the later stage (the intermediate plus the late), and it is not surprising that $R(k)$ obtained by forced–fit of the linear theory to it gives a κ inconsistent with de Gennes' theory. However, when looking at Figure 10–12, one would not readily agree to suppose that the "linear region" is off the early stage. Nonetheless, it does not seem for the moment that

---

[4] Though crude theoretically, eq 2.51 is expected to give the order of magnitude of κ correctly.

there is any other idea to explain why $R(k)$ from the "linear region" conflicts with de Gennes' prediction.

Because of their high internal viscosity, blends of high–molecular–weight polymers are supposed to undergo much slower phase separation than fluid mixtures of small molecules. Hence, the early stage of their spinodal decomposition could be measured by appropriate experimental means. However, if the "linear region" in Figure 10–13 were to be regarded as being already in the intermediate stage, we would have to consider that the early stage passed so fast that Hashimoto et al. failed to observe it. Probably, the early stage of spinodal decomposition for polymer blends is much shorter than usually imagined.

The parameter I_∞ in eq 2.48 is often associated with the thermal noise effect. In this consideration, I_b in $I_\infty = K S_T + I_b$ is tacitly ignored. Thus, when, on the basis of a recent study of Okada and Han [28] on a d– PS/PVME blend, Hashimoto [19] anticipated that the thermal noise effect should be insignificant except for shallow quench depths and for k near $k^\star$ at which $R(k)$ vanishes, he would not have minded the term I_b. In actuality, however, its contribution to I_∞ in the early stage of phase separation is by no means negligible, and a due analysis of spinodal decomposition data has to take the effect of I_b into account.

4. Reverse Quench Experiment   The parameter $R(k)$ discussed above refers to temperatures in the two–phase region. Its values in the one–phase state can be estimated by the reverse quench experiment, in which a spinodally decomposing mixture is quenched to outside the binodal, and the accompanying change in $I(k,t)$ is measured as a function of t. It can be shown that whatever the initial unmixed state of the mixture, $I(k,t)$ eventually obeys eq 2.48. Since $\partial^2 f / \partial u^2$ is positive in a stable mixture, D_{app} is negative (see eq 2.45) and hence $R(k)$ in eq 2.48 is negative for any k (see eq 2.44). Thus, $I(k,t)$ decays toward I_∞, and $\ln(I(k,t) - I_\infty)$ plotted against t should tend to follow a straight line with a slope $-|2R(k)|$. The reverse quench experiment done by Kumaki and Hashimoto [29] on a PS/PVME blend showed that $R(k)$ was nearly in direct proportion to k^2. This finding means that κ is very small and, interestingly, conflicts with the conclusion from the above–mentioned work of Hashimoto et al. on forward quench experiments. To check the finding of Kumaki and Hashimoto, Sato and Han [26] quenched their PS/PVME blend with $\phi = 0.81$ from $149.9°C$ back to various temperatures. The $R(k)$ data obtained are shown in the lower part of Figure 10–15. It is seen that, as in the upper part, $R(k)$ is in direct proportion to k^2, showing that κ is negligibly small in the one–phase region as well as in the two–phase one. Thus, differing from Hashimoto's group, Sato and Han found D_{app} which varies continuously as the blend is heated through the spinodal temperature, as should be the case physically.

2.7.3 Later Stage

1. Typical Data As cited by Hashimoto [19] and Nose [21], many experiments have ben done concerning the later stage (defined here as the intermediate plus the late) of spinodal decomposition of condensed binary polymer mixtures. Actually, much of them has focused on testing the dynamic scaling laws for $S(k,t)$, k_m, and $\bar{I}_m$ discussed in Section 2.6.

Figure 10–16 displays typical $I(k,t)$ data showing how the k dependence of I varies as phase separation of a critical polymer blend proceeds from the early to the later stage [25]. Except for a few at short times, each of the indicated curves shows a well–defined peak. As time goes on, the peak is markedly sharpened, and especially, in (a), its position k_m moves appreciably to small k. As explained in Section 2.5.4, the shift in k_m is caused by the increased spreading of the correlation function, while the peak rises owing to the increases in both spread and amplitude of this function. However, the amplitude will shortly come to a saturation point, and then the spreading of the correlation function will govern the kinetics toward phase equilibrium. The late stage of spinodal decomposition may be defined as one characterized by this dynamics. However, it is not immediately obvious from the curves of Figure 10–16 when the kinetics crosses over from the early stage to the late one.

2. Scaling for $S(k,t)$ Hashimoto et al. [24] examined the applicability of the scaling law given by eq 2.32 or 2.36 to their $I(k,t)$ data for a critical PS/PVME blend quenched at different T_q, with $k_m(t)$ taken as $\bar{k}(t)$. Figure 10–17 illustrates the results for $T_q = 98.2°C$. The abscissa represents the scaled scattering intensity $I(k,t)[k_m(t)]^3$, which, according to eq 2.32 and 2.34, is proportional to $\overline{u(t)^2}F$. After 59.1 min the data points fall approximately on a single curve. This fact implies that $\overline{u(t)^2}$ becomes constant and F depends only on η, i.e., the process crosses over to the late stage, about one hour after the start of a spinodal decomposition. Hashimoto et al.[24] found that the same process began occurring earlier as the quench depth $T_q - T_s$ was increased. Thus, as might be expected, spinodal decomposition enters the late stage earlier for deeper quenching. Hashimoto et al. [24] also found that $F(\eta)$ became somewhat broader for deeper quenching.

Takahashi et al.[30] examined spinodal decomposition of a liquid blend of polystyrene and poly(methylphenylsiloxane) ($M_w = 6400$ and $M_w/M_n < 1.04$ for PS and $M_w = 4000$ and $M_w/M_n = 1.43$ for PMPS), focusing on the late stage behavior. They found that the similarity postulate held for the critical composition (30 wt% PS) and $T_s - T_q = 0.05°C$, but only approximately for an off–critical composition (38.5 wt% PS) and $T_s - T_q = 1.95°C$. Furthermore, for both critical and off–critical blends, $F(\eta)$ became broader with increasing

353

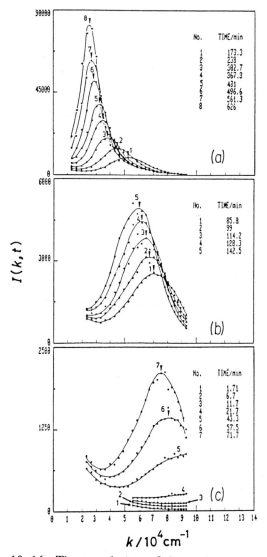

For plot (a):

No.	TIME/min
1	173.3
2	238
3	302.7
4	367.3
5	431
6	496.6
7	561.3
8	626

For plot (b):

No.	TIME/min
1	85.8
2	99
3	114.2
4	128.3
5	142.5

For plot (c):

No.	TIME/min
1	1.71
2	6.7
3	11.7
4	21.7
5	43.3
6	57.5
7	71.7

$k / 10^4 \mathrm{cm}^{-1}$

Fig. 10–16. Time evolution of the k dependence of scattering intensity $I(k, t)$ after quenching at $96.8°\mathrm{C}(T_q - T_s = 1°\mathrm{C})$ of a critical PS/PVME blend [25] (taken from Ref. [19]).

quench depth.

To check the similarity postulate it is more advantageous to use eq 2.36, which does not contain $\overline{u(t)^2}$. In fact, this equation is nothing but eq 2.32 corrected for the amplitude of the correlation function so that it reflects only the shape

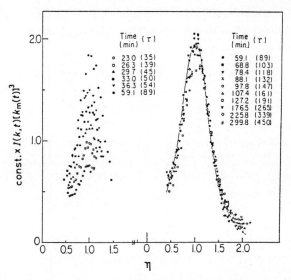

Fig. 10-17. Time change of scaled scattering intensity $I(k,t)[k_{\mathrm{m}}(t)]^3$ for a critical PS-1/PVME-1 blend quenched at $98.2°\mathrm{C}$, where $\eta = k/k_{\mathrm{m}}(t)$ [24].

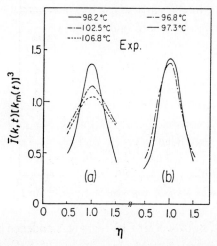

Fig. 10-18. Normalized scaled scattering intensity $\bar{I}(k,t)[k_{\mathrm{m}}(t)]^3$ $(= \overline{F}(\eta))$ at indicated T_{q} for a critical PS-1/PVME-1 blend [24].

of $H(r,t)$. Figure 10-18 illustrates the T_{q} dependence of $\bar{I}(k,t)[k_{\mathrm{m}}(t)]^3$(i.e., $\overline{F}(\eta)$ by eq 2.36) for a critical PS-1/PVME-1 blend [24]. We see that the $\overline{F}(\eta)$ for three T_{q} closer to $T_{\mathrm{s}}(95.8°\mathrm{C})$ agree well with one another, while

355

those for the higher T_q are distinctly broader than them. Hashimoto [19, 20] attributes this phenomenon to an asymmetry of the phase diagram about the critical composition.

3. Scaling for $k_m(t)$ and $\bar{I}_m(t)$ To determine the dimensionless wavenumber $q_m(\tau)$ and the dimensionless scattering intensity $\tilde{I}_m(\tau)$ defined, respectively, by eq 2.39 and 2.40 we need information about $\xi_c(T)$ and $D(T)$. In deriving the data points shown in Figures 10-9 and 10-10, Chou and Goldburg [17] used the data of Chu et al. (J. Am. Chem. Soc. **90**, 3042 (1968); $\xi_c(T) = (3.57 \pm 0.07)\epsilon^{-0.613\pm0.001}$ (in Å) and $D(T) = k_B T_c/[6\pi\xi_c(T)\eta^\star]$ (in cm^2/s) with $\eta^\star = 2.42 \times 10^{-2}$ (in poises), where $\epsilon = |T - T_c|/T_c$). Izumitani and Hashimoto [31] equated D to D_{app} defined by eq 2.45 and determined the latter by

$$D_{app} = \lim_{k \to 0} R(k)/k^2 \qquad (2.52)$$

which follows from eq 2.44. Snyder and Meakin [32] used

$$D_{app} = 2R(k_m^0)/(k_m^0)^2 \qquad (2.53)$$

which follows from eq 2.10, 2.44, and 2.45 and where k_m^0 is the k value for the maximum of $S(k,t)$ in the early stage of spinodal decomposition. These are reasonable maneuvers. Izumitani and Hashimoto [31] tried to estimate ξ_c by using

$$\xi_c = 1/k_m^0 \qquad (2.54)$$

which had been proposed by Snyder and Meakin [33]. However, when, as in Figure 10–16, the measured $I(k,t)$ vs. k curves in the early stage exhibit no peak, this proposal is of no use. In the following, we propose a new method which allows the time dependence of q_m and $\tilde{I}_m$ to be determined with no need for information on ξ_c (H. Fujita, T. Hashimoto, and M. Takenaka, Macromolecules **22**, 4663 (1989)).

Using eq 2.41 and 2.42, we can transform eq 2.39 into

$$X(t,T) \equiv k_m(t,T)[D(T)t]^{1/2} = (t/t_c)^{1/2}q_m(t/t_c) \qquad (2.55)$$

This indicates that $X(t,T)$ depends on a single variable t/t_c. Hence, plots of $X(t,T)$ (calculable from measured $k_m(t)$ and known $D(T)$) against $\ln t$ for different T should be superimposable on a master curve by horizontal shifting. The master curve gives us the form of $X(t,T')$ at a fixed temperature T' over a wider range of t. According to eq 2.55, if the master curve is divided by $t^{1/2}$, the resulting curve yields the values of $(t_c')^{-1/2}q_m$ as a function of t, where t_c' is the value of the time constant t_c at $T = T'$, i.e., $[\xi_c(T')]^2/D(T')$.

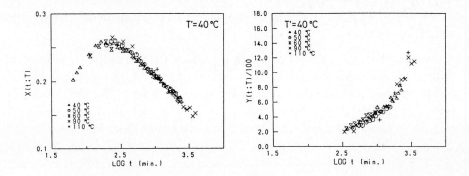

Fig. 10-19. Superposition of $X(t,T)$ and $Y(t,T)$ against $\log t$ for different T by horizontal shifting, for a critical SBR/PB blend (SBR: $M_w = 1.18 \times 10^5$, $M_w/M_n = 1.18$; styrene content = 20 wt%; *cis : trans* : vinyl = 0.16 : 0.23 : 0.61. PB: $M_w = 1.90 \times 10^5$, $M_w/M_n = 1.16$; *cis : trans* : vinyl = 0.19 : 0.35 : 0.46.

Next, with eq 2.41 and 2.42, we derive from eq 2.40

$$Y(t,T) \equiv \bar{I}_m(t,T)[D(T)t]^{-3/2} = (t/t_c)^{-3/2}\tilde{I}_m(t/t_c) \qquad (2.56)$$

This indicates that plots of $Y(t,T)$ against $\ln t$ for different T can be reduced to a master curve by horizontal shifting and that multiplication of the master curve $Y(t,T')$ by $t^{3/2}$ yields the values of $(t_c')^{3/2}\tilde{I}_m$ as a function of t.

Though incapable of determining the absolute values of q_m and $\tilde{I}_m$, the method described above gives us information that suffices to see how the exponents α and β defined by $q_m \sim t^{-\alpha}$ and $\tilde{I}_m \sim t^\beta$ vary with time. We emphasize that no data on $\xi_c(T)$ are needed for its application. If the value of $\xi_c(T)$ (along with that of D) at $T = T'$ is available, the curves for $(t_c')^{-1/2}q_m(t)$ and $(t_c')^{3/2}\tilde{I}_m(t)$ can be converted into those of $q_m(\tau)$ and $\tilde{I}_m(\tau)$ at $T = T'$, respectively.

Figure 10-19 illustrates with unpublished data of Hashimoto and Takenaka for a critical blend of styrene–butadiene copolymer (SBR) and polybutadiene (PB) that master curves can be obtained for X and Y, i.e., the scaling laws for k_m and $\bar{I}_m$ (eq 2.39 and 2.40) hold. This result is somewhat surprising, because these data were obtained at very deep quench depths (T_c for the blend was estimated to be about $400°C$ [31]). The solid curves in Figure 10-20 give $\log q_m(t,T')$ and $\log \tilde{I}_m(t,T')$ as functions of $\log t$ calculated from the smooth curves in Figure 10-19. In the range of timescales shown, the slope for q_m

357

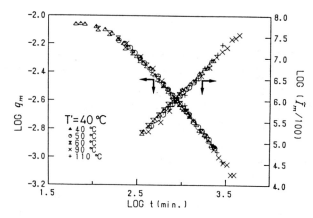

Fig. 10–20. Time dependence of q_m and $\tilde{I}_\mathrm{m}$ at $40°\mathrm{C}$ derived from Fig. 10–19.

gradually changes from −0.2 to −0.8, while that for $\tilde{I}_\mathrm{m}$ is constant, being equal to about 2.3. In this book, we do not go further than this.

References

1. P. J. Mills, J. W. Mayer, E. J. Kramer, G. Hadziioannou, P. Lutz, C. Strazielle, P. Rempp, and A. J. Kovacs, Macromolecules **20**, 513 (1987).
2. J. Klein, Macromolecules **19**, 105 (1986).
3. P.-G. de Gennes, J. Phys. (Paris) **36**, 1199 (1975).
4. K. Dodgson, D. J. Bannister, and J. A. Semlyen, Polymer **21**, 663 (1980); J. A. Semlyen, Pure & Appl. Chem. **53**, 1787 (1981).
5. J. Roovers, Macromolecules **18**, 1359 (1985).
6. J. D. Ferry, "Viscoelastic Properties of Polymers," Wiley, New York, 1980.
7. G. B. McKenna, G. Hadziioannou, P. Lutz, G. Hild, C. Strazielle, C. Starupe, P. Rempp, and A. J. Kovacs, Macromolecules **20**, 498 (1987).
8. G. B. McKenna and D. J. Plazek, Polym. Commun. **27**, 304 (1986).
9. J. Roovers, Macromolecules **21**, 1517 (1988).
10. J. D. Gunton, M. S. Miguel, and P. S. Sahni, in "Phase Transitions and Critical Phenomena," Ed. C. Domb and J. L. Lebowitz, Academic Press, New York, 1983, Chap. 3.
11. T. Nishi, T. T. Wang, and T. K. Kwei, Macromolecules **8**, 227 (1975).
12. J. W. Cahn and J. E. Hilliard, J. Chem. Phys. **29**, 258 (1958); **31**, 688 (1959).
13. J. W. Cahn, J. Chem. Phys. **42**, 93 (1965).
14. H. E. Cook, Acta Metall. **18**, 297 (1970).
15. J. S. Langer, M. Bar-on, and H. D. Miller, Phys. Rev. A **11**, 1417 (1975).
16. J. S. Langer, Ann. Phys. **65**, 53 (1971).
17. Y. C. Chou and W. I. Goldburg, Phys. Rev. A **20**, 2105 (1979).
18. K. Kawasaki and T. Ohta, Progr. Theor. Phys. **59**, 362 (1978).
19. T. Hashimoto, Phase Transitions **12**, 47 (1988).
20. T. Hashimoto, in "Dynamics of Ordering Processes in Condensed Matter," Ed. S. Kimura and H. Furukawa, Plenum, London, 1988, p. 421.
21. T. Nose, Phase Transitions **8**, 245 (987).
22. See, for example, T. Hashimoto, K. Sasaki, and H. Kawai, Macromolecules **17**, 2812 (1984).
23. H. L. Snyder, P. Meakin, and S. Reich, Macromolecules **16**, 757 (1983).
24. T. Hashimoto, M. Itakura, and H. Hasegawa, J. Chem. Phys. **85**, 6118 (1986).
25. T. Hashimoto, M. Itakura, and N. Shimidzu, J. Chem. Phys. **85**, 6773 (1986).
26. T. Sato and C. C. Han, J. Chem. Phys. **88**, 2057 (1988).
27. P.-G. de Gennes, J. Chem. Phys. **72**, 4756 (1980).
28. M. Okada and C. C. Han, J. Chem. Phys. **85**, 5317 (1986).
29. J. Kumaki and T. Hashimoto, Macromolecules **19**, 763 (1986).

30. M. Takahashi, H. Horiuchi, S. Kinoshita, Y. Ohyama, and T. Nose, J. Phys. Soc. Jpn 55, 2687 (1986).
31. T. Izumitani and T. Hashimoto, J. Chem. Phys. 83, 3694 (1985).
32. H. L. Snyder and P. Meakin, J. Chem. Phys. 79, 5588 (1983).
33. H. L. Snyder and P. Meakin, J. Polym. Sci., Polym. Symp. 73, 217 (1985).

Subject Index

mean–field approximation, 205, 316
mean–field theory of Flory,
   original, 20
   modified, 21
mean force, 5
mean force potential, 6
mean–square end–to–end distance,
   definition of, 10
   of Edwards continuous chains, 11,
     18
   of helical wormlike chains, 170
   of spring–bead chains, 10, 17
   of wormlike chains, 140, 141
   renormalization group calculation
     of, 77
mean–square radius of gyration,
   definition of, 23
   in concentrated solutions, 199
   of macrorings, 125
   of wormlike chains, 140, 162
   renormalization group calculation
     of, 80
measuring methods for
   self–diffusion, 247, 248
moderately concentrated solution,
   182
mother solution, 286
Murakami et al. method, 150
Muthukumar–Edwards theory, 228

N
Nagai theory for optical anisotropy,
   163
Ngai coupling theory, 277
Nies et al. approach, 296
non–draining limit, 55
normalized scattering intensity, 342
nucleus growth, 333
number–average relative chain
   length, 288

O
one–body distribution function, 185
one–point distribution functional,
   338
oneset of coil overlapping, 180
optical anisotropy,
   correction of light scattering data
     for, 165
   Nagai theory for, 163
Ornstein–Zernike form, 191
Ornstein–Zernike relation, 192
osmotic pressure,
   of semi–dilute solutions, 206
   virial expansion for, 8, 205
overlap concentration,
   apparent, 184
   definition of, 180
   estimation of, 183

P
Padé approximant method, 22
pair correlation function, 336
particle scattering function,
   Debye function for, 79, 192
   for Gaussian coils, 144
   for Gaussian rings, 125
   for rods, 144
   for wormlike chains, 144
   renormalization group theory for,
     78
paucidisperse quasi–binary solution,
   282
penetration function,
   definition of 37, 91
permeability, 222
persistence length,
   definition of, 140, 142
   determination of, 150, 153–155
perturbation theory,
   for end distance expansion factor,
     19

in d-benzene, 255
in dibutyl phthalate, 257, 258
in dioctyl phthalate, 108
in d-toluene, 126, 180
in ethyl acetate, 224
in ethyl benzene, 218
in methyl acetate, 106
in methylcyclohexane, 314, 317
in methyl ethyl ketone
  (2-butanone), 62, 63, 64, 184,
  212, 219
in n-butyl acrylate, 246
in poly(methyl methacrylate) +
  benzene, 258
in poly(vinylmethylether)
  + ortho-fluorotoluene, 259, 260
in poly(vinylmethylether)
  + toluene, 258, 260
in tetrahydrofuran, 51, 216, 254
in toluene, 43, 48, 51, 80, 126,
  184, 198, 209, 212, 220, 221,
  255
melt, 200, 249, 252, 328, 329
plasticized, 246
polystyrene (ring),
  blended with linear polystyrene,
    330
  in benzene, 129
  in cyclohexane, 123, 126, 127,
    128, 129
  in d-cyclohexane, 123, 126, 128
  in d-toluene, 126
  in linear d-polystyrene, 326
  in tetrahydrofuran, 128, 130
  in toluene, 128. 130
  melt, 328, 329
poly(terephthalamide-p-
  benzohydrazide), 166
  in dimethyl sulfoxide, 156, 160,
    165, 173
poly(terephthaloyl-trans-2,5-

dimethyl piperazine),
  in trifluoroethanol, 160, 173
poly(tetrahydrofuran), melt, 262
poly[trans-bis(tributyl-phosphine)
  platinum 1,4-butadiynediyl],
  in heptane, 173
poly(vinylmethylether), 258, 259,
  260, 344
porous plug, 222
preaveraging approximation to the
  Oseen tensor, 59
primary (or bare) chain, 14
primitive chain, 236
pseudo-network model, 183
pulse gradient spin echo, 248
pulse-induced critical scattering
  (PICS), 287

Q

quasi-binary solution, 235, 255, 282
quasi-elastic light scattering
  (QELS), 113
quench depth, 353

R

radius expansion factor,
  approximate closed expressions
    for, 24, 27
  asymptotic form of, 24
  below the theta temperature, 103
  definition of, 23
  Domb-Barrett interpolation for-
    mula for, 24
  for macrorings, 127
  perturbation theory for, 24
  renormalization group theory for,
    89
  Weill-des Cloizeaux theory for,
    96
random flight chain, 2

367

random phase approximation, 194, 351

reduced characteristic ratio, 172

reduced density correlation function, 188
  interchain contribution to, 188
  intrachain contribution to, 188

reduced osmotic pressure, 206

reduced pair correlation function, 336

reduced temperature, 40, 103

regular helix, 168

relative chain length, 283

relaxation time,
  maximum, 274
  primitive, 277
  Rouse, 265
  terminal, 245

renormalization, 71

renormalization constants, 72

renormalization group equation, 73

renormalization group theory,
  first-order, 75
  for characteristic frequency, 121
  for hydrodynamic factors, 60
  of Douglas and Freed, 82
  of Ohta and Oono, 211

renormalized excluded-volume strength, 228

reptation,
  definition of, 236
  Doi-Edwards theory for, 237
  model, 236

residual (irreducible) potential, 6

reverse quench experiment, 352

rigid-body approximation, 60

ring (or cyclic) polymers (see macrorings)

rotational isomeric state (RIS) chain model, 167

Rouse chain, 244

Rouse diffusion coefficient, 245

Rouse-like motion, 244

S

scaling equation of Phillies, 262

scaling law,
  for $k_m(t)$ and $\bar{I}_m(t)$, 342, 356
  for $S(k,t)$, 341, 353

scaling theory,
  for correlation length, 227
  for osmostic pressure, 226
  for radius of gyration, 226
  of Kosmas and Freed, 227

scattering vector, 78

Schäfer equation, 208

schizophyllan,
  in aqueous sodium hydroxide, 160, 173
  in water, 154, 173

screening effect, 179
  on chain dimensions, 198
  on density fluctuation correlation, 197
  on excluded-volume interaction, 195
  on hydrodynamic interaction, 200

second virial coefficient,
  approximate closed expressions for, 38
  below the theta temperature, 45
  definition of, 8
  for macrorings, 130
  for spring-bead chains, 37
  molecular weight dependence of, 44
  perturbation theory for, 38
  reduced, 37
  renormalization group theory for, 80, 90

sedimentation coefficient,

concentration dependence of, 183, 218, 223, 224
solvent–fixed, 215
volume–fixed, 215
segment, 2
self–avoiding chains, 29, 74, 87
self–avoiding limit, 29, 87
self–diffusion, 214
self–diffusion coefficient,
   computer simulations for, 268
   concentrated solutions, 253
   definition of, 214
   Doi–Edwards theory for, 239
   Hess theory, 244
   Philles scaling equation for, 262
   Skolnick et al. theory for, 245
self–knotted rings, 131
self–knotting, 131
self–similarity postulate,341, 342
semi–dilute regime, 182
semi–dilute solution, 182
semi–flexible polymer molecules, 4
separation factor, 308
shadow curve, 287
shift factor, 149, 173
Siegert relation, 115, 133
Skolnick et al. theory, 245
smooth–density sphere model, 110
solvent, good, 4
solvent, poor or bad, 4
solvent–mediated force, 5
spin–echo neutron scattering, 262
spinodal,
   definition of, 285
   of binary solutions, 296
spinodal decomposition, 333
spring–bead chain,
   definition of, 3
   end–to–end distance of, 10
   spring bond probability in, 8
spring bond vector, 6

spring length, 8
   mean, 17
steady gradient spin echo, 248
steady (-state) transport coefficients, 49
stages of spinodal decomposition,
   early, 333, 334, 337
   intermediate, 333, 338
   late, 333
   later, 353
stiff polymer molecules, 4
Stokes radius,
   definition of, 54
   of macrorings, 129
straight rods, 140, 147, 148, 170
stretched exponential form, 277
structure factor,
   dynamic, 116, 261
   excess, 189
   static, 189

T

ternary cluster, 6, 48, 102, 111
thermal blob, 211
thermal noise,
   effect, 335
   term, 346
thermodynamic factor, 213
thermodynamic parameter χ,
   definition of, 288
theta condition, 8
theta solvent, 8
theta temperature, 9
   for macrorings, 123
third–power type, 20
third virial coefficient,
   definition of, 8
   molecular weight dependence of, 47
   reduced, 47
   near the theta temperature, 113

renormalization group theory for, 47

three–phase separation, 313

threshold point, 303

topological constraint, 236

topological factor,
    concentration dependence of, 242
    definition, of, 235

torsional energy, 141

tracer component, 235

tracer diffusion coefficient, 235, 250
    in quasi–binary solutions, 255
    of macrorings, 325

triblock copolymer of h–, d–, and h–styrenes, 100

truly polydisperse solution, 282

tube model, 236

tube renewal, 242

two–body distribution function, 185

two–parameter approximation,
    definition of, 19
    tests of, 61

two–parameter theory, 19

two–point distribution functional, 336

U

universality assumption, 71

uphill diffusion, 332

V

van't Hoff solution, 299

very shallow critical quench, 342

virial coefficients, 8

virial regime, 182

viscosity (bulk)
    of ring polymer melts, 327
    theoretical predictions, 274
    3.4 power law for, 272

viscosity expansion factor,

approximate expressions for, 54, 55

definition of, 54

for macrorings, 129

W

Wegner expansion, 319

weight–average relative chain length, 289

weight–basis chain length distribution, 284

Weill–des Cloizeaux approximation, 95

Weill–des Cloizeaux theory, 96

Williams–Watts equation, 277

wire model, 141

wormlike chains,
    chain dimensions of, 140
    characteristic function for, 143
    definition of, 139
    distribution function for, 142
    isotropic scattering function for, 144

wormlike cylinder, 145

X

xanthan, in aqueous sodium chloride, 160, 173

Y

Yamakawa–Fujii theory, 145
    for friction coefficient, 147
    for intrinsic viscosity, 148

Yamakawa–Fujii–Yoshizaki theory, 149
    tests of, 155

Yamakawa operational method, 143

Yamakawa–Shimada theory, 161

Yoshizaki–Nitta–Yamakawa theory, 64